Signal Transduction Protocols

Methods in Molecular Biology™

John M. Walker, SERIES EDITOR

Earlier volumes are still available. Contact Humana for details.

Methods in Molecular Biology™ • 41

Signal Transduction Protocols

Edited by

David A. Kendall
and Stephen J. Hill

Queen's Medical Centre,
University of Nottingham, UK

Humana Press Totowa, New Jersey

© 1995 Humana Press Inc.
999 Riverview Drive, Suite 208
Totowa, New Jersey 07512

This publication is printed on acid-free paper. ∞
ANSI Z39.48-1984 (American Standards Institute)
Permanence of Paper for Printed Library Materials.

Photocopy Authorization Policy:
Authorization to photocopy items for internal or personal use, or the internal or personal use of specific clients, is granted by Humana Press Inc., provided that the base fee of US $4.00 per copy, plus US $00.20 per page, is paid directly to the Copyright Clearance Center at 222 Rosewood Drive, Danvers, MA 01923. For those organizations that have been granted a photocopy license from the CCC, a separate system of payment has been arranged and is acceptable to Humana Press Inc. The fee code for users of the Transactional Reporting Service is: [0-89603-298-1/95 $4.00 + $00.20].

Printed in the United States of America. 10 9 8 7 6 5 4 3 2

Library of Congress Cataloging in Publication Data

Main entry under title:

Methods in molecular biology™.

Signal transduction protocols / edited by David A. Kendall and Stephen J. Hill.
    p. cm. — (Methods in molecular biology™ ; 41)
    Includes bibliographical references (p.).
    ISBN 0-89603-298-1
    1. Cellular signal transduction—Research—Methodology. 2. G Proteins—Research—Methodology. 3. Cell receptors—Research—Methodology. I. Kendall, David A. II. Hill, Stephen J. III. Series: Methods in molecular biology™ (Totowa, NJ) ; 41.
QP517.C45S555   1995
574.87'6—dc20   95-2769
                                                                 CIP

Preface

As our understanding of the biological sciences expands, the boundaries between traditional disciplines tend to blur at the edges. Physiologists and pharmacologists, for instance, now need to embrace techniques that until recently were the strict preserves of biochemists and molecular biologists. However, the acquisition of new technologies can be a time-consuming and frustrating business, and unless an expert is on hand to give instruction, precious hours can be spent poring over half-described Methods sections with no guarantee of eventual success. The aim of *Signal Transduction Protocols* has been to get experts with "hands-on" experience in particular techniques to give detailed accounts of experimental protocols in a recipe-type format, which we hope will circumvent the problems of ambiguity often encountered when reading the literature.

The techniques described in *Signal Transduction Protocols* are those that we think will be most useful in addressing questions in the area of receptor-mediated cell signaling, with particular regard to those receptors that are part of the G-protein-linked superfamily. To keep it to a manageable size, we have omitted any reference to electrophysiology and have instead concentrated on more biochemical approaches.

We have tried to cover signal transduction processes at various levels, ranging from the identification and localization of recognition sites by radioligand binding and autoradiography through assessments of G-protein function, intracellular messenger-generating enzymes (phospholipases and cyclases), and catabolic enzymes (e.g., phosphodiesterases) to the measurement of the kinases that mediate many of the actions of G-protein-coupled receptors.

We do not pretend that the list of techniques is fully comprehensive, but we do hope that *Signal Transduction Protocols* will be of practical value and that it will find its way not only onto library bookshelves, but also to where it is really aimed—laboratory benches.

David A. Kendall
Stephen J. Hill

Contents

Contributors

STEPHEN P. H. ALEXANDER • *Department of Physiology and Pharmacology, University of Nottingham Medical School, Nottingham, UK*

PETER M. BAXENDALE • *Amersham International, Whitchurch, Cardiff, UK*

MICHAEL R. BOARDER • *Department of Cell Physiology and Pharmacology, University of Leicester, UK*

RONALD M. BURCH • *Department of Cellular Pathology, Armed Forces Institute of Pathology, Washington, DC*

R. A. JOHN CHALLISS • *Department of Cell Physiology and Pharmacology, University of Leicester, UK*

ENRIQUE CLARO • *Department of Biochemistry and Molecular Biology, University of Barcelona, Spain*

JOHN M. DICKENSON • *Department of Physiology and Pharmacology, Queen's Medical Centre, Nottingham, UK*

PETER R. DUNKLEY • *Department of Medical Biochemistry, University of Newcastle, Callaghan, Australia*

VANESSA FURST • *Wellcome Research Laboratories, Beckenham, Kent, UK*

TREVOR J. HALLAM • *Roche Research Centre, Hertfordshire, UK*

JEREMY M. HENLEY • *Department of Anatomy, School of Medical Sciences, University of Bristol, UK*

FÉLIX HERNÁNDEZ • *Department of Molecular Biology, University of Madrid, Spain*

MATTHEW HODGKIN • *Institute for Cancer Studies, University of Birmingham, UK*

JEFFREY K. HORTON • *Amersham International, Whitchurch, Cardiff, UK*

PHILIP A. IREDALE • *Department of Physiology and Pharmacology, Queen's Medical Centre, Nottingham, UK*

ix

PAULA E. JARVIE • *Department of Medical Biochemistry, University of Newcastle, Callaghan, Australia*

STEPHEN JENKINSON • *The Neurosciences Institute, La Jolla, CA*

MARY KEEN • *Department of Pharmacology, Medical School, University of Birmingham, UK*

EAMONN KELLY • *Department of Pharmacology, Medical School, University of Bristol, UK*

DAVID M. KIRKHAM • *Department of Pharmacology, Medical School, University of Birmingham, UK*

TAKAYOSHI KUNO • *Department of Pharmacology, Kobe University School of Medicine, Kobe, Japan*

ANNA M. LEONE • *Wellcome Research Laboratories, Beckenham, Kent, UK*

KATRINA A. MARSH • *Department of Physiology and Pharmacology, Queen's Medical Centre, Nottingham, UK*

ASHLEY MARTIN • *Institute for Cancer Studies, University of Birmingham, UK*

PETER MOLENAAR • *Department of Pharmacology, University of Melbourne, Parkville, Victoria, Australia*

SALVADOR MONCADA • *Wellcome Research Laboratories, Beckenham, Kent, UK*

BRIAN J. MORRIS • *Department of Pharmacology, University of Glasgow, Scotland*

HIDEYUKI MUKAI • *Department of Biochemistry, Kobe University, Kobe, Japan*

KENNETH J. MURRAY • *SmithKline Beecham, Hertfordshire, UK*

STEFAN R. NAHORSKI • *Department of Cell Physiology and Pharmacology, University of Leicester, UK*

C. DAVID NICHOLSON • *Organon International, Oss, The Netherlands*

FERNANDO PICATOSTE • *Department of Biochemistry and Molecular Biology, University of Barcelona, Spain*

JOHN R. PURKISS • *Department of Cell Physiology and Pharmacology, University of Leicester, UK*

PETER RHODES • *Department of Clinical Pharmacology, Ninewells Hospital, Dundee, Scotland*

ELISABET SARRI • *Department of Biochemistry and Molecular Biology, University of Barcelona, Spain*

MOHAMMED SHAHID • *Department of Pharmacology, Organon Laboratories, Ltd., Newhouse, Scotland*
JAMES STRUPISH • *Department of Anaesthesia, University of Leicester, UK*
ROGER J. SUMMERS • *Department of Pharmacology, University of Melbourne, Parkville, Victoria, Australia*
MARVA SWEENEY • *Department of Physiology, College of Medicine, University of Saskatchewan, Saskatoon, Canada*
MICHAEL J. O. WAKELAM • *Institute for Cancer Studies, University of Birmingham, UK*
ROBERT A. WILCOX • *Department of Cell Physiology and Pharmacology, University of Leicester, UK*
SANDRA E. WILKINSON • *Roche Research Centre, Hertfordshire, UK*
ROBERT J. WILLIAMS • *Department of Pharmacology, Medical School, University of Bristol, UK*

CHAPTER 1

The Problems and Pitfalls of Radioligand Binding

Mary Keen

1. Introduction

Radioligand binding is an extremely versatile technique that can be applied to a wide range of receptors in a variety of preparations, including purified and solubilized receptors, membrane preparations, whole cells, tissue slices, and even whole animals. The basic method is very easy to perform. It can even be automated, and the throughput of samples that can be achieved is very high. The data obtained are typically extremely "tight" and reproducible, allowing receptor number, ligand affinity, the existence of receptor subtypes, and allosteric interactions between binding sites and/or receptors and effector molecules to be determined with great precision and subtlety.

Perhaps it is the very ease of radioligand binding that presents the problem. It is extremely simple to produce data, feed the data into a computer, and generate numbers. The question of whether these numbers really mean what one hopes they mean is often overlooked. This chapter provides a brief overview of the potential problems and artifacts that may occur in radioligand binding experiments. A much more in-depth treatment of the subject can be found in ref. *1*.

2. Materials and Methods

2.1. Basic Method

The basic outline of all radioligand binding assays is very similar.

From: *Methods in Molecular Biology, Vol. 41: Signal Transduction Protocols*
Edited by: D. A. Kendall and S. J. Hill Copyright © 1995 Humana Press Inc., Totowa, NJ

1. Incubate the radioligand with the receptor preparation. In parallel preparations, incubate the radioligand with the receptor preparation in the presence of an unlabeled ligand to define nonspecific binding or to investigate
   the binding characteristics of that unlabeled ligand (*see* Section 2.3.).
2. Separate bound ligand from free ligand.
3. Quantify the amount of radioligand bound, using liquid scintillation or γ
   counting, depending on the radioisotope used. In tissue slices, autoradiography may be used.
4. Analyze the data.

Several different types of binding assay can be performed using this
basic technique. By investigating the amount of radioligand bound at
various times after its addition, the association kinetics of the radioligand can be investigated. Similarly, dissociation can be investigated by
allowing the radioligand and receptor to come to equilibrium, and then
measuring binding at various times following infinite dilution of the
sample or, more practically, following the addition of a saturating concentration of unlabeled ligand, so that the probability of radioligand binding to the receptors is reduced to close to zero. The affinity of the
radioligand for the receptor and the total number of receptor sites (B_{max})
can be determined by investigating the equilibrium binding of a range of
concentrations of the radioligand. The binding of unlabeled drugs to a
receptor can be measured by their ability to inhibit the specific binding
of the radioligand either in equilibrium "competition" binding assays or
in time-course experiments *(1,2)*.

The data from binding assays are best analyzed using computerized
"curve-fitting" techniques to find the best fit of an appropriate function
to the data. Thus, the fitting of mono- or biexponentials to time-course
data, or single- or multiple-site models to equilibrium data allows the
estimation of association and dissociation rate constants, affinity constants, and receptor number. The functions used relate the amount of
ligand bound to the free concentration, and thus, any errors in the
measurement of either of these concentrations will inevitably affect the
accuracy of the parameters obtained (*see* Section 3.2.).

2.2. Choice of Radioligand

The isotopes most commonly used to label ligands for use in binding
assays are [³H] and [¹²⁵I]. Of these, [³H] has the advantage of a long half-
life and is less hazardous than [¹²⁵I], so that it requires fewer handling

precautions, and larger quantities can be safely stored and disposed of. Labeling of ligands with [^{3}H] is difficult to do in-house because gaseous tritium presents a particular hazard, whereas [^{125}I]iodination of ligands is a relatively simple procedure, especially for peptides and proteins. However, the addition of an iodine atom to the ligand molecule is likely to affect its binding characteristics, so the "cold" iodinated compound should always be used for comparison, rather than the unaltered parent molecule. The energy produced by the decay of [^{125}I] is greater than that produced by [^{3}H], and so the specific activity of ligands labeled with 1 atom/mol of [^{125}I] (>2000 Ci/mmol) is greater than for ligands similarly labeled with [^{3}H] ($\approx$80 Ci/mmol). This means that [^{125}I]-labeled ligands are particularly suitable for autoradiography and in situations where receptor density is low.

High-affinity ligands, with dissociation constants in the range 10^{-10} to $10^{-8}M$, are preferred, because they can be used at low concentrations and they tend to remain bound to the receptor during the separation procedure (*see* Section 2.6.). However, an extremely high affinity for the receptor is not necessarily a good thing because these ligands only reach binding equilibrium very slowly, necessitating very prolonged incubation periods.

The more selective the radioligand is the better. It is possible to restrict the binding of rather nonselective radioligands to particular receptors by including saturating concentrations of unlabeled drugs selective for the unwanted receptors in the assay. However, this introduces an extra degree of complexity into the assay, and the high concentrations of unlabeled drugs required to suppress binding to other receptors means that these unlabeled drugs must be very selective indeed.

2.3. Definition of Nonspecific Binding

Even the best radioligands inevitably bind to all sorts of things apart from their receptors. Thus, it is necessary to differentiate this "nonspecific" binding from the "specific" binding of interest. This is achieved by measuring the "total" binding obtained when the radioligand alone is incubated with the receptor preparation and measuring nonspecific binding in a parallel incubation, in which the binding of the radioligand to the receptor, but hopefully not the nonreceptor sites, is suppressed by a concentration of unlabeled ligand sufficient to occupy all of the available receptors in the presence of the radioligand. Specific binding is then defined as the difference between total and nonspecific binding. This procedure is not infallible. The question of whether specific binding really represents binding to the receptor must always remain provisional (*see* Section 3.4.), and for this reason, it might be wiser to refer simply to "displaceable" binding. In order to reduce the chances of the unlabeled ligand displacing radioligand from saturable, nonreceptor sites, such as enzymes or uptake sites, it is best to use a displacing ligand that is structurally dissimilar from the radioligand. However, this is by no means always possible, especially since the relatively large concentrations of these ligands that are required may mean that cost becomes a prohibiting factor.

In competition experiments, where the binding of unlabeled ligands is studied by their ability to inhibit the binding of a labeled drug, all of the ligands that are known to bind to the receptor of interest should inhibit binding to the same extent, i.e., give the same estimate of nonspecific binding. If they do not, it tends to suggest that some of these compounds displace nonspecific binding as well as specific binding. Examination of data from these experiments is a good way to find the most appropriate drug to use to define nonspecific binding.

2.4. Choice of Receptor Preparation

Membrane preparations are most widely used and probably generate the most reproducible and reliable data. However, the radioligand binding technique can be applied to other preparations, such as purified receptors, solubilized receptors, whole cells, and tissue slices. The choice of preparation depends on the question to be addressed. However, this will affect the assay conditions and separation procedures that can be used, and these in turn will affect the reliability of the data that can be obtained.

2.5. Choice of Assay Conditions

Radioligand binding to membrane preparations can often be successfully achieved using very simple buffer solutions, such as unsupplemented Tris-HCl or phosphate buffers. However, in many cases, binding kinetics and/or affinity are affected by the presence of various ions, pH, temperature, and so forth, and choosing the best system presents something of a problem. There is often no alternative to trying out various ionic strengths, divalent cations, and so on in an attempt to optimize binding in a new system. When comparing results with data performed in other laboratories, it is particularly important to try and use the same assay conditions.

2.6. Choice of Separation Technique

There are several methods available, and the choice depends on the radioligand used and the type of preparation. The technique most widely used with membrane preparations and whole cells is filtration using a vacuum filtration manifold or a cell harvester. The bound ligand is retained on glass-fiber filters, and the free ligand passes through. Following the initial filtration, the filters are usually rinsed with assay buffer to reduce the level of nonspecific binding owing to free radioligand loosely associated with the membranes or the filter. This washing step can considerably improve the signal:noise ratio. The method is very reproducible and quick; 24 samples can be filtered and washed in <10 s using a cell harvester. However, the removal of free ligand necessarily promotes dissociation of the ligand from the receptor, and in the case of low-affinity ligands, which have fast off-rates, this loss of binding may be so pronounced as to render filtration useless. Filtration may also be unsuitable for some ligands that display a very high level of binding to filters, especially if this binding is displaceable.

The problem of dissociation owing to removal of free ligand can be minimized in a centrifugation assay, in which bound radioligand is pelleted with the membranes, allowing the free ligand to be removed with the supernatant. This method is useful for low-affinity ligands, but suffers from a relatively high level of nonspecific binding, because some of the free ligand is inevitably trapped within the pellet. Centrifugation is therefore best reserved for ligands that exhibit a low level of nonspecific binding.

Solubilized receptors can be separated from free ligand by gel filtration, polyethylene glycol precipitation, charcoal adsorption, or filtration onto glass-fiber filters pretreated with 0.3% polyethylenenimine, which then retain acidic receptors by an ion-exchange mechanism. In the case of binding to tissue slices immobilized on slides for autoradiography, free ligand is simply removed by several rinses in beakers of buffer. In all cases, dissociation from the receptor can be a problem. The extent to which dissociation occurs will depend on the time that the separation procedure takes, the affinity of the ligand, and the temperature. Dissociation is slowed at low temperatures and can be reduced by making sure that all buffers are ice-cold.

3. Problems and Pitfalls

3.1. Insufficient Specific Binding

Perhaps the most obvious problem in any binding assay is when the levels of specific binding turn out to be extremely low or even nonexistent. This can arise either because of problems with the ligand, problems with the preparation, or a combination of both.

Individual radioligands vary enormously in their capacity to bind to nonspecific sites. If a ligand exhibits a high degree of nonspecific binding, it may be that it is impossible to pick out a low level of specific binding over the high level of background noise. This problem is obviously compounded in a tissue with low receptor density. Furthermore, the properties of the preparation can influence nonspecific binding, so that a radioligand which can be used successfully in one tissue may be much less useful in another that contains, for example, very high levels of lipid into which lipophilic drugs partition.

By and large, the problem of a high level of nonspecific binding compared to specific binding should be minimized by using a low concentration of a ligand with a high affinity for the receptor of interest, because nonspecific binding is usually directly proportional to ligand concentration. However, very often the only approach to solving this problem is to try to use a different radioligand to see if this gives any better results. It may not, and there are many instances of preparations that must contain a given receptor, but in which radioligand binding cannot be used to study that receptor because of an unfavorable combination of a rather "sticky" radioligand and a low density of receptor sites. It is perhaps worth emphasizing that there is no hard and fast rule that relates the density of receptors in any given tissue to their physiological and pharmacological importance.

The situation can sometimes be improved by enriching the number of receptors in the preparation by, for example, preparing a purified plasma membrane preparation rather than binding to the whole-cell homogenate, which will reduce the amount of nonspecific binding while hopefully retaining all the receptors. However, it is, in the author's experience, possible to lose more than you gain by this type of procedure and, unless a high level of nonspecific binding is a big problem, it is usually better to avoid differential centrifugation protocols in favor of simple washed homogenates.

3.2. Problems Associated
with Quantifying "Bound"
and Free Concentrations of Ligand

3.2.1. Uncertain Specific Activity

An accurate assessment of the concentration of radioligand bound to the receptor or free in solution depends on knowing its specific activity, and any errors in the estimation of this quantity will have profound effects on the concentrations calculated from radioactive "counts." In the vast majority of cases, radioligands are obtained commercially, and the specific activity is quoted. The author would not suggest that it is worth the effort to check this routinely. However, it is probably as well to bear in mind that this estimate may not be 100% accurate, especially when the stock bottle has been hanging around for some time, and that any apparent changes in radioligand affinity or B_{max} over time, or with different batches of ligand, may well be attributable to problems with specific activity.

In the case of "homemade" radioligands, you will probably need to try to determine specific activity for yourself, although a method for determining B_{max} that does not require specific activity to be known has been reported *(3)*. Unfortunately, the concentration of radioactive ligand likely to be synthesized is usually too small to be assayed reliably using chemical methods. Furthermore, the use of, for example, mass spectrometry could result in heavy radioactive contamination of the equipment. The most suitable methods for determining specific activity involve the use of immunoassay, if an antibody which binds the ligand is available, or a similar technique using a preparation of the receptor. These methods (outlined in ref. 2) rely on the assumption that labeled and unlabeled versions of the ligand have identical binding characteristics; although this is usually the case, it is not inevitable.

3.2.2. Impure Ligands

Impurity of the radioligand can affect binding assays in several different ways. If the radioligand stock consists of both labeled and unlabeled molecules, and if these pools are differentially contaminated, specific activity may be affected. The use of racemic radioligands, in which both stereoisomers bind to the receptor, but with different affinities, can give rise to apparently biphasic binding kinetics. In cases where either the radioligand or an unlabeled competing ligand is contaminated with a compound that does not bind to the receptor, the free ligand concentration will be overestimated. If an unlabeled impurity competes for binding to the receptor, ligand affinities will be underestimated, because their binding curves will be shifted to the right owing to the presence of the competing ligand.

A major source of a "contaminating" ligand may be the receptor preparation itself, which may well contain substantial concentrations of the receptor's "natural" agonist. Extensive washing of membrane preparations reduces the concentration of any natural ligand remaining in the binding assay. In addition, some ligands can be removed by the addition of enzymes, such as acetylcholine esterase, to remove acetylcholine or adenosine deaminase to breakdown adenosine. The synthesis of prostaglandins can be inhibited by preparing the membranes and performing the assay in the presence of indomethacin.

A special case of contamination of the membrane preparation with endogenous ligand is the occurrence of "locked" agonist binding to G-protein-coupled receptors. In the absence of guanine nucleotides, agonist, receptor, and G-protein appear to remain "locked" together in a complex that dissociates only very slowly. Thus, during the preparation of membranes, agonist is not removed from this site by washing or enzyme treatments. If the subsequent binding assay is carried out in the absence of guanine nucleotide, the agonist will still not dissociate, and the occupied receptors will not be detected. However, the addition of guanine nucleotide disrupts the complex, allowing the agonist to dissociate and the radiolabeled antagonist to bind. This process thus gives rise to an artifactual increase in antagonist B_{max} in the presence of guanine nucleotides.

3.2.3. Ligand Instability

Instability of ligands can be a problem both during storage and during the assay, and leads to an overestimation of the free ligand concentration. It is relatively easy to check the purity of the radioligand from time to time, or at the end of the assay, using thin-layer chromatography (TLC) or high-performance liquid chromatography (HPLC); in the case of radioligands obtained commercially, suitable systems are usually described on the data sheet provided. To minimize breakdown of the radioligand during storage, the manufacturer's instructions should always be followed. However, some instability of radioligands is inevitable; the isotopes decay, and this radioactive decay can lead to lysis of the ligand, since the energy released during decay is absorbed within the sample.

It is much harder to check the stability of unlabeled ligands, unless there is a suitable sensitive method for detecting the ligand and its breakdown products following TLC or HPLC. It is, however, possible to compare samples of the suspect ligand with *bona fide* ligand in a radioreceptor assay to see if they inhibit binding of a radioligand to a preparation containing the receptor to the same extent.

Very unstable ligands are best avoided if at all possible, but in some cases, stability during the assay can be improved. For example, peptidase inhibitors will reduce breakdown of peptide ligands, and the stability of prostacyclin is greatly improved if the assay is carried out at pH 8.5. However, it is worth noting that changing the pH can affect the receptor by altering the ionization of various groups important for binding *(4)*. Sensible precautions to reduce breakdown of ligands include making up all competing ligands just before use, keeping the solutions on ice, and never storing diluted stocks.

3.2.4. Receptor Instability

Instability of the receptors may be a problem while making the receptor preparation and during the binding assay. Any loss of receptor sites will obviously lead to an underestimate of the true B_{max}. Instability during the binding assay may also lead to a premature plateauing of the association time-course, owing to the combination of continuing association with decreasing receptor number.

It is a wise precaution to keep everything cold while processing the receptor preparation. Receptor stability during the assay may be improved by performing the incubation at low temperatures. However, this may not provide an enormous advantage because the time required for equilibrium to be approached will be increased, so a longer incubation will be required. Inhibition of proteases by the inclusion of chelating agents, such as ethylenediamine tetraacetic acid (EDTA) in the buffers, has been shown to be very useful, but it will be necessary to supplement the assay buffer with Mg^{2+} in order to allow the interaction of receptors with G-proteins. A cocktail of specific protease inhibitors *(1)* may also be included if breakdown is particularly troublesome; this may be especially useful during lengthy procedures, such as receptor solubilization and purification.

Ligand binding may stabilize receptors, and the best way to check the stability of the receptor during the binding assay is to see if B_{max} goes down with prolonged preincubation. The receptor preparation should be preincubated for various lengths of time under the assay conditions, but in the absence of ligands, before the binding assay is performed as usual.

3.2.5. Dissociation

The problem of dissociation of radioligand from the receptor during the separation procedure was mentioned in Section 2.6. If the separation is slow in comparison with ligand off-rate, the loss can be considerable, but the extent of the problem is rather hard to calculate. It might be possible to compare the level of binding obtained using the suspect procedure with that obtained using equilibrium dialysis or a centrifugation assay, which does not affect the equilibrium between bound and free ligand to the same extent. Alternatively, it may be possible to prolong the separation procedure and extrapolate back to determine the original level of binding.

3.2.6. Incomplete Separation of Bound and Free Ligand

The separation of bound and free radioligand at the end of the binding assay is not always perfect. B_{max} will be underestimated if some of the bound component is collected with the free radioligand. This can occur when filtering large amounts of protein in a filtration assay, when the capacity of the filters may be exceeded, or during centrifugation of very dilute membrane preparations, when a cohesive pellet may not be formed. Free ligand may also be collected with the bound component owing to entrapment within the pellet in a centrifugation assay or filter binding, for instance. This free ligand may give the appearance of extra nonspecific binding or, if displacement occurs, an additional high-capacity, low-affinity binding site.

3.2.7. Depletion

Depletion occurs when so much drug binds to receptor or nonreceptor sites that the concentration that remains unbound (free) is significantly different from the total concentration added. Thus, the free concentration, which is generally assumed to be the same as the total concentration, is underestimated, which leads to an artifactual steepening of a saturation binding curve. In competition assays, depletion of the radioligand can produce a substantial rightward shift of the binding curve, in addition to the familiar "Cheng-Prusoff" effect *(5)*. This occurs because, as the competitor displaces bound radioligand, the free radioligand concentration is significantly increased, which allows the radioligand to rebind. Furthermore, if the competing ligand recognizes more than one population of sites, depletion will result in an underestimation of the proportion of high-affinity sites *(6)*. Both radioligands and unlabeled ligands in competition assays may be subject to depletion.

In order to minimize the problem of depletion, the receptor concentration in the binding assay should be kept as low as possible. Ideally, depletion of the radioligand should be <10%. If depletion is between 10 and 50%, it can be corrected, but the accuracy of the results of the experiment is reduced *(1)*. If depletion is >50%, the assay is invalid.

Depletion may be owing to binding to the receptor, but also to binding to other components of the receptor-containing preparation, the assay tube, and so forth. If a centrifugation assay is used, the real concentration of free radioligand can be measured very easily by sampling the supernatant. It is much harder to assess the degree of depletion if other separation techniques are used, and the best strategy is probably to carry out a parallel centrifugation assay.

Depletion of the ligands can also occur before the assay. Ligands can be lost during dilution and storage, because they stick to the tubes and pipet tips used. The error is greatest at the lowest ligand concentrations and again produces an artifactual steepening of the binding curve. The problem is fairly trivial for radioligands, because the real concentration in each dilution should be routinely determined by counting a sample. It is much harder to ascertain for unlabeled ligands, but the sort of radioreceptor assay just outlined could be tried. The problem can be reduced in a number of cases by, for example, diluting peptides in a buffer containing albumin and silanizing the tubes.

3.3. Nonequilibrium Binding

Assuming that equilibrium has been achieved in a binding assay when it has not been can have profound effects on the results obtained. Radioligand saturation curves are steepened and tend to be shifted to the right, because the lower concentrations approach equilibrium more slowly than the higher concentrations. The problem is compounded in competition assays, because it takes longer for two ligands competing for the same receptor to reach equilibrium than for either ligand alone. The effect of not reaching equilibrium in a competition assay depends on the relative rates at which the radioligand and the competing ligand bind; if the competitor binds more slowly, its affinity will be underestimated; if it binds more quickly, its affinity will be overestimated *(7)*. This can lead to rather dramatic effects. Preliminary experiments investigating the binding of prostanoid TP receptor antagonists to endothelial cell membranes suggested that the antagonists had a different order of potency when [^{125}I]PTA-OH was used as the radioligand than when [^{3}H]SQ29548 was used. However, following a more prolonged incubation period, the same order of potency was obtained with either radioligand *(8)*.

The period of incubation required for equilibrium to be achieved may be impracticably long for some very high-affinity ligands, which approach equilibrium very slowly. Incubation time may be particularly hard to gage in the case of ligands that influence each other's binding allosterically, because these ligands can profoundly affect each other's binding kinetics in an unpredictable fashion. Gallamine inhibits the binding of [^{3}H]*N*-methylscopolamine to muscarinic receptors via a negative cooperative interaction, but considerably slows the radioligand's binding kinetics *(9)*.

Problems with nonequilibrium binding may also occur if the ligands are able to bind to part of the receptor population faster than the rest. For example, when binding is measured using whole cells, the radioligand will be able to access those receptors on the cell surface fairly rapidly. However, in order for the ligand to bind to any receptors within intracellular compartments, it will have to partition into the cell, and this process may take some time. A similar phenomenon can occur in membrane preparations if some of the receptors are contained with sealed vesicles. In this case, the binding curve will appear flattened, because only the higher ligand concentrations will be starting to equilibrate with the occluded receptor sites. It is worth comparing the results of binding to whole cells with the binding obtained to membranes prepared from these cells and including a hypotonic lysis step in any membrane preparation procedure to try to disrupt any sealed vesicles that may have formed during homogenization.

3.4. Defining Receptor Sites

Defining nonspecific binding has been mentioned in Section 2.3., but how can one be sure that specific binding really represents receptor binding? The binding should fulfill the normal criteria expected for a receptor: being saturable, reversible, stereoselective, and having the same tissue distribution as the receptor under investigation. However, the definition of a receptor depends on function, which cannot be measured in a binding assay. Thus, the identification of any binding site as a receptor must always remain provisional and must rely on a correlation of the potencies of a variety of drugs in the binding assay with their potencies in a functional assay. The discovery of one substance whose potency differs in the two systems must question the validity of the binding data.

Displaceable, nonreceptor binding of radioligands is a fairly common phenomenon. How big a problem this nonreceptor binding presents depends on how easy it is to distinguish from receptor binding in the same system, i.e., the concentration range over which it occurs and the specificity of the nonreceptor binding. For example, the prostanoid IP receptor agonist [^{3}H]iloprost appears to bind to *bona fide* IP receptors producing half-maximal occupancy at concentrations around 10 nM. However, at concentrations >50 nM, [^{3}H]iloprost starts to bind to a high-capacity, low-affinity site from which it is displaced by iloprost and other prostanoids. This site becomes a problem in saturation experiments, since the specific binding of [^{3}H]iloprost does not reach a plateau when all the available IP receptors are occupied. However, the problem can largely be avoided by using self-competition experiments to study iloprost binding; a low (<10 nM) concentration of [^{3}H]iloprost is used to label receptor sites selectively and the ability of unlabeled iloprost to displace this binding is investigated *(10)*.

If nonreceptor binding occurs at similar concentrations to binding to the receptor and if the capacity of the nonreceptor sites is very large in comparison with the *bona fide* receptors, it may render the radioligand useless in that preparation. Furthermore, nonreceptor binding may be rather difficult to distinguish from true receptor binding. The radiolabeled adenosine A$_2$ receptor agonist, [^{3}H]N-ethylcarboxamidoadenosine ([^{3}H]NECA) labels a high density of specific binding sites in NG108-15 cell membranes. This binding is saturable and reversible, and there is a very good correlation between the affinities of five A_2 receptor ligands (three antagonists and two agonists) for inhibition of this specific binding and their functional affinities. However, other agonists known to interact with adenosine A_2 receptors do not displace [^{3}H]NECA binding at all, and the specific binding of [^{3}H]NECA does not represent binding to an A_2 receptor *(11)*. Thus it is important to compare the binding and functional characteristics of as wide a range of drugs as possible to check the identity of a binding site. However, this is not always possible. For many receptor systems, there is only one ligand available, the natural agonist, and in many cases, functional assays may be rather difficult to perform.

3.5. Analyzing and Interpreting the Data

Considerable errors can be introduced by incorrect analysis of the data. The problems associated with the use of the Scatchard plot have been well documented *(12,13)* and with the widespread availability of microcomputers allowing nonlinear regression analysis of the untransformed data, the continued use of Scatchard plots and other linearizations is unjustified.

Even with the best curve-fitting routine, the results are only as good as the original data, and some techniques inevitably incorporate greater errors than others; the data obtained by quantitative autoradiography are never going to be as reliable as those obtained in a membrane binding assay, for instance. The danger of using a computer to analyze the data is that the computer yields a number and it is very tempting to take that number at face value. The best curve-fitting routines provide not only an estimate of the parameters that give the best fit to the data, but also an estimate of the 95% confidence limits of these values—inspection of these limits can be a very sobering experience indeed!

It is important to remember that the results of curve-fitting routine depend on the model used, and each model makes considerable assumptions about the underlying binding process. As the number of variable parameters within the binding models increases, the number of combinations of these values which will fit the data more or less equally well increases enormously. It is therefore necessary to interpret the results of curve fitting with an element of caution and common sense, particularly when the more complex models are involved. The number and spread of the data points must be sufficient to justify the use of any particular model; limited spread of data points may be a particular problem in saturation binding assays, where cost and/or high levels of nonspecific binding may limit the maximum concentration of radioligand that can be used. Complex models, such as a multiple-site binding model, should only be used when the data clearly cannot be adequately described by a more simple model, such as a binding to a single site. It is also useful to examine the effect of constraining the values of various parameters on the values of the other parameters and the "goodness of fit"; this is particularly appropriate when an estimate of some parameter has been obtained independently.

4. Conclusions

Thus, despite its relative simplicity, the techniques of radioligand binding is subject to as wide an array of problems and potential artifacts as any other technique. The design of binding experiments has to reflect a compromise between conflicting demands, such as reducing receptor number to minimize depletion while still retaining sufficient specific binding to enable reliable data to be obtained, or having a radioligand with sufficiently high an affinity that it does not dissociate from the receptor during the separation procedure, but not so high an affinity that equilibrium can be achieved within a reasonable period of time. In the interpretation of binding experiments, it is important to be aware of the potential problems, so that any clues to the presence of artifacts are not missed or wrongly interpreted. Wherever possible, it is a good idea to attempt to verify binding parameters using different techniques to check for internal consistency, by, for example, using two different radioligands to determine B_{max} or determining affinity both by saturation and competition experiments. Never forget that a binding site is not a receptor. As long as these points are borne in mind, radioligand binding assays remain an extremely useful and powerful method for studying receptors.

References

1. Hulme, E. C. (ed.) (1992) *Receptor–Ligand Interactions. A Practical Approach.* IRL Press at Oxford University Press, Oxford.
2. Keen, M. and MacDermot, J. (1993) Analysis of receptors by radioligand binding, in *Receptor Autoradiography: Principles and Practice* (Wharton, J. and Polak, J. M., eds.), Oxford University Press, Oxford, UK, pp. 23–56.
3. Swillens, S. (1992) How to estimate the total receptor concentration when the specific activity of the ligand is unknown. *Trends Pharmacol. Sci.* **13,** 430–434.
4. Birdsall, N. J. M., Chan, S.-C., Eveleigh, P., Hulme, E. C., and Miller, K. W. (1989) The modes of binding of ligands to cardiac muscarinic receptors. *Trends Pharmacol. Sci.* **10(Suppl. Subtypes of Muscarinic Receptors IV),** 31–34.

5. Cheng, Y. C. and Prusoff, W. H. (1973) Relationship between the inhibition constant Ki and the concentration of inhibitor which caused 50% inhibition (I50) of an enzymic reaction. *Biochem. Pharmacol.* **22,** 3099–3108.

6. Wells, J. W., Birdsall, N. J. M., Burgen, A. S. V., and Hulme, E. C. (1980) Competitive binding studies with multiple sites: effects arising from depletion of free radioligand. *Biochim. Biophys. Acta* **632,** 464–469.

7. Motulsky, H. J. and Mahan, L. C. (1984) The kinetics of competitive radioligand binding predicted by the law of mass action. *Mol. Pharmacol.* **25,** 1–9.

8. Hunt, J. A., Merritt, J. E., MacDermot, J., and Keen, M. (1992) Characterization of the thromboxane receptor mediating prostacyclin release from cultured endothelial cells. *Biochem. Pharmacol.* **43,** 1747–1752.

9. Stockton, J. M., Birdsall, N. J. M., Burgen, A. S. V., and Hulme, E. C. (1983) Modification of the binding properties of muscarinic receptors by gallamine. *Mol. Pharmacol.* **23,** 551–557.

10. Krane, A., MacDermot, J., and Keen, M. (1994) Desensitization of adenylate cyclase responses following exposure to IP prostanoid receptor agonists: homologous and heterologous desensitization exhibit the same timecourse. *Biochem. Pharmacol.* **47,** 953–959.

11. Keen, M., Kelly, E. P., Nobbs, P., and MacDermot, J. (1989) A selective binding site for [^{3}H]-NECA that is not an adenosine A_2 receptor. *Biochem. Pharmacol.* **38,** 3827–3833.

12. Klotz, I. M. (1982) Numbers of receptor sites from Scatchard graphs: facts and factasies. *Science* **217,** 1247–1249.

13. Burgisser, E. (1984) Radioligand-receptor binding studies: what's wrong with the Scatchard analysis? *Trends Pharmacol. Sci.* **5,** 142–144.

CHAPTER 2

Solubilization and Purification of a Functional Ionotropic Excitatory Amino Acid Receptor

Jeremy M. Henley and David M. Kirkham

1. Introduction

Excitatory amino acid receptors are the predominant type of neurotransmitter receptor in the vertebrate central nervous system (CNS). They are membrane-spanning proteins that mediate the stimulatory actions of glutamate and possibly other related endogenous amino acids. Excitatory amino acid receptors are crucial for fast excitatory neurotransmission; they are believed to be involved in the neuropathology of ischemic damage; and they have been implicated in many disease states, including Alzheimer's disease and epilepsy (1,2).

Despite the cloning of numerous cDNAs encoding ionotropic glutamate receptor subunits (GluRs), the isolation and purification of the hetero-oligomeric native receptor complexes are still of undoubted value. Indeed, the availability of recombinant receptor assemblies in a variety of expression systems has not, and probably will not, successfully address crucial questions about the structures and hetero-oligomeric assemblies of these receptors as they occur in vivo. Such questions include aspects of posttranslational modification and receptor regulation by specific phosphorylation and/or glycosylation events.

Based on their pharmacological and physiological profiles, two main families of excitatory amino acid receptor have been classified: (1) the metabotropic receptors, which are G-protein-linked receptors associated with the stimulation of inositol phospholipid and/or cyclic nucleotide metabolism; and (2) the ionotropic receptors, which are membrane-span-

From: *Methods in Molecular Biology, Vol. 41: Signal Transduction Protocols*
Edited by: D. A. Kendall and S. J. Hill Copyright © 1995 Humana Press Inc., Totowa, NJ

ning ligand-gated channels. The latter family comprises the NMDA receptors and non-N-methyl-D-aspartate (NMDA) or kainate and α-amino-3-hydroxy-5-methyl-4-isoxazolepropionate (AMPA) types *(3,4)*. We describe the solubilization and purification of an example of an ionotropic glutamate receptor from *Xenopus* brain and spinal cord.

Successful solubilization protocols for [^{3}H]kainate binding sites include those from chicken cerebellum with Triton X-100 *(5)*, from rat and frog brain with Triton X-100/digitonin *(6,7)*, and from goldfish and *Xenopus* CNS with octylglucoside *(8,9)*. Among the lower vertebrate systems investigated, *Xenopus* brain has been shown to be an exceptionally rich source of both kainate and AMPA binding sites, and the relationship between these sites has been studied in detail *(10)*. The functional interaction between kainate and AMPA was correlated with a physical co-localization of the two types of sites in a single protein complex. The two sites coexist in a 1:1 ratio, and cannot be separated by physical or by chemical fractionations *(11)*. Furthermore, purification of the binding sites to apparent homogeneity by domoic acid affinity chromatography has yielded a unitary kainate/AMPA receptor. To our knowledge, this represents the only functional excitatory amino acid receptor/channel protein to have been completely purified from any source *(12)*.

Sodium dodecyl sulfate-polyacrylamide gel electrophoresis (SDS-PAGE) analysis of the specific eluent from domoate columns revealed a major band with an M_r value of $\approx 42,000$ and an additional band $\approx 100,000$ Dalton in some preparations. The pure binding protein shows high-affinity binding for both AMPA and kainate, and they are mutually and fully competitive with K_i values identical to the K_d values for the radioligand (AMPA 34 nM; kainate 15 nM) *(12)*.

Patch-clamp electrophysiological analysis of this pure protein after reconstitution into lipid bilayers has demonstrated that cation channels (with conductance states and lifetimes similar to those for non-NMDA receptors on neurons) can be opened by low levels of AMPA or kainate *(13)*. Furthermore, under these conditions, AMPA can act as an antagonist of kainate-induced conductance, indicating that both ligands bind to a single site. This purified and reconstituted protein complex shows additional interesting properties in that the same protein complex appears also to form an NMDA-activatable ligand-gated ion channel. It has been suggested that the NMDA component of the receptor is conferred by the presence of the $\approx 100,000$-Dalton polypeptide.

Thus, the ionotropic glutamate receptor from *Xenopus* CNS has been proposed as a single multi-subunit protein complex to which kainate or AMPA or NMDA can bind and open the same channel. The receptor may then be fixed in a conformation appropriate to the activating ligand. Agents interacting with the agonist binding site need not necessarily act in the same way for the different conformations. Indeed, an established parallel to this proposal is the difference in effects of different agonists acting on one site in the nicotinic acetylcholine receptor *(14)*.

2. Materials

2.1. Preparation of **Xenopus** *CNS Membranes*

1. Tris buffer with protease inhibitors (*see* Note 1): 50 mM Tris citrate, pH 7.4, at 0°C, 1 mM EGTA, 1 mM EDTA 0.015% (w/v) benzamidine, 0.001% (w/v) soybean trypsin inhibitor, 0.02% (w/v) bacitracin, 1 mM phenylmethylsulfonyl fluoride, and 0.4 κIU/mL aprotinin.
2. Homogenization buffer: Tris buffer with protease inhibitors containing 320 mM sucrose.

2.2. Receptor Solubilization **(see Note 3)**

1. 10% *N*-octyl-β-glucopyranoside in Tris buffer (50 mM Tris-citrate, pH 7.4, 0°C).
2. Solubilization buffer: 50 mM Tris-citrate, pH 7.4, at 0°C, 1 mM EDTA, 1 mM EGTA, 0.015% (w/v) benzamidine, 0.02% (w/v) bacitracin, 1 mM phenylmethylsulfonyl fluoride, and 1% *N*-octyl-β-glucopyranoside, but without soybean trypsin inhibitor and aprotinin.

2.3. Preparation of Domoate-Affinity Column

1. 3.5 g AH Sepharose 4B (powdered gel) or 10 mL EAH Sepharose 4B (preswollen gel).
2. 1 L 0.5M NaCl.
3. 10 mg Domoic acid (*see* Note 1).
4. 10 mg EDAC (1-ethyl-3-[3 dimethylaminopropyl] carbodiimide).
5. 1 L Bicarbonate buffer (0.8M NaCl and 0.1M NaHCO$_3$, pH 8.3).
6. 1 L Acetate buffer (0.8M NaCl and 0.1M Na acetate, pH 4.0).
7. 1 L Tris buffer plus EDTA and EGTA (50 mM Tris-citrate, pH 7.4, at 0°C, 1 mM EDTA, and 1 mM EGTA).
8. 1 mL 4% (w/v) Na azide.

2.4. Affinity Purification

1. Wash buffer 1: Solubilization buffer, but containing 0.8% *N*-octyl-β-glucopyranoside and 0.1M NaCl.

2. Wash buffer 2: Solubilization buffer, but containing 0.8% N-octyl-β-glucopyranoside and 0.25M NaCl.
3. Wash buffer 3: Solubilization buffer, but containing 0.8% N-octyl-β-glucopyranoside.
4. Elution buffer: Solubilization buffer, but containing 0.8% N-octyl-β-glucopyranoside and 100 μM kainate.

2.5. Regeneration of Domoate Affinity Column

1. 1M NaCl.
2. 6M Urea.
3. 50 mM Tris-citrate buffer, pH 7.4, at 0°C.
4. 4% (w/v) Na azide.

2.6. Reconstitution of Pure Receptor

1. 90 mg Azolectin.
2. 10 mg Cholesterol.
3. 1 mL Chloroform.
4. Dialysis membrane.
5. Dialysis buffer: 10 mM Tris citrate, pH 7.0, and 100 mM NaCl.

3. Methods (*see* Note 7)

3.1. Preparation of **Xenopus** *CNS Membranes*

1. Place *Xenopus* brains and spinal cords into a precooled glass homogenizer in an ice-filled container.
2. Add homogenization buffer to approx 1 mL/brain.
3. Homogenize on ice in a glass/Teflon™ homogenizer with six passes.
4. Transfer to 35-mL polycarbonate centrifuge tubes, and add homogenization buffer to fill tubes.
5. Spin at 4°C for 10 min at 1000g.
6. Collect and save supernatant, and discard the pellet (cell debris and nuclei).
7. Transfer supernatant to clean 35-mL polycarbonate centrifuge tubes, and spin at 4°C for 30 min at 48,000g.
8. Discard the resultant supernatant.
9. Resuspend this membrane pellet by vortexing in ≈1 vol of Tris buffer with protease inhibitors (no sucrose).
10. Rapidly freeze suspension in polycarbonate centrifuge tube in liquid N_2, dry ice, or in the –80°C freezer.
11. Allow to thaw either at room temperature or by placing tube in tepid water.
12. Fill the tube with Tris buffer with protease inhibitors, and mix and spin for 20 min at 48,000g.
13. Repeat freeze–thaw–wash steps twice more (three washes total) (*see* Note 4).

14. Resuspend final pellet in appropriate volume of Tris buffer with protease inhibitors, but without soybean trypsin inhibitor and aprotinin. Aliquot into labeled/dated microcentrifuge tubes, and store in –80°C freezer.
15. Determine protein concentration of final membrane preparation. (We use the Bio-Rad [Richmond, CA] protein assay based on the Bradford dye-binding method.)

3.2. Receptor Solubilization

1. Spin down membranes, and resuspend in solubilization buffer to give a final protein concentration of 0.3 mg/mL.
2. Incubate at 4°C for 60–90 min with gentle agitation on rotating wheel.
3. Spin the detergent extract at 100,000g for 1 h at 4°C.
4. By definition, the supernatant contains the soluble protein.

3.3. Preparation of Domoate Affinity Column

1. If AH Sepharose 4B is used, swell 3.5 g in 50 mL 0.5M NaCl for 2 h at room temperature.
2. Wash swollen gel on scintered glass filter with 250 mL 0.5M NaCl and then with 250 mL distilled water. Approximately 10 mL of caked gel are recovered.
3. Dissolve 10 mg of domoic acid in 7 mL distilled water, and add to 7.5 mL swollen gel (either EAH or AH Sepharose 4B).
4. Add 10 mg EDAC, and adjust to pH 6.1 with 1M NaOH ($\approx$100 μL).
5. Incubate gel slurry with gentle agitation on rotating wheel for 12–24 h at 4°C.
6. Wash gel on scintered glass filter with 100 mL bicarbonate buffer followed by 100 mL acetate buffer.
7. Repeat the bicarbonate buffer/acetate buffer wash cycle three times.
8. Wash with 500 mL Tris buffer.
9. For control gel, follow exactly the same coupling/blocking procedure, except omit step 3, i.e., do not add the domoic acid.
10. Pour 2 mL of the gel to form a column, and store gels in Tris buffer with 0.2% Na azide at 4°C.

3.4. Affinity Purification (see Note 6)

1. Apply 8 mL soluble receptor extract to washed domoate column.
2. Recirculate at a rate of 0.3–0.7 mL/min overnight (or at least 5 h).
3. Collect and save run off.
4. Wash 1: wash column ($\approx$0.7 mL/min) with 25 mL wash buffer 1—collect wash.
5. Wash 2: wash column ($\approx$0.7 mL/min) with 50 mL wash buffer 2—collect wash.
6. Wash 3: wash column ($\approx$0.7 mL/min) with 15 mL wash buffer 3—collect wash.
7. Elution 1: Incubate gel with 5 mL 100 μM kainate in Tris buffer with protease inhibitors containing 0.8% N-octyl-β-glucopyranoside at room temperature with vigorous shaking for 15 min—collect eluent.

8. Elution 2: Repeat elution with 5 mL 100 μ*M* kainate in Tris buffer with protease inhibitors containing 0.8% *N*-octyl-β-glucopyranoside—collect eluent, and keep eluents 1 and 2 separate.
9. Concentrate each eluent from 5 mL to ≤ 2 mL using an ultrafiltration membrane (e.g., Flowgen or Centricon filters).
10. Remove kainate from "pure receptor" using a commercial desalting column or a washed, precalibrated G-25 column.
11. Collect the void volume from the column on ice. This fraction will contain the pure receptor (*see* Note 5).
12. The specific activity and homogeneity of the purified fraction should be analyzed by both radioligand binding and SDS-PAGE.

3.5. Regeneration of Domoate Gel

1. Wash 1: 10 mL 1*M* NaCl at ≈0.7 mL/min.
2. Wash 2: 10 mL 6*M* urea at ≈0.7 mL/min.
3. Wash 3: 100 mL Tris buffer at ≈0.7 mL/min.
4. Store in Tris buffer with 0.2% Na azide at 4°C.
5. The column should be viable for at least 20 purifications.

3.6. Reconstitution of Pure Receptor

1. Dissolve azolectin and cholesterol (9:1) in chloroform to give 10% solution.
2. Evenly coat inside of a glass tube with lipids by swirling and allowing chloroform to evaporate.
3. A final lipid concentration of 0.025% in detergent/protein buffer is required. Thus, if 4 mL of pure receptor solution are to be reconstituted, 10 μL of 10% azolectin/cholesterol (9:1) in chloroform should be used to coat the tube.
4. Add soluble pure receptor to tube, and sonicate for 1 min.
5. Dialyze resultant mixture against at least five changes of dialysis buffer for 48 h to remove *N*-octyl-β-glucopyranoside.
6. Collect dialysate, aliquot, freeze in liquid nitrogen, and store at –80°C.
7. Subsequent uses include insertion into black lipid membranes at the tip of a patch pipet and the effect of lipids on the binding parameters of the receptor *(12)*.

4. Notes

The protocols given in this chapter are those used in our laboratory. However, it should be borne in mind that the techniques involved are far from trivial. Protein purification is often an involved and frustrating business. The following notes may be of some interest.

1. The ligand immobilized on the affinity column is domoic acid. This compound is purchased as the natural, purified toxin isolated from infected shellfish in Canada. As such, domoic acid is prone to differences between batches. It is advisable, therefore, to test individual batches in competition binding experiments using [³H]kainate as the radioligand.

2. A difficulty often encountered with protein purification is the partial proteolysis of labile proteins. In the procedures outlined here, great care has been taken to avoid this problem as far as possible. Thus, we include a broad spectrum of protease inhibitors, and probably more importantly, we chelate calcium, which is crucial for the activity of many proteases. In addition, we carry out as much of the manipulation at 0°C or 4°C as possible to restrict proteolytic enzyme activity.

3. The solubilization of the *Xenopus* AMPA receptor is optimal at 0.3 mg/mL protein with an *N*-octyl-β-glucopyranoside concentration of 1.0% as stated in the text. At this protein:detergent ratio, 80–90% solubilization of the binding activity is achieved. This percentage is altered dramatically by changing this ratio; an excess of protein will not be solubilized, and too much detergent will result in the loss of binding activity. The solubilization stage is highly dependent on the detergent used, for example, Triton X100 has been shown to result in the loss of kainate binding activity from *Xenopus* CNS membranes *(10)*. Solubilization will also be dependent on the biological membrane in which the receptor is found, since the presence and proportion of the lipid components of the membrane will affect solubilization.

4. It is important when using excitatory amino acid receptors to remove the endogenous glutamate. If this is not achieved, the information from binding assays will not be reliable. In the case given here, we have employed repeated freeze–thaw washing steps to release endogenous glutamate from the preparation.

5. Purification protocols are normally followed by a simple binding assay. We routinely use [³H]kainate to follow the purification and the unitary *Xenopus* AMPA/kainate receptor. In addition, we use SDS-PAGE followed by silver staining to demonstrate the purity of the receptor solution that is given.

6. For the protocol outlined here, the percentage of recovery is approx 10% of the initial binding activity. Although the minimum number of steps involved is few, considerable losses are inevitable and 10% represents a high recovery for protein purification. In our hands, 15% of the initial binding activity is found in the first NaCl wash of the column and 25–30% in the second NaCl wash. The remaining 50–55% of the binding activity is not recovered from the domoic acid column and may represent denaturation of the protein.

7. Generally, the whole protocol is completed in 3 d. The membranes are produced on the first day, and the solubilization is completed on the second, with the soluble receptor preparation being introduced onto the column overnight. The column washes and recovery of the pure protein from the column can be completed on the third day.

References

1. Barnes, J. M. and Henley, J. M. (1992) Molecular characteristics of excitatory amino acid receptors. *Prog. Neurobiol.* **39,** 113–133.
2. Choi, D. W. (1988) Glutamate neurotoxicity and diseases of the nervous system. *Neuron* **1,** 623–634.
3. Sommer, B. and Seeburg, P. H. (1992) Glutamate receptor channels—novel properties and new clones. *Trends Pharmacol. Sci.* **13,** 291–296.
4. Collingridge, G. L. and Lester, R. A. J. (1989) Excitatory amino acid receptors in the vertebrate central nervous system. *Pharmacol. Rev.* **40,** 143–208.
5. Gregor, P., Eshhar, N., Ortega, A., and Teichberg, V. I. (1988) Isolation, immunochemical characterisation and localisation of the kainate subclass of glutamate receptor from chick cerebellum. *EMBO J.* **7,** 2673–2679.
6. Hampson, D. R. and Wenthold, R. J. (1988) A kainic acid receptor from frog brain purified using domoic acid affinity chromatography. *J. Biol. Chem.* **263,** 2500–2505.
7. Hampson, D. R., Huie, D., and Wenthold, R. J. (1987) Solubilisation of kainic acid binding sites from rat brain. *J. Neurochem.* **49,** 1209–1215.
8. Henley, J. M. and Barnard, E. A. (1989) Solubilisation and characterisation of a putative quisqualate-type glutamate receptor from chick brain. *J. Neurochem.* **53,** 140–148.
9. Henley, J. M. and Oswald, R. E. (1988) Solubilisation and characterisation of kainate binding sites from goldfish brain. *Biochem. Biophys. Acta.* **937,** 102–111.
10. Henley, J. M. and Barnard, E. A. (1989) Kainate receptors in *Xenopus* central nervous system: solubilisation with *n*-octyl-β-D-glucopyranoside. *J. Neurochem.* **52,** 31–37.
11. Henley, J. M., Ambrosini, A., Krogsgaard-Larsen, P., and Barnard, E. A. (1989) Evidence for a single glutamate receptor of the ionotropic kainate/quisqualate type. *New Biologist.* **1,** 153-158.
12. Henley, J. M., Ambrosini, A., Rodriguez-Ithurralde, D., Sudan, H., Brackley, P., Kerry, C., et al. (1992) Purified unitary kainate/ alpha-amino-3-hydroxy-5-methylisooxazole-propionate (AMPA) and kainate/AMPA/*N*-methyl-D-aspartate receptors with interchangeable subunits. *Proc. Natl. Acad. Sci. USA* **89,** 4806–4810.
13. Barnard, E. A., Ambrosini, A., Sudan, H., Prestipino, G., Lu, Q., Rodriguez-Ithurralde, D. (1991) Purification and properties of a functional unitary non-NMDA receptor from *Xenopus* brain, in *Excitatory Amino Acids* (Meldrum, B. S., Moroni, F., Simon, R. P., and Woods, J. H., eds.), Raven, New York, pp. 135–143.
14. Changeux, J. P. (1991) The nicotinic acetylcholine receptor: an allosteric protein prototype of ligand-gated ion channels. *Trends Pharmacol. Sci.* **11,** 485–491.

CHAPTER 3

Autoradiography
of β_1- and β_2-Adrenoceptors

Roger J. Summers and Peter Molenaar

1. Introduction

Receptor autoradiography is derived from receptor binding techniques that were originally developed for use in homogenate membranes or isolated cells. The use of tissue sections has the advantage of retaining anatomical information that can provide useful insights into cell signaling systems involved in neurotransmitter receptor interactions. Like homogenate membrane or whole-cell binding techniques, receptor autoradiography can also be used to derive kinetic, saturation, competition binding, and receptor density data. This chapter describes methods for receptor autoradiography using specific ligands that can be used to localize, characterize, and quantitate β_1- and β_2-adrenoceptors in anatomically defined regions.

The major ligands used for autoradiographic localization of β_1- and β_2-adrenoceptor subtypes are ^{125}I-labeled cyanopindolol $((-)[^{125}$I]cyanopindolol [CYP]), [^{3}H]dihydroalprenolol, and $(-)[^{125}$I]pindolol. Of these, $(-)[^{125}$I]CYP has been the most widely used because of its high affinity, ease of preparation, and high specific activity (1). Delineation of β-adrenoceptor subtypes is achieved using this ligand in conjunction with highly selective compounds, such as CGP 20712A (β_1-adrenoceptor selective) or ICI 118,551 (β_2-adrenoceptor selective) at appropriate concentrations to inhibit binding to each of the β-adrenoceptor subtypes. Although selective ligands are available to label β_1- and β_2-adrenoceptor subtypes, such as [^{3}H]bisoprolol and [^{3}H]ICI 118,551, these are not recommended for use in autoradiography because of their relatively low

From: *Methods in Molecular Biology, Vol. 41: Signal Transduction Protocols*
Edited by: D. A. Kendall and S. J. Hill Copyright © 1995 Humana Press Inc., Totowa, NJ

affinity, low specific activity, and the low energy emission of [³H]. These properties lead to rather poor discrimination between receptors and also unacceptably long exposure times for autoradiographic studies.

2. Materials

2.1. Perfusion Solutions

1. Krebs phosphate: 118.4 mM NaCl (6.9 g/L), 10 mM NaH$_2$PO$_4$·2H$_2$O (1.56 g/L), 4.7 mM KCl (0.35 g/L), 1.2 mM MgSO$_4$ 7H$_2$O (0.29 g/L), and 1.3 mM CaCl$_2$ (0.19 g/L), adjust pH to 7.2 with 5–6M NaOH.
2. Krebs phosphate with sucrose: 1:1 dilution of Krebs phosphate and 0.32M sucrose (0.32M sucrose = 109 g sucrose/L water).
3. 10% Paraformaldehyde: 10 g formaldehyde and 100 mL of distilled water. Stir and heat (<60°C), pH to 7.4. Use 1 mL of concentrate/100 mL of perfusion solution.

2.2. Preparation of Subbed Slides

1. Cleaning slides: Add 120 mL Decon 90–880 mL of dH$_2$O, soak slides, then rinse copiously with dH$_2$O, and allow to dry.
2. Subbing solution: Add 2 g gelatin and 0.2 g chromic potassium sulfate to 400 mL dH$_2$O at room temperature, heat, and stir to just below boiling. Remove from heat when gelatin has dissolved completely, and cool before use. Dip slides and allow to dry in Peelaway® holders in a dust-free area.

2.3. Staining Techniques

1. Hematoxylin and eosin: Mayer's hematoxylin = 50 g potassium alum sulfate, 2 g hematoxylin, 0.2 g sodium iodate, 1 L of water, 1 g citric acid, 3–7 drops acetic acid, and 50 g chloral hydrate. Dissolve the hematoxylin in dH$_2$O at 30°C for 30 min. Add sodium iodate and alum, shaking at intervals. Leave overnight to dissolve, and dissolve citric acid, acetic acid, and chloral hydrate the following day. Boil for 3 min (use within 3 mo).
2. Scots tap water: 1000 mL dH$_2$O, 3.5 g sodium bicarbonate, and 20 g magnesium sulfate.
3. Eosin stock solution: 1 g eosin in 100 mL distilled water. For working solution, dilute 10 mL of stock solution in 90 mL of distilled water with 2 g of calcium chloride.
4. Staining procedure (for rat/guinea pig cardiac frozen tissue sections, 10-μm thickness): Stain in Mayer's hematoxylin for 2 min, rinse in tap water, blue in Scott's tap water for 1 min, rinse in tap water, stain in eosin working solution for 2 min, and rinse in tap water. Rinse in 70% alcohol and then in absolute alcohol. Wash in absolute alcohol for 2 x 1 min, then in xylene for 2 x 1 min, and mount in Depex.

2.4. Radioiodination

Stock solutions (must be made up fresh on day of iodination): 40 μL of 10 mM (–)CYP (14.3 mg/5 mL N/100 HCl), 40 μL 0.3M KH$_2$PO$_4$, pH 7.6 (408.3 mg/10 mL), 40 μL (4 mCi of sodium ^{125}I), 40 μL of aqueous chloramine T (0.34 mg/mL), 600 μL of Na$_2$S$_2$O$_5$ (1 mg/mL), 20 μL 1M NaOH (20 g/500mL), 300 μL ethyl acetate containing 0.01% phenol.

2.5. Incubation Solutions

1. Preincubation solution: a modified Krebs phosphate containing GTP 0.1 mM (8.5 mg), phenylmethylsulfonyl fluoride (PMSF) 10 μM (15 μL of 100-mM solution), and 150 mL of Krebs phosphate.
2. Incubation solution: a modified Krebs phosphate (*see* Section 2.1., item 1) containing PMSF 10 μM (15 μL of 100 mM solution) and (–)[^{125}I]CYP (50 pM).

2.6. Preparation of Emulsion-Coated Coverslips

1. Cleaning of coverslips: Wash coverslips in 12% Decon 90, rinse with dH$_2$O, dry in an oven at 150°C for 1 h or until dry, and store at room temperature in sealed boxes.
2. Coating with emulsion: In the dark, melt one pack of Kodak NTB 3 or Amersham LM-1 in a 40°C water bath for 1 h, pour emulsion into a container (Coplin jar), and if NTB 3 is used, add an equal volume of distilled water. Test the diluted emulsion with a coverslip to ensure that no bubbles are present. When bubble-free, immerse coverslips in solution once to obtain an even coating, hang up to dry overnight in the dark, and store in light-tight containers with silica gel at 4°C.

3. Methods

The autoradiographic studies fall into three categories: biochemical studies, which are performed to examine the integrity of the binding of the ligands for the receptor of interest; qualitative autoradiography to determine the sites of localization; and quantitative autoradiography to determine the density of receptors in anatomically defined areas. Qualitative autoradiography can be carried out at several levels of resolution from the standard technique using 10-μm tissue sections to high-resolution light and electron microscopic (EM) autoradiography. The tissue preparation and conditions used for these techniques vary somewhat, and the differences are indicated in the following sections.

3.1. Preparation of Tissues—Standard Method

1. Anesthetize the animals (40 mg/kg pentobarbitone sodium ip), and open the chest.
2. For tissues other than the heart, place a cannula in the left ventricle of the heart, excise the right atrial appendage, and perfuse the animal with equal parts 0.32M sucrose/Krebs buffer and then with the same solution containing 0.1% formaldehyde.
3. In experiments where the heart is the tissue of interest, cannulate the aorta, excise the heart, and perfuse retrogradely.
4. Dissect the tissues, trim to obtain blocks containing the areas of interest, mount in Tissue Tek™ (Miles, Elkhart, IN), and either use immediately or store at –80°C.

3.2. Preparation of Tissues—High-Resolution Studies

1. Rapidly remove tissue from animal, perfuse with 0.32M sucrose/Krebs buffer, and then with the same solution containing 0.1% glutaraldehyde/formaldehyde (1:1). If several tissues are required, perfuse entire animal.
2. Small blocks (1 × 1 × 2 mm) of tissue containing the area of interest are prepared using a tissue slicer and tissue chopper (McIlwains').
3. Incubate blocks with (–)[^{125}I]CYP for 3 h at 37°C.
4. Remove and fix with 4% glutaraldehyde/formaldehyde (1:1) in 0.32M sucrose/Krebs buffer (3 h, 4°C).
5. Wash with 0.32M sucrose/Krebs buffer (12 h, 4°C).
6. Postfix with 2.5% osmium tetroxide (1 h, 20°C).
7. Wash with 0.32M sucrose/Krebs buffer (3 x 15 min, 20°C).
8. Dehydrate with increasing concentrations of acetone from 10% to absolute dry acetone.
9. Infiltrate blocks with araldite/absolute dry acetone (1:1) for 30 min. Change araldite after 1, 2, and 12 h, and then bake at 55°C for 48 h.

3.3. Cutting of Sections

1. Section tissues on a cryostat at –20.5°C. Thaw mount onto gelatin chromic potassium sulfate-coated microscope slides.
2. If required, store the remaining tissue blocks at –80°C taking care to cover the exposed surface of the sections with frozen Tissue Tek.
3. Cut sections can be stored at –80°C for a few days in sealed slide boxes.
4. For high-resolution studies, cut semithin sections (0.7 µm) on a microtome with a glass knife and mount on microscope slides.
5. For EM autoradiographic studies, cut ultrathin sections (0.1 µm) on a microtome with a glass or diamond knife, collect sections on a nickel grid, and place on a glass microscope slide.

3.4. Preparation of Radioligand

1. Add in the following order: 40 µL of 10 mM (–)CYP, 40 µL 0.3M KH$_2$PO$_4$, pH 7.6, 40 µL (4 mCi of sodium ^{125}I), and 40 µL of aqueous chloramine T (0.34 mg/mL). Mix well and allow the reaction to proceed for 10 min.
2. Stop the reaction by adding 600 µL of Na$_2$S$_2$O$_5$ (1 mg/mL) and 20 µL 1M NaOH (20 g/500 mL).
3. Wash with 300 µL ethylacetate containing 0.01% phenol, mix well, and collect the top layer. Spot onto Whatman 3MM paper.
4. Chromatograph in a descending manner at room temperature with 0.1M ammonium formate pH 8.5, containing 0.01% phenol.
5. Cut paper into 1-cm horizontal strips, place each strip into 5 mL methanol, take a 20-µL sample, dilute 1:500 with water, and count a 20-µL sample in a γ counter, R_f (–)[^{125}I]CYP = 0.2, R_f (–)CYP = 0.7.

3.5. Preincubation and Labeling

1. Slide-mounted sections are slowly brought to room temperature.
2. Sections or tissue blocks are preincubated for 30 min in Krebs buffer containing PMSF and GTP.
3. Preincubated sections/blocks are transferred to Krebs buffer containing 10 µM PMSF with (–)[^{125}I]CYP (50 pM) (*see* Notes 3 and 4).
4. To delineate subtypes, sections are labeled in the absence or presence of the β_2-adrenoceptor-selective antagonist ICI 118,551 (70 nM), the β_1-adrenoceptor-selective antagonist CGP 20712A (100 nM), or (–)propranolol (1 µM) to define specific binding (*see* Notes 5 and 6).
5. Incubation takes place for 150 min (sections) or 3 h (blocks) at 25°C.
6. At the end of the incubation period, the labeled tissues are rinsed in buffer, followed by 2 x 15 min (sections), or 4 x 15 min (blocks) washes at 37°C in the same medium, and finally rinsed in distilled water at room temperature.
7. The sections are then dried in a stream of dehumidified air and stored at 4°C in sealed boxes containing silica gel. Labeled tissue blocks are processed as per Section 3.2. Note that for the high-resolution technique, the tissue is fixed and sectioned after labeling and not before as in the standard technique.

3.6. Biochemical Studies

Biochemical studies of the characteristics of binding of radioligands in slide-mounted tissue sections or in tissue blocks are an essential step that should be taken before receptor autoradiography to establish the integrity of the binding *(2–4)*. They should include an examination of the kinetics of binding, its saturation properties, and ways of determining nonspecific binding. Competition studies should also be carried out with a variety of competitors known to compete for the β-adrenoceptor. Stereoselectivity should be established using the stereoisomers of, e.g., propranolol. Wash conditions need to be established to optimize the ratio of specific to nonspecific binding, while losing little if any specific binding.

1. The rate of association of binding is studied using 50 pM (–)[^{125}I]CYP for 0, 30, 60, 90, 120, 150, and 180 min.
2. Dissociation is studied by incubating with (–)[^{125}I]CYP (50 pM) for 150 min at 25°C, then adding 1 µM (–)propranolol, measuring binding at 0, 15, 30, and 60 min, and then at hourly intervals up to 480 min.
3. Saturation properties of the binding are examined in slide-mounted sections incubated with 5, 10, 20, 30, 40, 60, 80, and 120 pM (–)[^{125}I]CYP using 1 µm (–)propranolol to determine nonspecific binding (*see* Note 1).
4. Competition studies to determine the characteristics of binding are conducted using (–)[^{125}I]CYP (50 pM) and appropriate concentrations of competitors and stereoselectivity using the stereoisomers of propranolol at concentrations of between 10^{-9} and $10^{-6}M$ for (–)propranolol and 10^{-7} to $10^{-4}M$ for (+) propranolol (*see* Notes 2 and 7).
5. Delineation of β-adrenoceptor subtypes is performed using the β$_1$-adrenoceptor-selective antagonist CGP 20712A (20 concentrations from 50 pM to 0.5 mM) and the β$_2$-adrenoceptor-selective antagonist ICI 118,551 (17 concentrations, 50 pM to 20 µM) (*2–4*).
6. Binding conditions are 25°C, and Krebs phosphate buffer incubation medium, pH 7.4.
7. Wash conditions are rinse in buffer at 25°C, followed by 2 x 15 min washes at 37°C, followed by a brief rinse in dH$_2$O at 25°C.

3.7. Qualitative Autoradiography—Standard Method

1. Qualitative autoradiography is performed using the coverslipping technique (*2–7*).
2. Slide-mounted tissue sections are incubated with (–)[^{125}I]CYP in the absence or presence of ICI 118,551 (70 nM), CGP 20712A (100 nM), or (–)propranolol (1 µM) to define nonspecific binding.
3. For cardiac tissues, emulsion-coated coverslips (Kodak NTB 3) are apposed to the labeled tissues for 2 d.
4. After exposure, coverslips are developed in Kodak Dektol, rinsed briefly in water, and fixed in Kodak Rapid Fix at the paper dilution.
5. Sections are stained with pyronine Y, dried, and mounted in Depex for light microscopy.
6. The stained sections can be compared with the serial sections stained with hematoxylin and eosin during the initial sectioning procedure.

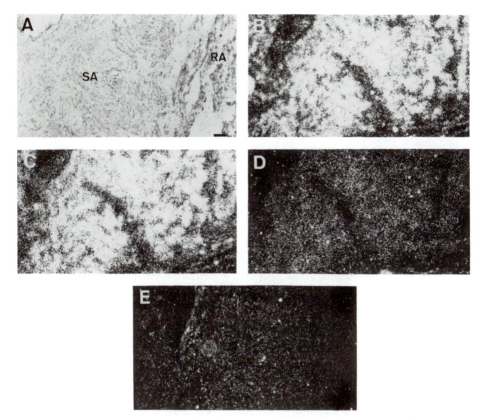

Fig. 1. Qualitative receptor autoradiography—localization of β-adrenoceptor subtypes in the human sino-atrial (SA) node and right atrium (RA)
using the coverslipping technique. Sections were labeled with (–)[^{125}I]CYP
(56 pM) in the absence or presence of ICI 118,551 (β_2-adrenoceptor antagonist, 70 nM); CGP 20712A (β_1-adrenoceptor antagonist, 100 nM); or with
propranolol (nonselective β-adrenoceptor antagonist, 1 μM), and apposed to
nuclear emulsion coated (Kodak NTB 3) coverslips. (**A**) Under bright-field
illumination, a hematoxylin-and-eosin-stained section of the SA node and
surrounding RA, bar represents 120 μm; (**B**) under dark-field illumination,
the location of the total β-adrenoceptor population; (**C**) a high density of β_1-
adrenoceptors in both SA node and RA; (**D**) β_2-adrenoceptors over SA node
and RA; and (**E**) nonspecific binding.

7. Location of receptors is determined using a combination of light- and dark-
   field microscopy with an Olympus (Tokyo, Japan) BH2 microscope as
   illustrated in Fig. 1 for human sino-atrial node.

3.8. Qualitative Autoradiography—High Resolution

1. For high-resolution light autoradiography, the semithin sections (*see* Section 3.3.) contain the radiolabel. Dip slides in nuclear emulsion (Kodak NTB 2 nuclear emulsion or Amersham equivalent).
2. Develop and fix after approx 2 mo.
3. For EM autoradiography, ultrathin sections are cut and placed on a nickel grid and then mounted on a microscope slide, which is dipped in Ilford L4 nuclear emulsion.
4. Develop and fix after approx 3–4 mo.
5. View under a transmission electron microscope.

3.9. Quantitative Autoradiography

1. Film images are produced by apposing labeled slide-mounted sections to Kodak Ektascan EC1 film (clear-base emulsion coated on one side).
2. A series of five slides with approx 25 x 76-mm pieces of film can be interleaved and clipped together with bulldog clips, or for a larger series, slides can be arranged in 18 x 24-cm X-ray cassettes and sheets of film applied.
3. After exposure (14 d at 4°C), film is developed in D19 at 18–20°C for 5 min and fixed in Rapid Fix (Kodak) for the same time.
4. Film images (Fig. 2 shows examples of β-adrenoceptors in human sinoatrial node) are quantitated using the MCID M1 computer-image analysis system (Imaging Research, St. Catherines, Ontario, Canada) utilizing a Northern light illuminator, a Mintron MTV 1801 CB CCD video camera, and an imaging technology FG100 frame grabber, mounted in an 80386 33 MHz computer equipped with an 80387 maths coprocessor.
5. After establishing shading error and linear calibration, the system is standardized using a series of [^{125}I]-labeled standards made up in heart paste.
6. Alternatively, appropriately calibrated [^{125}I] microscales (Amersham) can be used.
7. Further details and examples can be obtained from refs. *4,7–11*.

3.10. Protein Determination

1. Protein in discrete regions of tissue can be determined in the same sections used for autoradiography using a densitometric method based on coomassie brilliant blue *(8,12)*.
2. The sections are incubated with coomassie blue dissolved in Krebs-phosphate buffer (1.5% [w/v]) for 3 h at room temperature, followed by 2 x 10 min washes in buffer.
3. The degree of coomassie blue staining of the sections is proportional to the amount of protein in the regions and can be quantitated by densitometry.

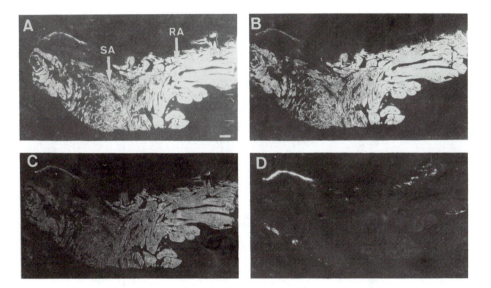

Fig. 2. Quantitative receptor autoradiography—measurement of β-adreno-ceptor subtypes in the human SA node and RA using the X-ray film technique. Sections were labeled with (–)[^{125}I]CYP (56 pM) in the absence or presence of ICI 118,551 (β$_2$-adrenoceptor antagonist, 70 nM); CGP 20712A (β$_1$-adreno-ceptor antagonist, 100 nM); or with propranolol (nonselective β-adrenoceptor antagonist, 1 μM) and apposed to X-ray film. **(A)** A print from film that had been apposed to a section which had been labeled for β-adrenoceptors with (–)[^{125}I]cyanopindolol; **(B)** localization of β$_1$-adrenoceptors; **(C)** β$_2$-adrenoceptors; and **(D)** nonspecific binding. Receptor density in anatomically defined areas is measured densitometrically using the MCID system and corrected for protein density. The density of β$_1$- and β$_2$-adreno-ceptors in the SA node was 7.9 for β$_1$ and 1.6 for β$_2$, and in RA was 39.6 and 7.2 fmol/mg protein. Bar represents 1.5 mm.

4. The micro computer imaging device (MCID) computer densitometry system is used in the same way as described for analysis of autoradio-grams.
5. Slide-mounted protein standards containing varying amounts of protein are prepared as mixtures of guinea pig heart paste and Tissue Tek.
6. There is an excellent correlation between the amount of protein in stan-dards determined using the Lowry method and values determined using the densitometric method.

4. Notes

4.1. Quantitative Autoradiography—Analysis of Results

1. The most rigorous analysis of β_1- and β_2-adrenoceptor density in discrete regions of tissue is done by performing full saturation experiments with $(-)[^{125}I]CYP$ (5–120 pM) and full competition binding experiments between $(-)[^{125}I]CYP$ (50 pM) and ICI 118,551 (50 pM–20 µM) or CGP 20712A (50 pM–0.5 mM). The affinities of $(-)[^{125}I]CYP$, CGP 20712A, ICI 118,551, and β_1- and β_2-adrenoceptor densities can then be determined by analysis of binding data with nonlinear curve-fitting programs, such as LIGAND *(13)*. However, sufficient quantities of tissue are not always available for this type of analysis, and if this is the case, then determination of the maximal density of β-adrenoceptor subtypes can be made using a single concentration of $(-)[^{125}I]CYP$, and the proportions of β_1- and β_2-adrenoceptors determined by the extent of inhibition of $(-)[^{125}I]CYP$ binding by either CGP 20712A or ICI 118,551 at a concentration chosen using the mass action equation to give selective blockade of β_1- or β_2-adrenoceptors.
2. Inhibition of binding of a radioligand by a selective competitor is described by the equation developed by Neve et al. *(14)* as used by Murphree and Saffitz *(15)* and ourselves *(8)* for autoradiography.

$$B = (B_{max1} \cdot L)/[L + K_{d1}(1 + I/K_{i1})] + (B_{max2} \cdot L)/[L + K_{d2}(1 + I/K_{i2})] \quad (1)$$

where B is the amount of radioligand bound, B_{max1} and B_{max2} are the densities of β_1- and β_2-adrenoceptors, L is radioligand concentration, K_{d1} and K_{d2} are dissociation constants for the binding of ligand to β_1- and β_2-adrenoceptors, K_{i1} and K_{i2} are dissociation constants for the competing agent at β_1- and β_2-adrenoceptors, and I is the concentration of competing agent. In the absence of competitors, this reduces to:

$$B = (B_{max1} \cdot L)/(L + K_{d1}) + (B_{max2} \cdot L)/(L + K_{d2}) \quad (2)$$

Equations (1) and (2) can then be used to solve for B_{max1} and B_{max2}.
3. Typical values for $(-)[^{125}I]CYP$ affinities at β_1- and β_2-adrenoceptors are 20.3 and 10.2 pM.
4. The concentration of $(-)[^{125}I]CYP$ *(L)* is determined in each experiment, but is usually approx 50 pM.
5. Competitor concentrations *(I)* are 70 nM (ICI 118,551) or 100 nM (CGP 20712A).
6. The affinity of ICI 118,551 at β_1-adrenoceptors is 0.132 µM and at β_2-adrenoceptors, 0.724 nM. Equivalent values for CGP 20712A are 0.933 nM at β_1-adrenoceptors and 6.31 µM at β_2-adrenoceptors.
7. Calculations involved in these equations are facilitated by a simple computer program, SIMUL *(16)*, which will run on any MS DOS computer.

4.2. Pitfalls and Artifacts
in Autoradiographic Studies of β-Adrenoceptors

8. High background can either be localized to the tissue section or be widespread over the slide. Generalized high background is usually owing to exposure of the emulsion-coated coverslips or film to light. This can occur because of poor sealing of the dark room or of cassettes used to store the specimens and the photographic medium. Another common cause of high background is contamination of the photographic emulsion or exposure of the film to a source of radiation. This can either be introduced as a result of poor techniques in the laboratory and a failure to separate the photographic process from the labeling process, or the presence of naturally occurring isotopes in materials used for autoradiography (e.g., ^{40}K in glass). There can also be a build-up of contamination in photographic emulsions if these are stored incorrectly or for long periods of time owing to cosmic radiation. However, these factors are unlikely to pose a problem in the autoradiography of β-adrenoceptors using $(-)[^{125}I]$CYP, since exposure times are relatively short.

9. High background over the tissue usually represents positive chemography, which usually results from the fixatives used to treat the tissues. In the case of β-adrenoceptor autoradiography with $(-)[^{125}I]$CYP, this is not usually a problem, since the tissues are only very lightly fixed with 0.1% formaldehyde (positive chemography is usually associated with glutaraldehyde).

10. False positives may occur. In certain tissues, $(-)[^{125}I]$CYP will label receptors other than β-adrenoceptors. Thus, in brain, it labels $5HT1_B$ receptors, although these sites can be readily recognized by their sensitivity to 5HT and their lower affinity for $(-)[^{125}I]$CYP (K_D 1 nM, cf. 20 pM at β-adrenoceptors).

11. In heart and in guinea pig trachea, $(-)[^{125}I]$CYP labels peroxidase-related sites (17). The binding is resistant to blockade by propranolol and is enhanced in the presence of ascorbic acid. These sites can be blocked by a range of diverse compounds, including 5HT, phentolamine, and (−)isoprenaline. $(-)[^{125}I]$CYP can also label two distinct (−)propranolol-resistant binding sites in rat hindlimb and skeletal muscle. One site is highly localized and can be blocked by (−)isoprenaline, 5HT, and phentolamine, whereas the other is evenly distributed (18). These sites are recognized by their relatively low affinity for $(-)[^{125}I]$CYP (K_D 500 pM) and their resistance to propranolol (19).

12. These examples illustrate that in any study localizing β_1- and β_2-adrenoceptors using $(-)[^{125}I]$CYP as the ligand, binding to other sites is commonly observed. However, with appropriate treatment, such as the removal of ascorbic acid from the incubation medium or the inclusion of 10 μM 5HT to inhibit binding to $5HT1_B$ receptors, it is possible to examine successfully the localization and distribution of β-adrenoceptor subtypes in the presence of these other sites.

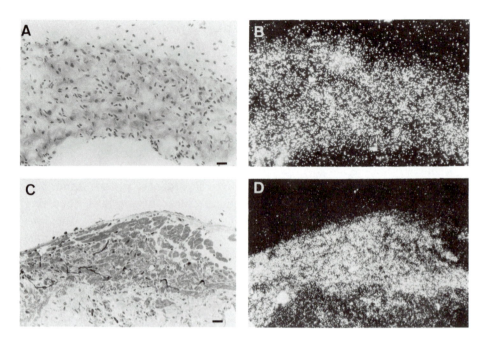

Fig. 3. Localization of β-adrenoceptors in the guinea pig SA node—comparison of the "standard" and high-resolution techniques. (**A**) A hematoxylin-and-eosin-stained section (10 μm) of SA node that underwent mild fixation (0.1% formaldehyde) before labeling; (**B**) the corresponding distribution of (–)[^{125}I]CYP binding; (**C**) a methylene blue-stained section (0.7 μm) of SA node that was labeled and then fixed with glutaraldehyde/formaldehyde and postfixed with osmium tetroxide. Note the marked improvement in tissue resolution compared to (A); (**D**) the distribution of silver grains corresponding to β-adrenoceptors in the nuclear emulsion coating the section—note the tighter definition of boundaries compared to (B). Bar = 25 μm.

13. Other false positives are common to all autoradiographic studies. These include positive chemography caused by a reaction of the photographic emulsion to histological stains and fixatives, and drying artifacts where the ligand collects in particular areas of the tissue and forms concentrations of label as they dry, often along histological features. Pressure artifacts are a problem particularly with emulsion-coated coverslips and with highly sensitive films designed to detect tritium, such as Ultrofilm ^{3}H or Hyperfilm ^{3}H. All of these artifacts can be expected to occur in slides set up to study nonspecific binding, as well as in those examining total receptor binding.

4.3. High-Resolution Qualitative Autoradiography

14. Improvements in the resolution of receptor autoradiographs can be made at the light microscopic level *(7)* (Fig. 3) and with electron microscopy

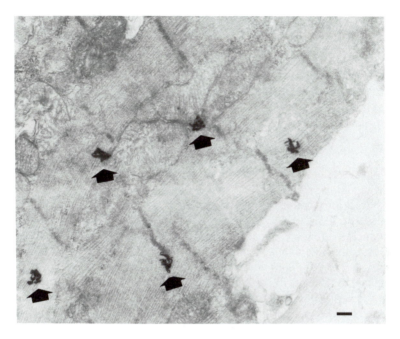

Fig. 4. Localization of cardiac β-adrenoceptors by EM autoradiography. Guinea pig heart was labeled with $(-)[^{125}I]$CYP fixed in formaldehyde/glutaraldehyde, postfixed with osmium tetroxide, embedded, and ultrathin (0.1 μm) sections cut and coated with nuclear emulsion. Silver grains localized to particular structures and the density of labeling can be determined after correcting for the area occupied by each structure. In cardiac sections, 90% of the receptors are associated with the sarcolemma or blood vessels. Bar = 0.2 μm.

(20) (Fig. 4). The standard technique uses lightly fixed frozen tissues to preserve the integrity of ligand binding, but is associated with poor preservation of tissue morphology. To overcome these problems, lightly fixed tissue blocks are incubated with radioligand followed by rigorous (4% glutaraldehyde/formaldehyde) fixation. The modification takes account of the ability of $(-)[^{125}I]$CYP to label lightly fixed blocks of tissue with characteristics similar to those observed in tissue sections. Blocks are then fixed in formaldehyde/glutaraldehyde at 4°C at which temperature the dissociation of $(-)[^{125}I]$CYP from the receptor is slow. Tissues can therefore be fully fixed and embedded for sectioning without the ligand diffusing appreciably from the receptor.

Acknowledgments

The authors acknowledge the contribution of many students and fellows, including Vivien Bonazzi, Joy Elnatan, Richard Jones, Andrew

Kompa, Robert Lew, Sue Lipe, Janine Matthews, Lynne McMartin, Sue Roberts, Fraser Russell, Jenny Stephenson, and Darren Williams. Without their help and enthusiasm, much of this work would not have been possible. Thanks also go to Maria Loucas for her excellent secretarial assistance and to the National Health and Medical Research Council (NH&MRC) and National Heart Foundation of Australia, who have supported this work. P. Molenaar is a Research Fellow of the NH&MRC.

References

1. Engel, G., Hoyer, D., Berthold, D., and Wagner, H. (1981) $(\pm)[^{125}\text{Iodo}]$-cyanopindolol, a new ligand for β-adrenoceptors: identification and quantitation of subclasses of β-adrenoceptors in guinea-pig. *Naunyn-Schmeideberg's Arch. Pharmacol.* **317,** 277–285.

2. Buxton, B. F., Jones, C. R., Molenaar, P., and Summers, R. J. (1987) Characterization and autoradiographic localization of β-adrenoceptor subtypes in human cardiac tissues. *Br. J. Pharmacol.* **92,** 299–310.

3. Molenaar, P., Canale, E., and Summers, R. J. (1987) Autoradiographic localization of β_1- and β_2-adrenoceptors in guinea-pig atrium and regions of the conducting system. *J. Pharmacol. Exp. Ther.* **241,** 1048–1064.

4. Molenaar, P., Jones, C. R., McMartin, L. R., and Summers, R. J. (1988) Autoradiographic localization and densitometric analysis of β_1- and β_2-adrenoceptors in the canine left anterior descending coronary artery. *J. Pharmacol. Exp. Ther.* **246,** 384–393.

5. Young, W. S. and Kuhar, M. J. (1979) A new method for receptor autoradiography: [^{3}H]-opioid receptors in rat brain. *Brain Res.* **179,** 255–270.

6. Molenaar, P., Malta, E., Jones, C. R., Buxton, B. F., and Summers, R. J. (1988) Autoradiographic localization and function of β-adrenoceptors on the human internal mammary artery and saphenous vein. *Br. J. Pharmacol.* **95,** 225–233.

7. Russell, F. D., Molenaar, P., Edyvane, N., Smolich, J. J., and Summers, R. J. (1991) Autoradiographic localization and quantitation of β-adrenoceptor subtypes in the guinea-pig sinoatrial node. *Mol. Neuropharmacol.* **1,** 141–147.

8. Molenaar, P., Russell, F. D., Shimada, T., and Summers, R. J. (1990) Densitometric analysis of β_1- and β_2-adrenoceptors in guinea-pig atrioventricular conducting system. *J. Mol. Cell. Cardiol.* **22,** 483–495.

9. Molenaar, P., Smolich, J. J., Russell, F. D., McMartin, L. R., and Summers, R. J. (1990) Differential regulation of *beta*-1 and *beta*-2 adrenoceptors in guinea-pig atrioventricular conducting system after chronic (–)-isoproterenol infusion. *J. Pharmacol. Exp. Ther.* **255,** 393–400.

10. Kompa, A., Molenaar, P., and Summers, R. J. (1991) Regulation of cardiac β-adrenoceptor subtypes in guinea-pig after (–)-adrenaline and (–)-noradrenaline treatment. *Mol. Neuropharmacol.* **1,** 203–210.

11. Elnatan, J., Molenaar, P., Rosenfeldt, F. L., and Summers, R. J. (1994) Autoradiographic localization and quantitation of β_1- and β_2-adrenoceptors in the human

atrioventricular conducting system: a comparison of patients with idiopathic dilated cardiomyopathy and ischemic heart disease. *J. Mol. Cell. Cardiol.* **26,** 313–323.

12. Miller, J. A., Curella, P., and Zahniser, N. R. (1988) A new densitometric procedure to measure protein levels in tissue slices used in quantitative autoradiography. *Brain Res.* **447,** 60–66.

13. Munson, P. J. and Rodbard, D. (1980) LIGAND: a versatile computerized approach for the characterization of ligand binding systems. *Anal. Biochem.* **107,** 220–239.

14. Neve, K. A., McGonigle, P., and Molinoff, P. B. (1986) Quantitative analysis of the selectivity of radioligands for subtypes of beta adrenergic receptors. *J. Pharmacol. Exp. Ther.* **238,** 46–53.

15. Murphree, S. S. and Saffitz, J. E. (1988) Delineation of the distribution of β-adrenergic receptor subtypes in canine myocardium. *Circ. Res.* **63,** 117–125.

16. Williams, D. W. and Summers, R. J. (1990) SIMUL: an accurate method for the determination of receptor subtype proportions using a personal computer. *Comp. Prog. Biomed.* **32,** 137–139.

17. Molenaar, P., Kompa, A., Roberts, S. J., Pak, H. S., and Summers, R. J. (1992) Localization of (–)-[^{125}I]-cyanopindolol binding in guinea-pig heart: characteristics of non-β-adrenoceptor related binding in cardiac pacemaker and conducting regions. *Neurosci. Lett.* **136,** 118–122.

18. Molenaar, P., Kim, Y. S., Roberts, S. J., Pak, H., Sainz, R. D., and Summers, R. J. (1991) Localization and characterization of two propranolol resistant (–)[^{125}I]cyanopindolol binding sites in rat skeletal muscle. *Eur. J. Pharmacol.* **209,** 257–262.

19. Roberts, S. J., Molenaar, P., and Summers, R. J. (1993) Characterisation of propranolol-resistant (–)-(^{125}I)-cyanopindolol binding sites in rat soleus muscle. *Br. J. Pharmacol.* **109,** 344–352.

20. Russell, F. D., Molenaar, P., and Summers, R. J. (1992) Absence of mitochondrial β-adrenoceptors in guinea-pig myocardium: evidence for tissue disparity. *Gen. Pharmacol.* **23,** 827–832.

CHAPTER 4

In Situ Hybridization

Brian J. Morris

1. Introduction

In situ hybridization allows a particular mRNA species to be localized in its normal anatomical environment (i.e., in a tissue section). A large number of different potocols have been used with success, all similar in terms of their basic approach, but differing widely in their practical details. This chapter describes a procedure that has the double advantage of being very simple and also highly sensitive.

Most studies that have used *in situ* hybridization have employed either short synthetic oligodeoxyribonucleotides (oligos) or longer cRNA transcripts synthesized from cDNA clones. The cRNA probes may have a slight advantage in terms of sensitivity, but are markedly inferior to oligos in terms of the reproducibility of their signal and of their ability to distinguish between closely related mRNA sequences. Because receptor mRNAs are almost invariably highly homologous to other members of their gene family, oligos are generally a better choice than cRNA probes for receptor studies. In addition, there is currently great interest in studying the extent to which receptor gene expression can change at the mRNA level following alterations in neuronal activity *(1)*. The reproducibility of the *in situ* hybridization signal with oligos makes them ideal for such semiquantitative studies. Consequently, it is their use that is considered here.

The oligos should be designed to hybridize to regions of the mRNA that are not homologous to other mRNA species. When the target mRNAs are those encoding receptors or receptor subunits, the choice of

From: *Methods in Molecular Biology, Vol. 41: Signal Transduction Protocols*
Edited by: D. A. Kendall and S. J. Hill Copyright © 1995 Humana Press Inc., Totowa, NJ

region is particulary important, since certain parts of the sequences are highly conserved between different members of the same gene family. With both the ligand-gated ion channel families and the G-protein-coupled receptor families, the extreme 5' end of the sequence is generally a safe choice. With the ligand-gated ion channel subunits, the second intracellular loop region is suitable *(2)* since it tends to have low homology to related subunits (although beware in the case of the glutamate receptors), whereas with the G-protein-coupled receptors, the third intracellular loop region is normally a good region to choose *(3)*. Oligos have a unique ability to distinguish between sequences varying only by a few bases. This means that judicious selection of the region to which the oligo is synthesized *(4)* will allow the experimenter to detect selectively mRNA species as homologous as the flip and flop forms of the glutamate subunits GluR-A to GluR-D (Fig. 1).

Since the correct stringency for hybridization will depend on the length of the oligo, it is convenient to stick to a given length for all the different oligos used in the laboratory for *in situ* hybridization. In my laboratory, I have settled on 45 bases as a convenient length—longer oligos require higher hybridization temperatures, which are detrimental to tissue morphology, whereas shorter probes appear to lose sensitivity and selectivity.

The oligos can be labeled with radioactive nucleotides, the most commonly used isotopes being ^{35}S and ^{32}P. A faster result is obtained with ^{32}P because of the higher energy of its emissions, but this means that there is a consequent substantial drop in resolution. The lower energy of ^{35}S emissions gives exposure times that are still in the manageable range, but with resolution that is very good. This isotope has been found to be preferable for most purposes. Recently, ^{33}P-labeled nucleotides have become available, with properties in between ^{35}S and ^{32}P. Although the resolution they give is satisfactory, I have been disappointed by a tendency they show to give higher backgrounds with extended exposure times. I would therefore not recommend their use at present to detect low-abundance mRNAs.

The enzyme terminal transferase is used to add a string of radioactive nucleotides onto the the 3' end of the oligo. Unfortunately, this reaction is highly variable, and a combination of luck and skill is required to obtain high-specific-activity probes.

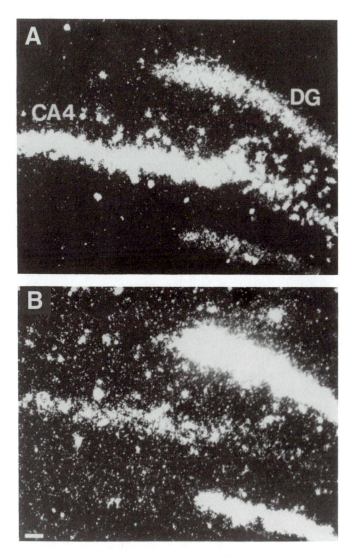

Fig. 1. Localization of glutamate receptor subunit mRNAs in rat hippoc-ampus. Darkfield photomicrographs showing (**A**) GluR-A flip mRNA and (**B**) GluR-A flop mRNA distribution. Oligonucleotides (39 mers) were syn-thesized as in Sommer et al. *(4)*. Note that the levels of the flip isoform mRNA are higher in the CA4 region than in the dentate gyrus (DG), whereas the flop isoform mRNA has a reciprocal distribution. The distinct distribution is readily discernible, despite the fact that the proteins derived from these tran-scripts differ by only eight amino acids *(4)*.

2. Materials

2.1. Tissue Preparation

1. Diethyl pyrocarbonate (DEPC): All solutions that need to be RNase-free—basically all those used prior to and during hybridization—should have 0.1% (v/v) DEPC added in a fume hood. The solutions are then shaken and autoclaved. Note that Tris solutions cannot be treated with DEPC.
2. Poly-L-lysine hydrobromide (mol wt > 350,000)—Sigma (St. Louis, MO) P-1254: Aliquot at a concentration of 5 mg/mL in DEPC-treated water and store at –20°C.
3. Microscope slides: Bake at 180°C for 3h (wrapped in aluminum foil) to destroy RNase.
4. 10X Phosphate-buffered saline (PBS) solution: $1.3M$ NaCl, 70 mM NaH_2PO_4, 30 mM Na_2HPO_4. Filter, treat with DEPC, and autoclave.
5. Paraformaldehyde solution (4%): Add 20 g paraformaldehyde (reagent grade) to 50 mL 10X PBS with 450 mL deionized water. Heat in a fume hood to about 65°C (do not allow to boil) when the powder will dissolve. Then cool on ice. This solution should be prepared fresh each day.
6. Prepare 1X PBS by diluting 10X PBS in DEPC-treated water.
7. Absolute ethanol (reagent grade).
8. Plastic containers with sealing lids for storing ethanol at 4°C. For safety reasons, containers full of ethanol should clearly not be placed in a normal refrigerator. A spark-free refrigerator or cold room licensed for storage of inflammable liquids is required.
9. Glass slide racks and staining jars—these should be baked at 180°C before use.

2.2. Probe Preparation

1. Oligonucleotide: Purchase from supplier, or design and synthesize on locally available machine. Deprotect oligo as directed by supplier if this has not already been done, but purification (gel, HPLC) is not normally necessary. Dilute a small amount of oligo in DEPC water to a concentration of around 0.3 pmol/μL. Store both the original stock solution and the diluted working solution at –20°C—oligos are stable if the solutions are sterile, and repeated feezing and thawing is not a problem.
2. Terminal deoxynucleotidyl transferase (i.e., Gibco, Paisley, UK, or Pharmacia, Milton Keynes, UK)—store at –20°C and never allow to warm to room temperature.
3. Buffer for terminal transferase (supplied free of charge with the Gibco enzyme): Aliquot and store at –20°C.
4. α-^{35}S dATP (NEG 034H—New England Nuclear)—store at –20°C.

5. Tris-EDTA (TE) buffer: DEPC-treated water to which has been added Tris (10 mM) and EDTA (1 mM) and the pH restored to 7.5.
6. Sephadex G25 equilibrated with TE buffer.
7. 1-mL Sterile plastic syringes.
8. Dithiothreitol (DTT): 1M stock in DEPC-treated acetate buffer.

2.3. Hybridization

1. Deionized formamide: In a sterile 50-mL centrifuge tube, shake 30 mL formamide with some Amberlite mono-bed resin beads. Filter to remove beads.
2. 20X SSC: 3M NaCl, 0.3M Na citrate, pH 7.0 with HCl, DEPC-treated and autoclaved.
3. Hybridization buffer: In a sterile 50-mL plastic centrifuge tube, add 10 mL 20X SSC to 25 mL deionized formamide. Add 2.5 mL DEPC-treated 0.5M Na phosphate buffer, pH 7.0, 1 mL aliquot of 5 mg/mL polyadenylic acid (Sigma), and 5 g dextran sulfate. After the dextran sulfate has dissolved, adjust to 50 mL with DEPC water.
4. Parafilm and 6-in. plastic Petri dishes.
5. Incubator or oven able to hold a steady temperature. For an oligo of 40–45 bases, the temperature required is 42°C.
6. Double-sided tape, X-ray cassettes, Kodak XAR-5 X-ray film, plus Ilford K5 liquid emulsion and dipping chamber if cellular resolution is required.

3. Methods
3.1. Tissue Preparation

1. Remove tissue and freeze on dry ice or isopentane/liquid nitrogen.
2. Cut sections in a cryostat—the tissue penetration of ^{35}S emissions is sufficiently great that a 20-μm section will give almost twice the signal of a 10-μm section, although with slightly less resolution. In practice, I select section thickness according to the abundance of the mRNA to be studied and the number of sections required from that region of tissue. The sections should be mounted onto baked slides that have been coated with 100 μg/mL Poly-L-lysine (diluted in DEPC water) and allowed to dry. The coated slides can be stored indefinitely at 4°C.
3. Allow the sections to dry, then place in a slide rack, and fix in fresh, ice-cold 4% paraformaldehyde for 5 min *(5)*.
4. Transfer rack to 1X PBS (dilute 10X PBS with DEPC water) in a baked staining jar.
5. After approx 5 min, transfer rack to 70% ethanol, 95% ethanol, and finally 100% ethanol (about 2 min. each, all in baked staining jars). Store rack of slides in sealed plastic container full of ethanol at 4°C. The mRNA in the sections is preserved in this manner for a number of years.

3.2. Probe Preparation (see Note 2)

1. In an Eppendorf tube, add sequentially 4.5 µL DEPC water, 2 µL (0.6 pmol) oligonucleotide, 2.5 µL 5X buffer, 2.5 µL ^{35}S-dATP, and finally 1 µL approx 20 U) terminal transferase. Mix gently by moving liquid up and down the pipet tip (do not vortex—air bubbles inhibit the enzyme activity dramatically). Leave at 37°C for about 2 h (or longer).

2. Prepare a spin column using the Sephadex G25 and the plastic syringe, as described by Maniatis et al. *(6).* Load the probe onto the column after diluting in 40 µL TE buffer, and spin in a bench centrifuge for sufficient time to bring down the labeled oligo but not the free nucleotides (determine this empirically). Add 2 µL 1*M* DTT to the eluate.

3. Count 2 µL by liquid scintillation. As a general rule, probes labeled to <100,000 cpm/µL are unlikely to be of any use unless the mRNA to be detected is incredibly abundant. Probes labeled to more than 800,000 cpm/µL are likely to give high nonspecific binding, since the polyA tail is too long. Anything in between is probably okay to use.

4. Dilute the probe 200-fold in hybridization buffer, adding in addition 4% (v/v) 1*M* DTT.

3.3. Hybridization (see Notes 1, 4, and 5)

1. Remove the microscope slides with the sections of interest from ethanol storage and allow to air-dry. Lay them in a 6-in. diameter plastic Petri dish. Pipet 150 µL of hybridization buffer containing the radiolabeled probe onto each slide, and gently lower a piece of parafilm, cut to a size slightly smaller than the slide surface, onto the sections, ensuring that no air bubbles are trapped as the buffer spreads out.

2. Put a piece of rolled tissue paper, soaked in 4X SSC, onto the side of the Petri dish. Put on the lid, and place in the hybridization oven at 42°C (if a 45 mer is being used) overnight. The soaked tissue prevents the fluid under the parafilm from evaporating.

3. On the next morning, place the slides in a rack in a staining jar containing 1X SSC at 55°C (if probe is a 45 mer) in a shaking water bath. Gently remove the parafilm with forceps, and leave shaking.

4. After 30 min, transfer rack to another jar of 1X SSC at 55°C.

5. After a further 30 min, transfer rack to 0.1X SSC at room temperature for 2 min, then to 70% ethanol for 1 min, and then to 100% ethanol for 2 min.

6. Allow slides to air-dry, then stick them onto a piece of card with double-sided tape, and expose to X-ray film in the cassette. Exposure times generally vary from 6 h to 3 wk, according to the abundance of the mRNA under investigation. Develop with Kodak D-19 developer and standard photographic fixer.

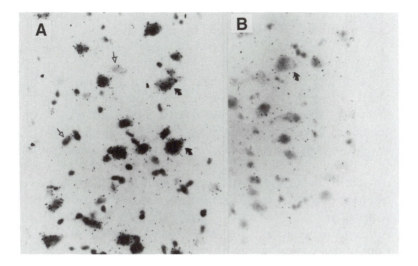

Fig. 2. Localization of GluR-B flop mRNA in rat cerebral cortex. (**A**) Brightfield photomicrograph of tissue hybridized with a 36-mer oligonucleotide, constructed as in ref. *4*, selective for the GluR-B flop isoform mRNA. Note that the layer V pyramidal cells (curled arrows) are strongly labeled, whereas some smaller cells (open arrows) do not express this mRNA. (**B**) Section adjacent to that shown in (A), hybridized with the same probe under identical conditions, except that a 25-fold excess of unlabeled GluR-B flop oligo was also included in the hybridization buffer. Note that the clustering of silver grains over the counterstained cell bodies in (A) is completely abolished, and only a low level of background labeling remains.

7. If cellular resolution is required *(7)*, dip the slides in liquid photographic emulsion (i.e., Ilford K5, diluted 1:1 in 0.5% glycerol) under safelight conditions, and leave in a light-tight slide box with silica gel at 4°C for roughly five times as long as the X-ray film exposure. Develop with D-19, and 30% sodium thiosulfate as fixer. The slides can be counterstained, after developing, with cresyl violet or toluidine blue, when the autoradiographic signal appears as black silver grains overlying the nissl-stained cell bodies (Fig. 2).

4. Notes

1. Controls: RNase pretreatment of sections has been used to demonstrate that the *in situ* hybridization signal represents binding to RNA, rather than to some other cellular constituent. This control is generally not worth the effort involved, because most of the nonspecific hybridization that is

encountered is also to RNA, and it is always slightly dangerous to have large quantities of RNase around in the laboratory.

2. A "sense" control probe can be synthesized, that has the *same* sequence as the tissue mRNA, rather than the complementary sequence. This probe will therefore have identical characteristics to the antisense oligo, except that it will not be complementary to any tissue mRNA. When it is labeled to the same specific activity as the antisense oligo and used under the same conditions, any signal that is observed must be nonspecific. This is a reasonable control, but it is risky because it will prove difficult to label the sense probe to the identical specific activity, and therefore the amount of nonspecific hybridization will vary. In addition, a likely source of artifactual hybridization of the antisense probe is to some mRNA species with a sequence related to the desired target. The sense probe will not hybridize to this related sequence, and so the signal will appear as specific.

3. Competition control, where for some sections, a 25- or 50-fold excess of unlabeled oligo is put in the hybridization buffer along with the labeled oligo. This control relies on the fact that hybridization to the target mRNA is saturated with the labeled oligo. If unlabeled oligo is added as well, then it will displace only the signal owing to the specific hybridization (Fig. 2). Many of the potential sources of artifactual hybridization are of much greater capacity, so that hybridization to them will not be displaced. However, this control also fails to pick out nonspecific hybridization to related sequences.

4. Northern analysis: It is very useful, if not essential, to hybridize the antisense oligonucleotide to a Northern blot under identical stringency conditions to those used with tissue sections. Any tendency for the probe to hybridize to ribosomal RNA or to a related mRNA sequence should show up as additional bands on the Northern blot. Although there are many examples in the literature of groups conducting *in situ* hybridization studies without characterizing the probe by Northern analysis, this is a highly suspect approach.

5. The separate use of two oligonucleotides complementary to different regions of the mRNA sequence is a useful control that reveals whether there is a tendency for one probe to hybridize to a related, but distinct mRNA sequence. It is unlikely that both probes would happen to hybridize to the same, unknown mRNA. Only an *in situ* hybridization signal common to both probes can be taken as representing the desired target mRNA.

Acknowledgments

I am grateful to the Wellcome Trust for financial support, and to Dr. W. Wisden for the oligonucleotides used in this study.

References

1. Morris, B. J. (1993) Control of receptor sensitivity at the mRNA level. *Mol. Neurobiol.* **7,** 1–17.
2. Morris, B. J., Hicks, A. A., Wisden, W., Darlison, M. G., Hunt, S. P., and Barnard, E. A. (1990) Distinct regional expression of nicotinic acetylcholine receptor genes in chick brain. *Mol. Brain Res.* **7,** 305–315.
3. Weiner, D. M. and Brann, M. R. (1989) The distribution of a dopamine D2 receptor mRNA in rat brain. *FEBS Lett.* **253,** 207–213.
4. Sommer, B., Keinanen, K., Verdoorn, T. A., Wisden, W., Burnashev, N., Herb, A., Koehler, M., Takagi, T., Sakmann, B., and Seeburg, P. H. (1990) Flip and flop: a cell-specific functional switch in glutamate-operated channels of the CNS. *Science* **249,** 1580–1585.
5. Wisden, W., Morris, B. J., and Hunt, S. P. (1990) In situ hybridisation with synthetic DNA probes, in *Molecular Neurobiology—A Practical Approach* (Chad, J. and Wheal, H., eds.), IRL, Oxford, UK, pp. 205–228.
6. Maniatis, T., Fritsch, E. P., and Sambrook, J. (1982) *Molecular Cloning. A Laboratory Manual.* Cold Spring Harbor Laboratory Press, Cold Spring Harbor, New York.
7. Morris, B. J. (1989) Neuronal localisation of Neuropeptide Y gene expression in rat brain. *J. Comp Neurol.* **290,** 358–368.

CHAPTER 5

Measurement of the GTPase Activity of Signal-Transducing G-Proteins in Neuronal Membranes

Marva Sweeney

1. Introduction

The methods described in this chapter are designed to measure the hydrolysis of guanosine triphosphate (GTP) to inorganic phosphate (P_i) and guanosine diphosphate (GDP), a reaction catalyzed by the GTPase enzymes (EC 3.6.1.–). The theory behind the experimental design involves using [γ-^{32}P]GTP as a marker, whereby any GTPase-induced hydrolysis will result in the ^{32}P label appearing as ^{32}P$_i$ and as unhydrolyzed [γ-^{32}P]GTP. It was originally described by Cassel and Selinger *(1)*. ^{32}P$_i$ must be separated from [γ-^{32}P]GTP, and then can be easily quantitated by using a liquid scintillation counter. The amount of ^{32}P$_i$ is directly proportional to the amount of [γ-^{32}P]GTP hydrolyzed, and, therefore, proportional to the activity of GTPases in the preparation.

There are a number of membrane-bound enzymes that are capable of hydrolyzing GTP to GDP and P_i. The specific interest in GTPase, when attempting to study the signal-transducing, regulatory guanine nucleotide binding proteins (G-proteins), is the GTPase intrinsic to the α-subunit of these hetero-trimeric G-proteins (Gα). GTP hydrolysis is an intrinsic part of all signal-transducing Gα subunits and is responsible for termination of the activated state of Gα, in which GTP is bound to the high-affinity nucleotide binding site, into the inactive GDP-bound state *(2–4)*. This action recycles the G-protein to its ground state, which is then capable of interacting with the neurotransmitter receptor and being reactivated. This

From: *Methods in Molecular Biology, Vol. 41: Signal Transduction Protocols*
Edited by: D. A. Kendall and S. J. Hill Copyright © 1995 Humana Press Inc., Totowa, NJ

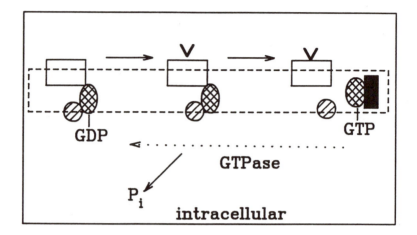

Fig. 1. The cycling of G-proteins from inactive to active states. (□) receptor;
(∨) agonist; (⊕) Gα; (⊘) Gβγ; (■) effector; (⌐_⌐) membrane.

cycling of G-proteins from inactive to active states has been described in
numerous review articles (*see* refs. *3,5,6*) and is depicted in Fig. 1.

Receptors that interact with G-proteins produce an increase in GTP
hydrolysis, ultimately via an increase in the GTPase activity of Gα, but
initially by stimulating the binding of GTP to Gα *(4)*. Thus, a study of
the ability of various receptor agonists to stimulate GTPase activity is
useful to determine:

1. The possible involvement of a G-protein(s) in the signal transduction of a
   particular pathway;
2. The identity of this G-protein, by using various specific G-protein toxins
   and antisera, or antisense oligonucleotides that are complementary to various G-protein subunits; and
3. The nature of the interaction between the G-protein and receptor.

It must be emphasized that in the presence of receptor agonists, an
increase in GTP binding likely precedes hydrolysis; this can be determined by measuring the ability of the agonist to increase the binding of
a nonhydrolyzable analog of GTP, such as guanosine 5'-(γ-[^{35}S]thio)-
triphosphate (GTPγ[^{35}S]) *(7,8)*. However, effectors such as the dihydropyridine-sensitive L-type Ca^{2+} channel can modulate the activity
(maximal rate of hydrolysis, V_{max}) of GTPase in its own right without
affecting nucleotide exchange *(8)*.

The measurement of a specific, high-affinity GTPase is accomplished by using the following experimental conditions to reduce or eliminate the possible activity of nonspecific nucleotidases:

1. A low concentration of nonradiolabeled GTP ($[GTP]_{cold}$) is used, usually <1 μM, since the affinity of the GTPase of Gα for binding GTP is approx 0.1–0.5 μM.
2. Nucleotide triphosphatases are inhibited by using ouabain and adenosine 5'-(β,γ- imino)triphosphate (p[NH]ppA or App[NH]p), although the presence of this latter agent did not significantly reduce nonspecific GTP hydrolysis in our preparation of rat frontal cortical membranes (Sweeney and Dolphin, unpublished observations);
3. The transfer of γ-^{32}P from GTP to ADP is suppressed with an ATP-regenerating system (creatine phosphate plus creatine phosphokinase), and the additional inclusion of ATP and ouabain in the assay conditions; and
4. Low-affinity (high K_M) GTPase activity is assessed following the addition of a high concentration of GTP, usually 100 μM, or alternatively, in the presence of a nonhydrolyzable analog of GTP, such as 10 μM GTPγS. These nucleotides compete with [^{32}P]GTP for the high-affinity nucleotide binding site, leaving primarily low-affinity GTPases available for hydrolyzing [^{32}P]GTP. Such low-affinity GTPase activity can then be subtracted from total basal or agonist-stimulated GTPase activity to yield a value for high-affinity GTPase activity.

The protocol that is used to determine the GTPase activity of G-protein is outlined in Sections 3.1. and 3.2. This method is based on that first described by Cassel and Selinger *(1)* and subsequently modified several times (e.g., refs. *8–11*).

2. Materials

1. Neuronal membranes: Prepare as indicated in Section 3.1.
2. Homogenizing buffer: Make up in 100-mL quantities, and store as 10-mL aliquots at –20°C: 10 mM Tris-HCl, pH 7.4, 1 mM EDTA, 5 mM dl-dithiothreitol (DTT), 2 mM benzamidine, 50 μM chlorpromazine, 50 μM leupeptin, and soybean trypsin inhibitor (1 trypsin inhibitor unit [TIU]/mL). Daily, add aprotinin (0.25 TIU/mL) and phenylmethylsulfonyl-fluoride (10 μM final concentration).
3. GTPase reaction buffer: Make fresh daily using the volumes of stock solutions (aliquots stored at –20°C) shown in Table 1. Daily, add 50 U of creatine phosphokinase/mL of GTPase reaction buffer. The inclusion of 1 mM p(NH)ppA may also be utilized, as mentioned in Section 1. Also daily, add enough [γ-^{32}P]GTP to yield 50,000–100,000 cpm/assay, using the follow-

Table 1
The Volumes of Stock Solutions Required to Make GTPase Reaction Buffer

| Stock solution | Final concentration | | Volume (μL) of stock required to make 5 mL buffer |
	Buffer	Assay	
Deionized water	—	—	2200
10 mM ouabain	2 mM	1 mM	1000
2M NaCl	200 mM	100 mM	500
400 mM Tris-HCl +	40 mM (Tris-HCl)	20 mM (Tris-HCl)	500
40mM EDTA	4 mM (EDTA)	2 mM (EDTA)	
100 mM dl-DTT	4 mM	2 mM	200
40 mM ATP	2 mM	1 mM	250
40 mM creatine phosphate (phosphocreatine)	2 mM	1 mM	250
1M MgCl$_2$	10 mM	5 mM	50
100 μM GTP	1 μM	0.5 μM	50

ing approximate guide: Order [γ-^{32}P]GTP from New England Nuclear (NEN) with a specific activity of 2 mCi/mL. On arrival, dilute this (or an aliquot) 1:10 to yield a solution of 0.2 mCi/mL. 50,000 cpm/assay are equivalent to 2.27×10^{-5} mCi/100 μL = 2.27×10^{-4} mCi/mL. Add 1.14 μL of diluted [γ-^{32}P]GTP/mL of GTPase reaction buffer to obtain a GTPase reaction buffer with 2.27×10^{-4} mCi/mL; increase accordingly to account for decay ($T_{1/2} = 14$ d).

4. Two percent (w/v) activated charcoal suspension: Make in ice-cold 20 mM phosphoric acid, pH 2.2, containing 5% ethanol. Ideally, this suspension should be sonicated at 0–4°C for 30 min, and then stirred constantly on ice until and during use.

5. ADP-ribosylating buffer: Make up in 100-mL quantities, and store as 10-mL aliquots at –20°C: 25 mM Tris-HCl, pH 7.4, 1 mM EDTA, 1 mM dl-DTT, 1 mM MgCl$_2$, 1 mM ATP, 10 mM thymidine, 10 mM NAD$^+$, 10 mM nicotinamide, and 100 μM GTP.

3. Methods

3.1. Preparation of Membranes

1. Collect neuronal tissue on ice.

2. Homogenize tissue in 10 vol of ice-cold homogenization buffer (*see* Section 2., item 2) using eight strokes, glass/Teflon™. Typically the amount of tissue necessary is approx 200 mg rat frontal cortex (wet wt) homog-

enized in 2 mL of buffer, or 4 plates of cerebellar granule cells grown in primary culture (total of 2×10^7 cells) pooled into 0.5 mL.

3. Centrifuge homogenates at 1000g for 10 min at 4°C.
4. Centrifuge the supernatant at 40,000 or 50,000g for 10 min at 4°C.
5. Wash the resulting membrane pellet in 2 mL of ice-cold homogenizing buffer, and then resuspend in 30 vol of homogenization buffer (or just 10 mM Tris-HCl, pH 7.4, 1 mM EDTA) using 4 strokes, glass/Teflon. Thus, if 200 mg of cortex were used, the final volume of membranes would be 6 mL. This yields a suspension of approx 1 mg protein/mL.

3.2. GTPase Assay

1. Add 20 µL of membrane suspension (1 mg protein/mL) to disposable 5-mL glass test tubes on ice containing 30 µL of test drug(s) (*see* Note 2) and 50 µL of GTPase reaction buffer (*see* Section 2., item 3). This gives a final assay volume of 100 µL. Perform all treatments in triplicate, and always include blank tubes that contain no membranes (i.e., 20 µL homogenization buffer + 30 µL H_2O + 50 µL GTPase reaction buffer).
2. To determine the amount of GTP hydrolyzed by low-affinity GTPases, include tubes containing a high concentration of GTP (100 µM) in basal and drug-stimulated conditions.
3. Vortex-mix tubes while still being kept on ice, and then place all tubes simultaneously into a 37°C water bath.
4. Let GTP hydrolysis occur for 10 min. At this time, terminate the reaction by removing all tubes to ice, and agitate the tubes for 1 min.
5. Immediately add 900 µL of ice-cold 2% activated charcoal suspension (Section 2., item 4) to each tube (*see* Note 3). The assay volume is now 1 mL.
6. Spin tubes at 1500g for 10 min at 4°C. Carefully remove 200 µL of the supernatant, and count this aliquot for ^{32}P content in a liquid scintillation counter. Always count 50 µL of the GTPase reaction buffer in triplicate, and determine the protein content of the membrane suspension, both values of which are used to calculate the rate of GTP hydrolysis (*see* Section 3.6.).

3.3. Lineweaver-Burk Analysis of GTPase Activity

In order to determine the V_{max}, and K_M for GTPase, the Lineweaver-Burk analysis can be used. The Lineweaver-Burk analysis plots the reciprocal of the rate of GTP hydrolysis vs the reciprocal of [GTP]$_{cold}$ (Section 3.6., Fig. 2B).

1. Perform GTPase assays exactly as outlined in Section 3.2., except vary the [GTP]$_{cold}$ from 0.5 µM. Thus, we have used [GTP]$_{cold}$ of 0.1, 0.25, 0.5, and

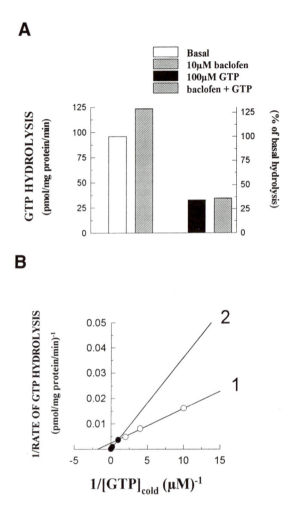

Fig. 2. Graphic representation of results calculated in Table 2 (**A**) and Table 3 (**B**).

1.0 μM to determine the activity of our high-affinity GTPase, and 1, 5, 10, 50, and 100 μM to study the low-affinity GTPase (Section 3.6., step 3, Fig. 2B). In subsequent experiments, $[GTP]_{cold} = 0.04, 0.08, 0.2,$ and $2.0 \mu M$ in order to determine the effect of various drugs on the activity of high-affinity GTPase *(8)*.

2. Plot the reciprocal of the rate of GTP hydrolysis vs the reciprocal of $[GTP]_{cold}$ (Section 3.6., step 3, Fig. 2B).

3.4. Pretreatment with Anti-G-Protein Toxins

1. Suspend membranes in ADP-ribosylating buffer (Section 2., item 5).
2. Activate small volumes of pertussis toxin or cholera toxin by incubating them for 10 min at 37°C with 40 mM DTT.
3. Add aliquots (1 mL) of membranes to 1.5-mL centrifuge tubes on ice containing activated pertussis or cholera toxins. The final concentrations of the toxins are 3 μg pertussis toxin/mg protein and 10 μg cholera toxin/mg protein, and in previous experiments, they were dissolved in 50% glycerol in phosphate-buffered saline (PBS).
4. Vortex-mix all tubes and then allow ADP ribosylation to occur for 20 min at 30°C.
5. At this time, centrifuge down the membranes at 50,000g for 15 min at 4°C, and then resuspend them in homogenizing buffer (four strokes, glass/Teflon) to be used in GTPase assays.

3.5. Pretreatment with Anti-G-Protein Antisera

1. Suspend membranes in homogenizing buffer.
2. Add 980 μL of membranes to 20 μL of nonimmune serum or anti-G-protein antisera in 1.5-mL centrifuge tubes on ice.
3. Vortex-mix all tubes, and then allow antisera to bind to the membranes for 1 h at 30°C. Mix occasionally. The membranes are used directly in the GTPase assay.

3.6. Calculation of GTP Hydrolysis

1. Calculate the rate of GTP hydrolyzed in pmol/mg protein/min as described in Eq. (4):

$$\text{Total cpm } {}^{32}P_i \text{ produced in each assay (in 1 mL volume at the end)} = \atop (\text{# of cpm in 200 μL} - \text{# of cpm in blank}) \times (1000 \text{ μL/mL}/200 \text{ μL}) \quad (1)$$

$$\text{Total mg protein in each assay} = (\text{mg protein/mL} \atop \text{membrane suspension}) \times (20 \text{ μL}/1000 \text{ μL/mL}) \quad (2)$$

$$\text{pmol GTP hydrolyzed/min/mg protein} = (\text{Total cpm } {}^{32}P_i \times \text{# of pmol} \atop \text{GTP added})/(\text{cpm in GTPase reaction buffer} \times \atop 10 \text{ min} \times \text{mg protein in assay}) \quad (3)$$

Eq. (3) simplified to include Eq. (1) and (2):

$$\text{Rate of GTP hydrolysis (pmol/mg protein/min)} = \atop [(\text{cpm counted} - \text{cpm in blank}) \times 5 \times \text{pmol GTP} \times 50]/ \atop (\text{cpm in GTPase reaction buffer} \times 10 \text{ min} \times \text{mg protein/mL}) \quad (4)$$

Table 2
Sample Calculations of Rate of GTP Hydrolyzed/min/mg Protein[a]

Sample counted	cpm in sample	cpm in sample— cpm in blank	pmol/mg protein/min
50 μL GTPase reaction buffer	69,394	—	—
Blank	514	0	0
Basal tube	5565	5051	95.77
Basal + 100 μM GTP	2208	1694	32.12
Basal + 10 μM (–)-baclofen	7019	6505	123.34
Basal + 10 μM (–)-baclofen + 100 μM GTP	2311	1797	34.07

[a]Membrane protein = 0.95 mg/mL. [GTP]$_{cold}$ = 0.5 μM → 50 pmol GTP/100 μL assay. Basal rate of GTP hydrolysis = $(5051 \times 5 \times 50 \times 50)/(69{,}394 \times 10 \times 0.95) = 95.77$ pmol/mg protein/min.

2. Sample calculations of the rate of GTP hydrolysis are depicted in Table 2. These values can be represented graphically as shown in Fig. 2A.
3. To determine the V_{max} and K_M for GTPase, follow the Lineweaver-Burk calculations given in Table 3. These values can then be graphically represented as shown in Fig. 2B. (Data from Tables 2 and 3 are depicted in Fig. 2.)

4. Notes

1. The stock solutions used to make GTPase reaction buffer in Section 2., item 3 and Table 1, should be stored at –20°C as small aliquots to prevent thawing and refreezing.
2. Typically the 30 μL of drug(s) used in the GTPase assay (Section 3.2., steps 1 and 2) consists of only 10 μL of each test drug at 10X the desired concentration. The balance of the volume is deionized water. For example, to determine the amount of GTP hydrolyzed by low-affinity GTPases (*see* Section 3.2., step 2), tubes would contain 50 μL of GTPase reaction buffer + 20 μL of membranes + 10 μL of 1 mM GTP + 20 μL water.
3. It is imperative that the charcoal suspension be stirred constantly on ice during the pipeting stage (*see* Section 3.2., step 5). This yields a homogeneous suspension, and aids quick and accurate pipeting.
4. As has been noted previously *(2,4)*, not all preparations display a receptor-stimulated high-affinity GTPase. In addition, only a few examples of a receptor-activated stimulation of the GTPase activity of G_s have been published, as opposed to several demonstrations for the pertussis toxin-sensitive G-proteins G_i and G_o. This observation may stem from the fact that the amount of G_s in membranes tends to be very low (<0.5%), whereas G_i and G_o account for 1–2% of total membrane protein (*see* ref. 2).

Table 3
Sample Calculations Used for Generating Lineweaver-Burk Plots[a]

Sample counted, $[GTP]_{cold}$	cpm in sample	cpm in sample— cpm in blank	cpm in GTPase reaction buffer	Rate of GTP hydrolysis	1/rate	1/ $[GTP]_{cold}$
Blank	624	0	—	—	—	—
Basal, 0.1 µM[b]	17,528	16,904	100,060	62.11	0.0161	10
Basal, 0.25 µM	13,841	13,217	98,364	123.5	0.0081	4
Basal, 0.5 µM	12,077	11,453	101,070	208.3	0.0048	2
Basal, 1 µM	9140.5	8516.5	112,710	277.8	0.0036	1
Basal, 5 µM	6360	5736	101,213	1041.7	0.00096	0.2
Basal, 10 µM	6339	5715	92,462	2272.72	0.00044	0.1
Basal, 50 µM	5641	5017	101,449	9090.9	0.00011	0.02
Basal, 100 µM	3641	3017	98,777	11,230.7	0.000089	0.01

[a]Membrane protein = 0.68 mg/mL.
[b]Basal rate of GTP hydrolysis at 0.1 µM GTP = $(16,904 \times 5 \times 10 \text{ pmol} \times 50) / (100,060 \times 10 \text{ min} \times 0.68) = 62.11$ pmol/mg protein/min.

Using a computer-assisted regression analysis to generate straight lines for 1/rate vs 1/$[GTP]_{cold}$, the K_M and V_{max} for high- (Fig. 2B, curve 1) and low-affinity (Fig. 2B, curve 2) GTPases can be determined:

Curve 1: $R = 0.99$ ($p < 0.001$)
              X intercept = -2.1
              $K_M = -1/X$ intercept = 0.48 µM
              Y intercept = 0.0024
              $V_{max} = 1/Y$ intercept = 416.67 pmol/mg protein/min

Curve 2: $R = 0.98$ ($p < 0.001$)
              X intercept = -0.0195
              $K_M = -1/0.0195 = 51.28$ µM
              Y intercept = 0.000673
              $V_{max} = 1/0.000673 = 1485.88$ pmol/mg protein/min

5. In some circumstances, the basal activity of GTPase may be very high relative to agonist-stimulated activity. This may be because of high levels of nonspecific nucleotidases. In this case, it is critical that the low-affinity GTPase activity, observed in the presence of 100 µM GTP (*see* Section 3.2., step 2), be subtracted away from the total GTPase activity to yield an accurate value for high-affinity GTPase activity.

6. As described in Section 1., there are several applications of the technique described in this chapter, including determining the involvement of a G-protein in a signal-transduction pathway or its interaction with a particular receptor, and the identity of such a G-protein using the protocol described in Sections 3.4. and 3.5. In addition, reconstitution experiments

in which purified receptors and G-proteins are combined, and additive experiments in which the effect of more than one agonist (or perhaps even antagonist; ref. *11*) on GTPase is examined, can provide information on whether more than one G-protein (or G-protein pool) is involved in the modulation by a receptor, or whether positive or negative cooperativity exists. Recently, in this laboratory it was demonstrated that the 1,4-dihydropyridine Ca^{2+} channel agonist Bay K 8644 increases the rate of GTP hydrolysis by $G_o\alpha$, which is already activated *(8)*, and this effect is profoundly synergistic with the adenosine receptor agonist (R)-phenylisopropyladenosine *(12)*. These observations *(8,12)* led myself and A. C. Dolphin to speculate that the positive cooperative effect of neurotransmitter receptor agonists and effectors on the activity of G-proteins may represent a widespread mechanism whereby the effector molecule, having interacted with the G-protein, which modulates its activity, limits the temporal effectiveness of the modulating signal. Thus, the lifetime of the effector/$G\alpha$ complex is reduced because the intrinsic GTPase activity of $G\alpha$ has been increased and the G-protein is turned off. The measurement of receptor-stimulated GTPase represents an important and yet simple tool that can be used, together with several other techniques described in this volume, to assist in defining the increasingly important role of regulatory G-proteins in signal transduction, as well as the emerging role of GTPases in protein synthesis and translocation, as well as cell proliferation and differentiation.

Acknowledgments

The author gratefully acknowledges Professor A. C. Dolphin, Department of Pharmacology, Royal Free Hospital School of Medicine, London, UK, in whose laboratory all of the GTPase experiments were conducted. That project was funded by the Medical Research Council (UK) and Wellcome Trust in grants to Professor Dolphin, as well as the Medical Research Council of Canada in a postdoctoral fellowship to Dr. Sweeney.

References

1. Cassel, D. and Selinger, Z. (1976) Catecholamine-stimulated GTPase activity in turkey erythrocyte membranes. *Biochim. Biophys. Acta* **452,** 538–551.
2. Milligan, G. (1988) Techniques used in the identification and analysis of function of pertussis toxin-sensitive guanine nucleotide binding proteins. *Biochem. J.* **255,** 1–13.
3. Bourne, H. R., Sanders, D. A., and McCormick, F. (1990) The GTPase superfamily: a conserved switch for diverse cell functions. *Nature* **348,** 125–132.

4. Gierschik, P. and Jakobs, K. H. (1990) Receptor-stimulated GTPase activity of G-proteins, in *G-Proteins as Mediators of Cellular Signalling Processes* (Houslay, M. D. and Milligan, G., eds.), Wiley, London, pp. 67–82.

5. Gilman, A. G. (1987) G proteins: transducers of receptor-generated signals. *Annu. Rev. Biochem.* **56,** 615–649.

6. Kaziro, Y., Itoh, H., Kozasa, T., Nakafuku, M., and Satoh, T. (1991) Structure and function of signal-transducing GTP-binding proteins. *Annu. Rev. Biochem.* **60,** 349–400.

7. Avissar, S., Schreiber, G., Danon, A., and Belmaker, R. H. (1988) Lithium inhibits adrenergic and cholinergic increases in GTP binding in rat cortex. *Nature* **331,** 440–442.

8. Sweeney, M. I. and Dolphin, A. C. (1992) 1,4-Dihydropyridines modulate GTP hydrolysis by G_o in neuronal membranes. *FEBS Lett.* **310,** 66–70.

9. Koski, G. and Klee, W. A. (1981) Opiates inhibit adenylate cyclase by stimulating GTP hydrolysis. *Proc. Natl. Acad. Sci. USA* **78,** 4185–4189.

10. Vachon, L., Costa, T., and Herz, A. (1987) GTPase and adenylate cyclase desensitize at different rates in NG 108-15 cells. *Mol. Pharmacol.* **31,** 159–168.

11. Costa, T. and Herz A. (1989) Antagonists with negative intrinsic activity at δ opioid receptors coupled to GTP-binding proteins. *Proc. Natl. Acad. Sci. USA* **86,** 7321–7325.

12. Sweeney, M. I. and Dolphin, A. C. (1993) Adenosine A1 agonists stimulate the GTPase activity of G_i- and G_o-type G-proteins in cortical membranes: synergism with Bay K 8644. *J. Neurochem.* **61(Suppl.),** S78D.

CHAPTER 6

Measurement
of Adenylyl Cyclase Activity
in Cell Membranes

Robert J. Williams and Eamonn Kelly

1. Introduction

In mammals, adenylyl cyclase is a family of membrane-bound enzymes that catalyze the conversion of adenosine triphosphate (ATP) to adenosine 3':5'-cyclic monophosphate (cAMP). cAMP is an ubiquitous intracellular signaling molecule that modifies cell function by activating cAMP-dependent protein kinase A (PKA). The adenylyl cyclase enzyme is itself regulated by various proteins, including the G-proteins (guanine nucleotide binding regulatory proteins) G_s and G_i, and Ca^{2+}-calmodulin *(1)*. Binding of a hormone to a stimulatory receptor induces coupling of the guanosine diphosphate (GDP) liganded heterotrimeric G_s protein to the receptor. Under these conditions, guanosine triphosphate (GTP) replaces GDP on the G_s protein, which then dissociates into the $G_s\alpha$-GTP subunit and the $\beta\gamma$ complex *(2)*. The $G_s\alpha$-GTP subunit then interacts with and activates the adenylyl cyclase enzyme. Activation of this enzyme ceases when the GTP is hydrolyzed back to GDP by GTPase activity intrinsic to the α-subunit of G-proteins. Binding of a hormone to an inhibitory receptor leads to a similar train of events involving the G_i class of proteins, except that it is unclear exactly how inhibition of the adenylyl cyclase enzyme occurs. Evidence exists that $G_i\alpha$-GTP directly inhibits adenylyl cyclase, whereas other studies implicate a role for $\beta\gamma$ complexes liberated from G_i in combining with and functionally inactivating free $G_s\alpha$-GTP *(1,3)*. Both mechanisms probably operate under different conditions.

From: *Methods in Molecular Biology, Vol. 41: Signal Transduction Protocols*
Edited by: D. A. Kendall and S. J. Hill Copyright © 1995 Humana Press Inc., Totowa, NJ

The first mammalian adenylyl cyclase to be cloned was from bovine brain *(4)*, and the 120-kDa protein encoded by this clone was proposed to have a somewhat unexpected topography, including a short intracellular amino terminal, two large hydrophobic domains each hypothesized to contain six transmembrane-spanning regions, and two ≈40-kDa cytoplasmic domains (C1 and C2), which are essential for catalytic activity. The proposed structure was unexpected because adenylyl cyclase was not thought to be a membrane-spanning protein, and also because of the striking structural resemblance between adenylyl cyclase and some ion channels (e.g., dihydropyridine-sensitive Ca^{2+} channels and K^+ channels) and transporter proteins (e.g., the P glycoprotein), implying that adenylyl cyclase may subserve other functions in addition to cAMP production. Since then, at least five other types of mammalian adenylyl cyclase have been cloned (types I–VI; reviewed in ref. *5*), and others are sure to follow. Although the different types are localized to different tissues (e.g., type I in brain and type III in olfactory tissue), all are likely to share the same basic structure just described, all are activated by GTP liganded $G_s\alpha$, and all are activated by the diterpine drug forskolin. In other respects, however, there are important differences in the regulation of these types of adenylyl cyclase. First, types I and III are activated by Ca^{2+}-calmodulin, implying that agents that elevate cytosolic Ca^{2+} will also increase cAMP levels in cells where these types are expressed, and second, βγ subunits inhibit the activity of type I, have little effect on types III, V, and VI, but can activate type II and IV adenylyl cyclase *(5)*. This latter stimulatory effect of βγ subunits is, however, conditional, in that stimulation only occurs in the presence of coactivation of the adenylyl cyclase by $G_s\alpha$-GTP *(6)*. Because this effect requires high concentrations of βγ, it is thought that the necessary βγ complexes are liberated upon activation of other G-proteins, such as G_i and G_o, which in brain at least are present in much higher concentrations than G_s.

As just mentioned, the C1 and C2 regions of adenylyl cyclase are necessary for catalytic activity, and interestingly, these regions are highly homologous to the catalytic domains of membrane-bound guanylyl cyclases. The substrate for adenylyl cyclase is Mg^{2+}-ATP rather than ATP alone, and Mg^{2+} ions also serve as a cofactor for the enzyme. Furthermore, Mn^{2+} ions can substitute for Mg^{2+} in this reaction. Once formed, the cAMP is rapidly degraded to 5'-AMP by a family of enzymes called cyclic nucleotide phosphodiesterases *(7)*, effective inhibition of which is vital if reliable estimates of adenylyl cyclase activity are to be made.

A number of techniques are available to measure the activity of adenylyl cyclase in tissues. Perhaps the simplest involves the assay of endogenous cAMP levels in membranes, whole cells, or tissue slices using a bovine adrenal cortex protein-binding technique *(8)* or a commercially available radioimmunoassay. However, it is often difficult to obtain near-complete inhibition of phosphodiesterase activity in whole tissues or cells. Another method involves the prelabeling of the ATP pool by preincubating cells or tissue slices with [^{3}H]adenine *(9)*, which then requires sequential chromatography to separate [^{3}H]cAMP from other tritiated precursors and products. This again requires careful attention to the method and extent of phosphodiesterase inhibition.

Another widely used method, and the one to be described in detail here, was originally described by Salomon et al. in 1974 *(10)*, and involves the use of [α^{32}P]ATP as the substrate for adenylyl cyclase. Cell homogenates or membranes are incubated with [α^{32}P]ATP in a suitable buffer normally in the presence of an ATP-regenerating system, Mg^{2+} ions, GTP, a phosphodiesterase inhibitor, and unlabeled cAMP (to prevent access of phosphodiesterase to the freshly synthesized [^{32}P]cAMP). Following production of [^{32}P]cAMP from the labeled substrate, the former is isolated from the latter and other labeled products by a simple sequential chromatography technique using Dowex 50W-X4 cation-exchange resin followed by activated alumina. This technique is based on earlier work by Krishna et al. *(11)*, who showed that cAMP could be separated from ATP, ADP, AMP, and Pi on Dowex cation-exchange resin, because cAMP binds to the Dowex, whereas ATP and other products pass through; and White and Zenser *(12)* and Ramanchandran *(13)*, who were able to separate cAMP from ATP and other products on activated aluminum oxide, which retains ATP, whereas cAMP can be readily eluted off. Although both of these methods were able to separate cAMP from ATP and other products almost completely, Salomon and colleagues demonstrated that the sequential use of both steps could achieve much greater separation. This is vitally important because the amount of [^{32}P]cAMP produced represents only a tiny fraction (usually around 0.05%) of the total [^{32}P]-labeled products (mainly ATP) in the reaction mixture. Because the recovery of [^{32}P]cAMP can vary from column to column, this is carefully quantified by the addition of [^{3}H]cAMP to the incubation mix before addition to the Dowex resin. This technique for quantifying adenylyl cyclase activity has the advantages of high sensitivity and reproducibility, coupled with the possibility of obtaining almost complete inhibition of phosphodiesterase activity by using a combination of phosphodiesterase inhibitor and a high concentration of unlabeled cAMP to saturate the phosphodiesterase enzyme.

2. Materials

1. Thermostatically controlled water bath.
2. Perspex shielding (at least 10-mm thick).
3. Plastic disposable test tubes (e.g., LP3 or LP4 type).
4. Teflon™-glass (or equivalent) hand-held homogenizer.
5. Plastic Konte Disposaflex polypropylene columns (overall length, 265 mm, internal diameter 8 mm, full stop). Available from Burkard (Uxbridge, Middlesex, UK).
6. Glass wool.
7. Dowex AG 50W-X4 (200–400) analytical-grade cation-exchange resin.
8. Chromatographic Alumina, activity-grade Super 1, type WN-6:neutral (Sigma [St. Louis, MO] no. A-1522).
9. Homogenization buffer: $0.3M$ sucrose, 25 mM Tris-HCl, pH 7.5.
10. Assay buffer (10X): $0.05M$ MgCl$_2$, $0.5M$ Tris-HCl, pH 7.5.
11. Assay premix (20X): $0.02M$ ATP (disodium salt, e.g., Sigma no. A2383), 20 μM GTP (sodium salt, e.g., Sigma no. G5756), and $0.02M$ cAMP (sodium salt, e.g., Sigma no. A6885) in 50 mM Tris-HCl, pH 7.5. This assay premix is stored in ≈0.5 mL aliquots at –20°C.
12. Creatine phosphate (20X): $0.4M$ creatine phosphate (disodium salt, e.g., Sigma no. P6502) in 50 mM Tris-HCl, pH 7.5. This is stored in ≈0.5-mL aliquots at –20°C. Aliquots of assay premix and creatine phosphate can be refrozen if partially used in an experiment.
13. Creatine phosphokinase (type I from rabbit muscle, ≈200 U/mg protein, e.g., Sigma no. C3755) weighed out fresh for each experiment; final concentration in incubation ≈130 U/mL.
14. Phosphodiesterase inhibitor, e.g., use a stock solution of 5 mM Ro 20-1724 {[4-(3-butoxy-4-methoxybenzyl)-2-imidazolidinone]; made up in 5% dimethyl sulfoxide}, to give a final concentration in the incubation of 250 μM.
15. [α^{32}P]-ATP (e.g., Amersham [Amersham, UK] PB171, 2 mCi/mL; ≈30 Ci/mmol) (*see* Note 1).
16. [8-^{3}H]-cAMP (e.g., Amersham TRK304, 1 mCi/mL; 26 Ci/mmol). Stock [^{3}H]-cAMP (≈10,000–20,000 cpm/100 μL) in water is stored in 10-mL aliquots at –20°C.
17. Trichloroacetic acid (6.25% w/v).
18. Imidazole buffer (10X): $1M$ imidazole, pH 7.5.
19. Disposable plastic scintillation vials (20 mL).

3. Methods

3.1. Column Preparation

1. The plastic Konte columns we employ in this procedure are supplied in three separate sections that can be glued together prior to use. A small plug of glass wool is inserted into the lower end of the column, with the aid of a glass rod, to support the resin or alumina bed. Gloves should be worn when handling glass wool. The plug of glass wool should be large enough not to allow resin or alumina to escape from the column, but not so large as to impair the flow rate (which in our setup is ≈0.5 mL/min).
2. Prepare a slurry of Dowex AG 50W-X4 by mixing the resin with an equal volume of 1M HCl. Two milliliters of the constantly stirred slurry are pipeted into each column to obtain a resin bed volume of approx 1 mL.
3. Add 0.8 g of neutral alumina to an equal number of columns as have been prepared with Dowex. This is easily achieved by weighing 0.8 g of alumina into a 1.5-mL plastic microcentrifuge tube, and then cutting off the top of the tube at the upper level of the alumina powder. With the aid of a pair of forceps, this can then be used as a scoop to add the correct amount of alumina to the required number of columns.
4. For the sequential chromatography steps, the Dowex ion-exchange columns need to be arranged directly above the alumina columns, and the alumina columns in turn drip directly into scintillation vials (*see* Fig. 1). For large numbers of columns, purpose built perspex racks are convenient (*see* Note 7).

3.2. Incubation Procedure

1. The total incubation volume per sample is 100 µL. This is composed of 30 µL of assay premix, 40 µL of membrane/homogenate, and the remaining 30 µL are used for addition of drugs as required. The first step therefore is to prepare sufficient premix for the number of incubation tubes set up, and a table giving relative proportions of the constituents of the premix in relation to assay size is given later (*see* Table 1). The final assay concentrations of premix constituents that we normally use are: 50 mM Tris-HCl, pH 7.5, 5 mM MgCl$_2$ (*see* Note 6), 1 µM GTP, 1 mM cAMP, 250 µM Ro 20-7240 (*see* Note 3), 20 mM creatine phosphate, ≈130 U/mL creatine phosphokinase, and 1 mM ATP (*see* Note 2), containing ≈2 µCi/tube [α^{32}P]ATP. Once the premix is prepared, it can be stored in an LP4 tube, in an ice-filled lead pig. Thirty microliters of premix can then be aliquoted into each incubation tube on ice.
2. Drugs can then be added to each tube as required in a maximum volume of 30 µL of distilled water or appropriate vehicle (*see* Note 10).

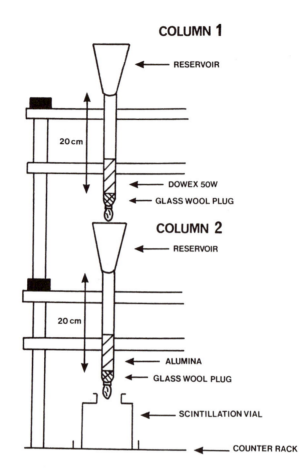

Fig. 1. Arrangement of columns for sequential chromatography.

3. If a crude cell homogenate is to be used, cells are lysed (e.g., 25 full strokes in a hand-held homogenizer) immediately prior to use in ice-cold homogenization buffer to give a homogenate concentration of ≈2–5 mg/mL of protein (*see* Note 4). If purified cell membranes are to be used in the assay, they can be prepared in advance by standard methods and stored in e.g., 25 m*M* Tris-HCl buffer, pH 7.5, at –80°C until required. For NG108-15 neuroblastoma × glioma hybrid cells, we prepare membranes by lysing washed cell pellets in homogenization buffer as described in step 3, and then centrifuging the lysate at 500*g* for 10 min at 4°C to remove nuclei and unbroken cells. The supernatant is then centrifuged at 16,000*g* for 30 min at 4°C, and the resultant pellet washed in ice-cold Tris buffer. The final washed

Table 1
Premix Constituents

Vol (µL)	Number of incubation tubes					
	30	40	50	60	70	80
Assay buffer	300	400	500	600	700	800
Assay premix	150	200	250	300	350	400
Creatine PO$_4$	150	200	250	300	350	400
CP kinase (mg)	2	2.7	3.3	4	4.7	5.3
[α^{32}P]ATP[a]			As required			
Ro 20-1724	150	200	250	300	350	400
H$_2$O			As required			
Final volume (µL)	900	1200	1500	1800	2100	2400

[a]To add 2 µCi [α^{32}P]ATP to each tube, the decay of the radiolabel must be taken into account. For example, if the experiment is performed 3 d after the reference date of the radiolabel, then in order to add 2 µCi/tube to the premix, it is necessary to divide the volume that would be added on the reference date by the decay factor (0.865) If performing the experiment 3 d before the reference date, you would multiply the volume by the same factor (i.e., use less radiolabel). For tissues with very low or very high enzyme activity, more (e.g., 3 µCi) or less (e.g., 1 µCi) [α^{32}P]ATP may be employed, respectively.

membrane pellet is resuspended in 25 mM Tris-HCl, pH 7.5, to a protein concentration of ≈1–2 mg/mL.

4. To begin the incubation (normally 10 min at 37°C in the water bath; *see* Note 5), vortex the homogenate/membrane preparation, and add 40 µL (≈100–200 µg of protein; *see* Note 4) of this to each incubation tube already containing 60 µL of premix and drugs to give the final incubation volume of 100 µL before placing in the water bath. For convenience, we normally stagger the addition of protein to each tube by 10 s, being careful to vortex briefly the membrane/homogenate preparation between each addition to prevent settling. In each experiment, it is also necessary to include two incubation tubes containing 30 µL premix, 30 µL distilled water, and 40 µL of homogenization buffer (i.e., enzyme blanks) to quantify background [^{32}P] levels in the absence of protein (enzyme blanks). Remember to retain a small amount of homogenate for protein determination.

5. To terminate the incubation, add 800 µL of ice-cold 6.25% (w/v) trichloroacetic acid to each tube, vortex, and place on ice for at least 10 min.

6. Add 100 µL of the [8-^{3}H]cAMP solution to each tube.

7. Centrifuge at 1500g, 4°C for 20 min to pellet precipitated protein.

3.3. Sequential Chromatography

1. Dowex columns are primed prior to use by addition of 6 mL $1M$ HCl, followed by 6 mL distilled water. Alumina columns are primed by addition of 8 mL $0.1M$ imidazole, pH 7.5, followed by 8 mL distilled water. If solutions do not pass through the columns, this will almost certainly be because of an air-block, which can be removed by agitating the column.
2. Place the Dowex columns over a suitable container to collect waste eluate, and carefully decant the contents of the incubation tubes onto the columns (*see* Note 1).
3. After allowing this to pass through the resin bed, add 2.5 mL distilled water to each column, and allow this to pass through. Most of the original [^{32}P]ATP will be present in the waste eluate.
4. Locate the Dowex columns directly above the alumina columns, and add 3 mL distilled water to the former to elute cAMP onto the alumina (*see* Fig. 1). This elution profile can be easily checked by monitoring the elution of a small amount of [8-^{3}H]cAMP from the Dowex column.
5. When the water has passed through both columns, the Dowex columns are removed and the alumina columns are then placed above 20-mL scintillation vials.
6. Add 4 mL $0.1M$ imidazole to the alumina columns to elute cAMP into the vials, ATP and other nucleotides being retained in the alumina.
7. Add sufficient scintillant to create a homogeneous sample for counting (this varies with type of scintillant, but we use Packard Emulsifier-Safe, which requires $\approx$16 mL).
8. Count the samples using a dual ^{3}H/^{32}P counting program (*see* Note 9).
9. For the purposes of the calculation of adenylyl cyclase described in Section 3.4., it is necessary also to prepare the following samples for scintillation counting:
   a. Total [α^{32}P]ATP. Add 30 µL of a 1:1000 dilution of premix to two vials followed by 4 mL imidazole and scintillant as necessary.
   b. Total [8-^{3}H]cAMP. Add 100 µL of the stock solution to two vials followed by 4 mL imidazole and scintillant as necessary.
   c. Imidazole buffer blanks. Add 4 mL imidazole to two vials followed by scintillant.

3.4. Calculation of Adenylyl Cyclase Activity

1. The following data are required for the calculation:
   a. ^{3}H counts = cpm in ^{3}H channel for a particular sample.
   b. ^{32}P counts = cpm in ^{32}P channel for the same sample.
   c. ^{3}H blank = cpm in ^{3}H channel for imidazole buffer blank.
   d. ^{32}P blank = cpm in ^{32}P channel for imidazole buffer blank.
   e. ^{3}H counts total = cpm in the total [8-^{3}H]cAMP standard.

f. Enzyme blank = cpm in the ^{32}P channel from incubation with no protein.

g. cpm/pmol of [^{32}P]cAMP (which is in fact calculated by dividing cpm in ^{32}P channel of the total [α^{32}P]ATP counts by 1000, i.e., the concentration of ATP in the premix is 3.33 mM, 10 µL of this are diluted by 100, and 30 µL of this counted, so you are counting the ^{32}P cpm equivalent to 1 nmol ATP or in effect the ^{32}P cpm equivalent to 1 nmol cAMP).

h. Crossover = cpm in ^{3}H channel/cpm in ^{32}P channel for the total [α^{32}P]ATP standard.

i. Protein (in mg) in sample.

j. Reaction time (in min).

2. First calculate the recovery of cAMP (%) from each column using the following calculation:

$$\text{Recovery} = \{(^3\text{H cpm} - {}^3\text{H blank cpm}) - [(^{32}\text{P cpm} - {}^{32}\text{P blank cpm}) \times \text{crossover}]/(^3\text{H cpm total} - {}^3\text{H blank cpm})\} \times 100 \quad (1)$$

3. Adenylyl cyclase activity (pmol/min/mg protein) is then calculated as follows:

$$[100/\text{recovery (\%)}] \times (^{32}\text{P cpm} - \text{enzyme blank cpm})/$$
$$[(^{32}\text{P})\text{cAMP, cpm/pmol}] \times (\text{mg protein/incubation}) \quad (2)$$
$$\times (\text{incubation time, min})$$

4. Example: To calculate basal adenylyl cyclase activity in an NG108-15 cell homogenate, the following values were obtained:

a. ^{3}H counts in sample = 9370.

b. ^{32}P counts in sample = 638.

c. ^{3}H blank = 20.

d. ^{32}P blank = 24.

e. ^{3}H counts total = 15,475.

f. Enzyme blank = 53.

g. cpm in ^{32}P channel for total [α^{32}P]counts was 36,000, so cpm/pmol ^{32}P = 36.

h. Crossover = 0.02.

i. Protein = 0.21 mg/incubation.

j. Time = 10 min

So,

$$\text{Recovery} = \{(9370 - 20) - [(638 - 24) \times 0.02]/ \quad (3)$$
$$(15,475 - 20)\} \times 100 = 60.4\%$$

Therefore,

$$\text{Activity} = [(100/60.4) \times (638 - 53)]/(36 \times 0.21 \times 10) = \quad (4)$$
$$12.8 \text{ pmol/min/mg protein}$$

A simple program containing this calculation and suitable for an IBM-PC is given in Note 11.

4. Notes

1. Safety precautions. [^{32}P] is a high-energy β-emitter, and the relevant precautions must therefore be taken. Once the radiolabel is present in the assay, we perform all manipulations behind 1-cm thick perspex screens (e.g., 60-cm high by 45-cm wide), and monitor constantly for emissions using a hand-held detector. The laboratory bench and water bath will also aid in screening. To minimize exposure to the premix, we rapidly aliquot the premix into the incubation tubes using an Eppendorf repeating pipet. It should be noted that by far the greatest proportion of radioactive waste appears in the primary eluate from the Dowex columns, and great care must be taken with disposal of this. The easiest method is to pour it carefully down the sink, along with copious amounts of water, but local regulations for your institution should be consulted before disposal. We store our contaminated dry waste (e.g., pipet tips) in plastic bags inside a perspex box The half-life of ^{32}P is 14.3 d, and so dry waste will decontaminate fairly rapidly.

2. ATP levels in the incubation are maintained by the creatine phosphate/ creatine phosphokinase ATP-regenerating system. This counteracts the conversion of ATP to ADP by membrane-bound ATPases, and concentrations of these constituents may need to be increased if the ATPase activity of a particular tissue is very high.

3. To inhibit phosphodiesterase activity during the incubation, we use the inhibitor Ro 20-1724 (250 μM) in combination with a high concentration of unlabeled cAMP (1 mM), which saturates the enzyme. The concentration of cAMP used for this purpose should not exceed the concentration of ATP in the assay, since cAMP is a weak competitive inhibitor of ATP at the catalytic subunit of adenylyl cyclase. It is also important to appreciate that different phosphodiesterase inhibitors may be required to inhibit the enzyme effectively in other tissues and cell lines *(7)*, and in some cases, a cocktail of inhibitors may be required. A time-course of adenylyl cyclase activity in the presence of the chosen phosphodiesterase inhibitor should be performed prior to any detailed experiments, in order to establish that there is a linear accumulation of cyclic AMP under these conditions.

4. Before detailed experiments are undertaken, it is important to determine that the protein concentration is in the linear range for adenylyl cyclase activation. This is readily checked by performing incubations with varying protein concentrations.

5. For NG108-15 cell homogenates, we perform incubations at 37°C, but for other systems, lower temperatures may be preferable. For further discussion of incubation conditions, refer to ref. *(14)*.

6. Under certain circumstances, it may be necessary to alter the Mg^{2+} concentration. For example, in studies of β_2-adrenoceptor-mediated desensitization of adenylyl cyclase activation in S49 lymphoma cells, lowering the Mg^{2+} concentration to submillimolar levels markedly enhances the observed desensitization *(15)*.

7. Rack dimensions: Each rack we use contains holes (diameter 1 cm) for 100 columns, and is made from two 36-cm^2 perspex sheets, one sheet positioned ≈10 cm below the other and joined by four connecting rods at the corners. The connecting rods protrude another 15 cm below the bottom perspex sheet. The whole arrangement is designed so that one rack (containing Dowex columns) can sit on top of the other (containing alumina columns). It is also convenient for the top rack to locate into recesses drilled at the corners of the top sheet of perspex of the second rack. The collecting trays that we use below the columns are large plastic food containers with the lids removed.

8. Column maintenance. Columns can be stored on the bench for long periods with no loss of recovery. Ensure, however, that the tops of the columns are covered to prevent dust from entering. If the columns dry out, stand the bottoms of the columns in distilled water to rehydrate. In our setup, recovery of ^3H is initially 60–80% with fresh columns, and can gradually diminish to 25–40% with use, but without affecting the reproducibility of the assay. We find, however, that if recoveries diminish further, it is better to make fresh columns. Otherwise, the quality of data deteriorates. In our hands, columns can be reused successfully over 100 times.

9. Scintillation counting: Modern counters usually contain preset programs to perform $^3H/^{32}P$ dual counting, and counts per minute values are sufficiently accurate for the purpose of the calculations.

10. Certain agents can be employed to manipulate adenylyl cyclase activity and render useful information about the function of various components of this signal-transduction pathway. Forskolin directly activates adenylyl cyclase *(16)* and, in NG108-15 cell homogenates, produces a very large activation of this enzyme *(17)*. However, forskolin is also able to interact with G_s and amplify hormonal stimulation of adenylyl cyclase *(16)*. Mn^{2+} ions can also directly activate the enzyme *(18)*. Under our incubation conditions, GTP is present at 1 μM, which will support both hormonal activation and inhibition of adenylyl cyclase by G_s and G_i, respectively. Nonhydrolyzable GTP analogs, such as GppNHp (5'-Guanylylimidodiphosphate), can also activate adenylyl cyclase or produce G_i-mediated cyclase inhibition under conditions where the adenylyl cyclase enzyme is already activated *(17)*. Sodium fluoride, which in solution forms fluoroaluminate complexes with trace quantities of aluminium in buffers, can also

activate adenylyl cyclase *(19)* by interacting with GDP-liganded $G_s\alpha$ and mimicking the γ-phosphate of GTP *(20)*. As with guanine nucleotides, sodium fluoride will also produce G_i-mediated inhibition of adenylyl cyclase when the enzyme is already activated *(17)*.

11. Program for the calculation of recovery and activity in adenylyl cyclase assays:

```
10 CLS
20 PRINT DATE$,TIME$
30 PRINT
40 PRINT "ADENYLYL CYCLASE RECOVERY AND ACTIVITY
   CALCULATION"
50 PRINT "................................................................"
60 PRINT
70 FOR N=1 TO 3:PRINT "TURN ON THE PRINTER !!!!!!":NEXT
80 PRINT
90 PRINT
100 REM INPUT CONSTANTS
110 INPUT "COUNTS FOR TRITIUM BLANK";A
120 INPUT "COUNTS FOR PHOSPHORUS BLANK";B
130 INPUT "CROSSOVER";C
140 INPUT "TOTAL TRITIUM COUNTS";D
150 INPUT "BUFFER ONLY PHOSPHORUS (E BLANK)";E
160 INPUT "PHOSPHORUS counts/pMOL cAMP";F
170 INPUT "mg PROTEIN per reaction";G
180 INPUT "REACTION TIME IN MINUTES";H
190 INPUT "HOW MANY REPLICATES";R
200 PRINT
210 PRINT
220 INPUT "IS THIS CORRECT ? (Y/N)";Q$
230 IF Q$ = "N" THEN GOTO 110
240 IF Q$ = "Y" THEN GOTO 250 ELSE GOTO 220
250 LPRINT TIME$,DATE$
260 LPRINT "............................................................"
270 LPRINT "3H BLANK = ";A
280 LPRINT "32P BLANK = ";B
290 LPRINT "CROSSOVER = ";C
300 LPRINT "TOTAL 3H = ";D
310 LPRINT "E BLANK = ";E
320 LPRINT "CPM / pMOL 32P cAMP = ";F
330 LPRINT "REACTION TIME (MINS) = ";H
```

```
340 LPRINT "ACTIVITY EXPRESSED AS pMOL cAMP/min/mg
protein
350 LPRINT
360 LPRINT "................................................................"
370 LPRINT
380 LPRINT "TUBE'','3H'',"32P'',"%RECOVERY","ACTIVITY"
390 LPRINT "................................................................"
400 LPRINT "mg PROTEIN/REACTION = ";G
410 COUNT=1
420 INC=1
430 PRINT
440 PRINT "TRITIUM COUNT OF 1 ENDS"
450 PRINT "TRITIUM COUNT OF 2 CHANGES PROTEIN"
455 PRINT "TRITIUM COUNT OF 3 EJECTS PAGE"
460 PRINT
470 INPUT "TRITIUM FOR THIS SAMPLE";TRI
480 IF TRI=1 THEN LPRINT CHR$(12):SYSTEM
490 IF TRI=2 THEN INPUT "NEW PROTEIN ? ";G:LPRINT
    "mgPROTEIN/REACTION =";G:INC=1 :REP=0:GOTO 470
495 IF TRI=3 THEN LPRINT CHR$(12):GOTO 470
500 INPUT "PHOS. FOR THIS SAMPLE ";PHO
510 TOP=(TRI-A)-((PHO-B)*C)
520 BOTTOM=D-A
530 REC (TOP/BOTTOM)*100
540 PRINT "RECOVERY = ";REC;"%"
550 TOP = (100/REC)*(PHO-E)
560 BOTTOM=F*G*H
570 ACTIV=TOP/BOTTOM
580 PRINT "ACTIVITY = ";ACTIV; "pMOL cAMP/MIN/mg
    PROTEIN"
590 LPRINT COUNT,TRI,PHO,REC,ACTIV
600 REP=REP+ACTIV
610 IF INC=R THEN LPRINT ,,,,"AVGE= ";REP/R:
    LPRINT,,,,"........................":INC=0:REP=0
620 COUNT = COUNT+1
630 INC=INC+1
640 GOTO 460
```

The above BASIC program calculates recovery and activity as described in the text. Originally written in GW-BASIC for the IBM-PC, it should run with little or no modification on most machines. Points to note:

a. Line 480 contains the GW-BASIC system command to return to the operating system when finished. Other basic dialects may use a different command.

b. The printer must be connected and turned on, or the program may crash.

c. Answers to "is this correct? (Y/N)" must be in upper case.

d. It is important to differentiate between commas, colons, and semicolons.

References

1. Gilman, A. G. (1989) G-proteins and regulation of adenylyl cyclase. *JAMA* **262**, 1819–1825.

2. Hepler, J. R. and Gilman, A. G. (1992) G proteins. *Trends Biochem. Sci.* **17**, 383–387.

3. Birnbaumer, L., Abramowitz, J., and Brown, A. M. (1990) Receptor-effector coupling by G-proteins. *Biochim. Biophys. Acta* **1031**, 163–224.

4. Krupinski, J., Coussen, F., Bakalyar, H. A., Tang, W.-J., Feinstein, P. G., Orth, K., Slaughter, C., Reed, R. R., and Gilman, A. G. (1989) Adenylyl cyclase amino acid sequence: possible channel- or transporter-like structure. *Science* **244**, 1558–1564.

5. Tan W.-J. and Gilman, A. G. (1992) Adenylyl cyclases. *Cell* **70**, 869–872.

6. Federman, A. D., Conklin, B. R., Schrader, K. A., Reed, R. R., and Bourne, H. R. (1992) Hormonal stimulation of adenylyl cyclase through G_i-protein βγ subunits. *Nature (Lond.)* **356**, 159–161.

7. Nicholson, C. D., Challiss, R. A., and Sahid, M. (1991) Differential modulation of tissue function and therapeutic potential of selective inhibitors of cyclic phosphodiesterase isoenzymes. *Trends Pharm. Sci.* **12**, 19–27.

8. Brown, E. L., Albano, J. D. M., Ekins, R. P., Sgherzi, A. M., and Tampion, W. (1971) A simple and sensitive saturation assay method for the measurement of adenosine 3':5'cyclic monophosphate. *Biochem. J.* **121**, 561,562.

9. Shimizu, H., Crevelling, C. R., and Daly, J. W. (1969) A radioisotopic method for measuring the formation of adenosine 3',5' monophosphate in incubated slices of brain. *J. Neurochem.* **16**, 1609–1616.

10. Salomon, Y., Londos, C., and Rodbell, M. (1974) A highly sensitive adenylate cyclase assay. *Anal. Biochem.* **58**, 541–548.

11. Krishna, G., Weiss, B., and Brodie, B. B. (1968) A simple sensitive method for the assay of adenyl cyclase. *J. Pharmacol. Exp. Ther.* **163**, 379–385.

12. White, A. A. and Zenser, T. V. (1971) Separation of cyclic 3',5' nucleoside monophosphates from other nucleotides on aluminum oxide columns. *Anal. Biochem.* **41**, 372–396.

13. Ramanchandran, J. (1971) A new simple method for separation of adenosine 3',5'-cyclic monophosphate from other nucleotides and its use in the assay of adenyl cyclase. *Anal. Biochem.* **43**, 227–239.

14. Johnson, R. A. and Salomon, Y. (1991) Assay of adenylyl cyclase catalytic activity. *Methods Enzymol.* **195**, 3–21.

15. Clark, R. B., Friedman, J., Johnson, J. A., and Kunkel, M. W. (1987) β-adrenergic receptor desensitization of wild-type but not cyc-lymphoma cells unmasked by submillimolar Mg^{2+}. *FASEB. J.* **1**, 289–297.

16. Seamon, K. B. and Daly, J. W. (1986) Forskolin: its biological and chemical properties. *Adv. Cyclic Nucleotide Protein Phos. Res.* **20,** 1–150.
17. Kelly, E., Keen, M., Nobbs, P., and MacDermot, J. (1990) NaF and guanine nucleotides modulate adenylate cyclase activity in NG108-15 cells by interacting with both G_s and G_i. *Br. J. Pharmacol.* **100,** 223–230.
18. Bender, J. L., Wolf, L. G., and Neer, E. J. (1984) Interaction of forskolin with resolved adenylate cyclase components. *Adv. Cyclic Nucleotide Res.* **17,** 101–109.
19. Kelly, E., Keen, M., Nobbs, P., and MacDermot, J. (1990) Segregation of discrete $G_s\alpha$ mediated responses that accompany homologous or heterologous desensitization in two related somatic hybrids. *Br. J. Pharmacol.* **99,** 309–316.
20. Bigay, J., Deterre, P., Pfister, C., and Chabre, M. (1985) Fluoroaluminates activate transducin-GTP by mimicking the γ-phosphate of GTP in its binding site. *FEBS Lett.* **191,** 181–185.

CHAPTER 7

The Measurement of Cyclic AMP Levels in Biological Preparations

Stephen P. H. Alexander

1. Introduction

The generation of cyclic AMP (cAMP) is probably the most scrutinized of signal transduction pathways. Receptors may be linked to the generation of cAMP by one of two routes. One group of receptors (such as β-adrenoceptors, A_2 adenosine receptors, D_1 dopamine receptors, histamine H_2 receptors, and some of the prostanoid receptors) is associated with an increased adenylyl cyclase activity and elevated cAMP levels mediated through a guanine nucleotide binding protein, G_s. Activation of the second group leads to inhibition of adenylyl cyclase and a reduction in cAMP generation. Examples of the latter group include α_2-adrenoceptors, A_1 adenosine receptors, D_2 dopamine receptors, 5-HT_1-type receptors, metabotropic glutamate receptors, and μ opioid receptors. The mechanism of inhibition of adenylyl cyclase activity appears to be mediated through a distinct guanine nucleotide binding protein, G_i. These two G-proteins act on adenylyl cyclase either to stimulate or inhibit the conversion of ATP to cAMP and PP_i.

This chapter deals with methods of measuring cAMP generation in tissue slices, cells in culture, and (briefly) cell-free preparations. The chapter is broadly divided in two sections: the use of radiolabeled adenine as a precursor of ATP and hence cAMP, and competition for [^{3}H]cAMP binding to the regulatory subunit of cAMP-dependent protein kinase—the radioreceptor method. Both these methods are currently in use in the author's laboratories, and are adaptations of published procedures (1–3).

From: *Methods in Molecular Biology, Vol. 41: Signal Transduction Protocols*
Edited by: D. A. Kendall and S. J. Hill Copyright © 1995 Humana Press Inc., Totowa, NJ

The former method is used on a daily basis for the measurement of cAMP generation in brain slices and for many cells in culture. The latter method is more applicable under circumstances where it is either undesirable or unworkable to use [^{3}H]adenine (*see* Notes 1–3).

2. Materials

2.1. Radioligands

1. [^{3}H]Adenine: The specific activity of [^{3}H]adenine used in the assay for [^{3}H]cAMP generation is approx 900 GBq/mmol (Amersham International, Amersham, UK, or NEN DuPont, Boston, MA). It is stored at 4°C undiluted prior to use.
2. [^{14}C]cAMP: The specific activity of [^{14}C]cAMP used in the assay for [^{3}H]cAMP generation is approx 1.6 GBq/mmol (Amersham International or NEN DuPont). It is stored at –20°C, diluted 1:30 in distilled water as aliquots of 100 µL ($\approx$1.3 kBq). Immediately prior to use, the [^{14}C]cAMP is diluted with 1M HCl to provide sufficient volume for halting the incubation (i.e., for 48 sample tubes, add 5 mL to include 1/2 "shots" for counting directly to estimate chromatographic recovery).
3. [^{3}H]cAMP: The specific activity of [^{3}H]cAMP used in the assay for measuring endogenous cAMP generation is approx 1600 GBq/mmol (Amersham International or NEN DuPont). It is stored at –20°C, diluted 1:10 with 50% ethyl alcohol (ETOH) as aliquots of 10 µL (37 kBq). Immediately prior to use, the [^{3}H]cAMP is diluted to 370 Bq/25 µL with 2.5 mL distilled water.

2.2. Assay Media

1. Krebs-Ringer-Bicarbonate (KRB): 118 mM NaCl, 4.7 mM KCl, 1.2 mM MgSO$_4$ · 7H$_2$O, 1.2 mM KH$_2$PO$_4$, 25 mM NaHCO$_3$, 11.7 mM glucose, 1.3 mM CaCl$_2$ · 2H$_2$O pregassed at 37°C with O$_2$/CO$_2$ 95/5. This medium must be made up fresh. It is also possible to prepare 10X concentrated solutions and keep these at 4°C for up to 2 wk, mixing and diluting them on the day of assay.
2. Hank's HEPES Buffer (HHB): 136 mM NaCl, 5.4 mM KCl, 0.83 mM MgSO$_4$ · 7H$_2$O, 0.45 mM KH$_2$PO$_4$, 0.13 mM Na$_2$HPO$_4$ · 12H$_2$O, 5.6 mM glucose, 1.3 mM CaCl$_2$ · 2H$_2$O, 20 mM HEPES. This medium may be prepared in advance and kept at 4°C, but should be used within 1 wk.
3. TENT Buffer: 0.1M Tris buffer, pH 7.4, containing 20 mM ethylenediamine tetraacetic acid (EDTA), 400 mM NaCl, and 20 mM theophylline. It may be

necessary to warm the solution to dissolve the theophylline. This buffer may be made up in advance and stored as aliquots (10–20 mL) at –20°C. Solutions should not be refrozen after use.

4. Wash buffer: 25 mM Tris buffer, pH 7.4, containing 5 mM EDTA. This buffer should be made up relatively fresh and used within 7 d.

2.3. Chromatographic Resins

1. Dowex 50: This cation-exchange resin (Bio-Rad, Richmond, CA) is prepared by washing with cycles of 1M HCl and distilled water. Dispense aliquots of the 50% slurry in water into disposable columns (Bio-Rad Econo-columns or equivalent) to approx 1 mL of resin bed volume. Regenerate the resin between uses with a single cycle of washing with 5 mL of 1M HCl and 20 mL of distilled water. Occasional elution with 2 mL of 1M NaOH followed by water and an HCl/water cycle will also help to regenerate the columns, and reduce elevated basal levels of [^{3}H]cAMP.

2. Alumina: This resin (Sigma, St. Louis, MO) is dispensed as 0.8-g aliquots of the dry powder into disposable columns (Bio-Rad Econo-columns or equivalent) and prepared by washing with 0.1M imidazole. Regenerate the alumina between uses by washing with 10 mL of 0.1M imidazole. Occasional washing with 2 mL of 1M NaOH followed by 20 mL of 0.1M imidazole is useful to reduce elevated basals.

3. Rack construction: The columns are mounted in two sets of aluminum racks that are constructed such that one rack allows elution from one set of columns (containing the Dowex 50 resin) directly to the second (containing the alumina), which in turn may be positioned over a set of 20-mL scintillation vials for collecting the eluate.

2.4. Preparation of the Radioreceptor Binding Protein

1. Obtain 12 bovine adrenals from the slaughterhouse, free from adipose tissue, and slice in two longitudinally.

2. Remove the lighter medulla, and scrape the cortex from the capsule.

3. Homogenize the cortex in 6 vol of ice-cold 20 mM NaHCO$_3$ and 1 mM dithiothreitol (at own pH) for 10 s using a Polytron (Kinematica, Luzern, Switzerland).

4. Centrifuge the homogenate at 5000g for 10 min, and then remove the supernatant layer for storage at 4°C.

5. Rehomogenize the pellet in 3 vol of the same solution, and again centrifuge at 5000g for 10 min.

6. Combine the supernatant fractions, and centrifuge at 38,000g for 30 min.

7. Store the supernatant fraction for isolation of the cAMP binding protein.

8. Rehomogenize the pellet in the same solution and recentrifuge.

9. This washing may be repeated once more.
10. Resuspend the final pellet (bovine adrenal cortical membranes [BACM]) to 15–20 mg protein/mL, and freeze as 1-mL aliquots (–20°C) for use in the measurement of inositol 1,4,5-trisphosphate (*see* Chapter 14).
11. Precipitate the cAMP binding protein from the high-speed supernatant fraction by the addition of ammonium sulfate to a final concentration of 400 g/L, stirring for 60 min on ice.
12. Centrifuge the suspension at 20,000g for 40 min, dissolve the resulting pellet in 50 mM Tris, pH 7.4, and dialyze overnight against 50 mM Tris and 2 mM dithiothreitol, pH 7.4.
13. Centrifuge the suspension again at 20,000g for 40 min, and freeze the supernatant layer (bovine adrenal cortical cytosol [BACC]) as aliquots at –20°C (*see* Note 4).

3. Methods
3.1. Tissue Preparation
3.1.1. Preparation of Tissue Slices

1. Dissect out particular brain regions, or dissect tissues on ice and then cut into miniprisms (350 × 350 μm) using a McIlwain tissue chopper, turning the tissue through 90° after the first series of cuts.
2. Disperse the slices into oxygenated KRB in a stoppered conical flask.
3. After 1 h of equilibration at 37°C with three to four changes of buffer (allowing slices to sediment and decanting the supernatant, then replacing with KRB), wash the miniprisms with KRB, and allow to settle under gravity.
4. Slices are then usable for assays of cAMP accumulation using either a radioactive precursor or the radioreceptor assay for measuring endogenous cAMP mass.

3.1.2. Preparation of Cells in Culture

1. Wash the cells twice with HHB to remove traces of cell-growth medium.
2. Cells are then usable for assays of cAMP accumulation.

3.2. The Use of a Radiolabeled Precursor to Follow cAMP Generation
3.2.1. [³H]cAMP Generation in Tissue Slices

1. Incubate the miniprisms with [³H]adenine to a concentration of 74 kBq/mL in a stoppered, gassed flask containing KRB (200 μL tissue/mL KRB).
2. After a 40–60-min incubation, wash the slices with two to three changes of KRB, and then allow to settle.

3. Distribute aliquots (25–50 μL, using an Oxford MicroDoser, Lancer, Kildare, Ireland) into flat-bottomed vials (Hughes & Hughes, Somerset, UK) containing KRB to maintain a final vol of 300 μL.

4. Antagonists/modulators (e.g., degradative enzymes) may be added to the slices at this point.

5. After a brief equilibration period (usually 10 min), add agonist in 10–20 μL and continue the incubation (usually for 10 min), before termination with 200 μL of ice-cold $1M$ HCl containing 25–30 Bq [^{14}C]cAMP.

6. Add 750 μL of cold water, and then centrifuge the suspension at >2000g for 5–10 min.

3.2.2. [^{3}H]cAMP Generation in Cells in Culture

1. Add [^{3}H]adenine to a final concentration of 74 kBq/mL, and continue the incubation at 37°C for 1–2 h.

2. Wash the [^{3}H]adenine out with two changes of HHB, and then incubate the cells in HHB (to a total vol of 1 mL in 24-well plates) in the presence of antagonists/degradative enzymes where necessary.

3. Add agonist in 10 μL, and incubate the cells for the required time period (usually 10 min) at 37°C before stopping the reaction with 50 μL of concentrated HCl.

4. The stopped incubation mixture may be stored at –20°C until required for resolution of [^{3}H]cAMP.

5. The mixture is removed and mixed with 25–30 Bq [^{14}C]cAMP for determination of recovery during resolution of [^{3}H]cAMP.

3.2.3. Resolution of [^{3}H]cAMP

1. Apply an aliquot of the supernatant (usually 0.8–1 mL) to Dowex 50 (Bio-Rad) columns (1 mL resin prepared as described in Section 2.3.1.).

2. Add an additional 2 mL of distilled water to elute [^{3}H]ATP and [^{3}H]ADP to waste.

3. Desorb the [^{3}H]cAMP from the Dowex resin by applying 4 mL water and allowing the effluent to drip directly onto the alumina columns (prepared as described in Section 2.3.1.).

4. Elute the [^{3}H]cAMP from the alumina columns with 5 mL of $0.1M$ imidazole directly into scintillation vials. ([^{3}H]Adenine and [^{3}H]adenosine remain on the Dowex resin, and may be desorbed during the regeneration cycle.)

5. Add 10 mL of a suitable scintillation cocktail (e.g., Scintillator 299-Fisons plc, for mixing with aqueous samples) to the effluent, and mix the suspension prior to counting in a liquid scintillation counter, using a dual-channel [^{3}H]/[^{14}C] program.

3.2.4. Correcting for Variable Tissue Distribution

It is possible to normalize data when using small amounts of tissue or cells that show variability in their distribution. Thus, after termination of the incubation and removal of an aliquot for chromatography, an aliquot of the supernatant may be removed for liquid scintillation counting. Alternatively, it is possible to add scintillation fluid directly to the remaining mixture including tissue slices, and count directly. Because the termination process involves the addition of recovery marker [^{14}C]cAMP, it is easy to correct for total [^{3}H]adenine nucleotides in the individual tubes.

3.2.5. Data Analysis

3.2.5.1. Calculation of [^{3}H]cAMP Generation

Dual-channel counting will provide two series of figures, [^{3}H] and [^{14}C] for each sample, together with a total [^{14}C]cAMP fraction. These data are best assessed with the aid of a computer spreadsheet program (Lotus 1-2-3, Quattro Pro, Microsoft Excel) that permits rapid, multiple calculations of large groups of figures. Thus, recovery is expressed as a fraction of the total [^{14}C]cAMP added to the sample. The [^{3}H]cAMP figure is thus multiplied by the reciprocal of the recovery to derive total [^{3}H]cAMP content of the sample.

$$[^{3}H]cAMP = [([^{3}H]cAMP \text{ eluted}) \times (\text{total } [^{14}C] \text{ added to sample})]/ [^{14}C]cAMP \text{ eluted} \qquad (1)$$

Typical basal values of [^{3}H]cAMP generations are between 600 and 1500 dpm.

3.2.5.2. A Fraction Converted from Total [^{3}H]Adenine Nucleotides

This allows comparison between samples of varying tissue content, and also between experiments in which tissue uptake of [^{3}H]adenine varies. An additional two sets of figures may be obtained when using this method, with [^{3}H] and [^{14}C] channels showing [^{3}H]adenine nucleotides and [^{14}C]cAMP remaining in the sample total:

$$[^{3}H]\text{adenine nucleotides} = [([^{3}H]\text{adenine nucleotides counted}) \times (\text{total } [^{14}C] \text{ added to sample})]/[^{14}C] \text{ counted} \qquad (2)$$

Alternatively, it is possible to count the remaining fluid on the [^{3}H] channel, because the contribution of [^{14}C]cAMP will be relatively minor (1000 dpm [^{14}C] compared to >60,000 dpm [^{3}H]). Thus, one then expresses the total [³H]cAMP produced as a percentage of the total [³H]adenine nucleotides present in the tissue. Typical basal values of cAMP formation are 0.01%.

3.3. The Radioreceptor Method for Measuring cAMP

3.3.1. cAMP Generation in Tissue Slices

1. Incubate pre-equilibrated aliquots (25 μL) of gravity-packed slices in flat-bottomed vials with KRB for 15 min in the presence of antagonists or modulators.
2. Add agonist (in 10–20 μL) to make a final vol of 300 μL, and continue the incubation for a given time period (usually 10–15 min).
3. Terminate the incubation by the addition of 100 μL of 7.5 % (w/v) ice-cold perchloric acid, and allow the tubes to stand on ice for 15 min, before neutralization with 1.2M KHCO$_3$ (100 μL).
4. Centrifuge the suspension at >2000g for 10 min, and remove an aliquot of the supernatant layer for storage (hours at 4°C or months at –20°C).
5. Add NaOH to the pellet (to a final concentration of 0.5–1.0M in a vol of 1 mL) in order to digest them for subsequent assay of protein (25 μL gravity-packed slices is approx 1 mg protein).

3.3.2. cAMP Generation in Cells in Culture

1. Prepare the cells as before, and preincubate in HHB (to a final vol of 1 mL/well in 24-well plates) in the presence of antagonist/degradative enzymes for at least 10 min at 37°C.
2. Add agonist in 10/20 μL, and continue the incubation for the required time (usually 10 min), before halting the incubation by aspirating the medium and replacing it with 400 μL of ice-cold 1.75% perchloric acid.
3. After standing for at least 15 min on ice, transfer the medium to a conical centrifuge tube, where it is neutralized by the addition of 1.2M KHCO$_3$ (100 μL).
4. Centrifuge the suspension at >2000g for 10 min, and remove an aliquot of the supernatant layer for storage (hours at 4°C or months at –20°C).
5. Add NaOH to the pellet (to a final concentration of 0.5–1M in 200 μL) in order to digest it for subsequent assay of protein (a confluent layer of cells in one well of a 24-well plate is approx 0.2 mg protein).

3.3.3. Estimation of cAMP

1. Distribute 25-μL aliquots of TENT buffer into 3.5-mL polypropylene tubes.
2. Add 25 μL of sample or standard to the tubes (standard curve 10^{-4} to 10^{-10} mol cAMP).
3. Distribute [³H]cAMP (300–400 Bq; 500 fmol) in 25 μL to the tubes.

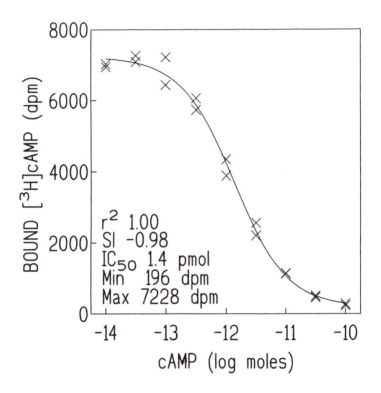

Fig. 1. A standard curve for cAMP competition at the radioreceptor [³H]cAMP binding protein.

4. Initiate the incubation by the addition of 25 μL binding protein, and mix thoroughly by vortex mixer.
5. After incubation at room temperature for 60 min, add 3 mL ice-cold wash buffer, and filter the samples rapidly through Whatman GF/B filters, washing with 2 × 3 mL ice-cold wash buffer.
6. It is very important that the standards are treated in the same way as the samples, i.e., diluted in prefrozen neutralized perchloric acid (the use of water as a diluent leads to overestimates of cAMP mass).

3.3.4. Data Analysis

Computer programs are available for radioimmunoassays (RIAs), and the transfer to the cAMP assay is straightforward. Alternatively, it is possible to use other computer programs (e.g., GraphPad, GraphPad, San Diego, CA) to generate estimates for binding maximae and minimae, EC_{50} and slope values (*see* Fig. 1). Such programs may also allow one to estimate unknowns or these figures may be used in a spreadsheet. Pico-moles/tube are calculated using the equation:

$$pmol = (total\ volume/aliquot) \times EC_{50} \times F^H$$

where F is the inverse of the fractional displacement:

$$F = (maximum - minimum) / (sample\ dpm - minimum) \qquad (3)$$

where ^{H} is the negative inverse of the slope. cAMP accumulation is usually expressed as pmol/mg protein. Basal levels in the brain are between 2 and 5 pmol/mg protein.

3.4. Measuring Adenylyl Cyclase Actvity in Cell-Free Preparations

Using a similar protocol and incubation conditions to those used for estimating GTPase activity (*see* Chapter 5), one can measure adenylyl cyclase activity with α-[^{32}P]ATP or [^{3}H]ATP as substrates *(4)* (*see* Chapter 6). Stop the incubation with sodium dodecyl sulfate, trichloroacetic acid, or perchloric acid (possibly with boiling to destroy ATP) containing either [^{3}H]cAMP or [^{14}C]cAMP to allow estimation of recovery in the subsequent double-chromatography techniques (*see* Section 3.2.3.).

4. Notes

1. Uptake/incorporation of radiolabel: The radioreceptor assay avoids the problems associated with radiolabel uptake and incorporation artifacts. Thus, in many cells and tissues there is insufficient [^{3}H]adenine uptake to allow the use of the prelabeling assay. However, in those tissues that exhibit sufficient [^{3}H]adenine uptake, the incorporation of [^{3}H]adenine into [^{3}H]ATP that is rapidly turning over may be advantageous, providing better signal-to-noise ratios.

2. Simultaneous measurement of cAMP and other factors: It is possible to measure simultaneously both cAMP levels and other intracellular messengers, e.g., inositol 1,4,5-trisphosphate (*see* Chapter 14) with some ease using the mass assay.

3. In vivo measurements: It is possible to measure in vivo levels (using microwave fixation or microdialysis) by the radioreceptor method without prior treatment of animals with radioactive precursors.

4. Stability of cAMP binding protein: The frozen aliquots of cAMP binding protein are stable for at least 1 yr without obvious loss of binding activity.

5. Magnitude of responses: No qualitative differences in prelabeling vs assay of endogenous cAMP are apparent. However, differences in the fold stimulation over basal may be evident.

6. Specificity and selectivity: High concentrations of ATP may lead to false positives or increased basals when using the mass assay. Heating to 100°C for 2–4 min will destroy remaining ATP.

7. Rapidity of cAMP responses: cAMP responses to agonists are comparatively slow, so both methods are applicable.

8. Inhibition of cAMP catabolism: Phosphodiesterase inhibitors are freely usable in either assay.

9. Stopping the incubation: Termination of incubations for analysis of cAMP mass with perchloric acid is less messy and quicker than the use of trichloroacetic acid:diethyl ether or trichloroacetic acid:freon:*n*-octylamine.

10. Radioreceptor vs antibody: The RIA using an antibody raised against a conjugated cAMP derivative requires derivatization of the sample, e.g., succinimidylation. The RIA is preferable when amounts of cAMP are small, owing to its increased sensitivity. The threshold for the radioreceptor assay is 0.2 pmol/25 μL of assay sample (8 n*M* cAMP), and is preferable when cAMP amounts are greater than this, because of the reduced sample handling involved. It is possible to increase the sensitivity of the assay by increasing the volume of tissue sample used and diluting the standards accordingly. That is, rather than a total assay volume of 100 μL with 25 μL of sample, increasing the sample volume to 250 μL and buffer volume to 100 μL to give a total volume of 400 μL, the assay sensitivity can be increased approx 10-fold (to 0.8 n*M* cAMP). In general, the radioreceptor assay is acceptable for use with the majority of tissue samples and cells in culture.

11. The use of filtration in the radioreceptor assay: As noted by Nordstedt and Fredholm *(1)*, the filtration of bound [^{3}H]cAMP over glass-fiber filters is more rapid than the use of charcoal *(5)*, polyethylene glycol, or ammonium sulfate precipitation—typical means for separating soluble proteins. However, filtration of soluble receptors is usually carried out over polyethyleneimine-treated filters *(6)*. In the case of the endogenous cAMP binding protein, polyethyleneimine treatment actually reduces the amount of bound [^{3}H]cAMP *(1)*, possibly through ionic repulsion mechanisms.

5. Summary

The radioreceptor method is sensitive, simple, and specific, requiring little, or no purification of sample prior to analysis. It is possible to process hundreds of samples in a matter of hours (sample throughput may be enhanced through the use of a cell harvester for rapid filtration of bound [^{3}H]cAMP). Assays for endogenous cAMP and cAMP measured using a radioactive precursor give similar results in terms of fold stimulation. The main advantage of the former technique lies in the use of cells or tissues with poor uptake of [^{3}H]adenine.

References

1. Nordstedt, C. and Fredholm, B. B. (1990) A modification of a protein-binding method for rapid quantification of cAMP in cell-culture supernatants and body fluid. *Anal. Biochem.* **189,** 231–234.
2. Shimizu, H., Daly, J. W., and Creveling, C. R. (1969) A radioisotopic method for measuring the accumulation of adenosine 3':5'-cyclic monophosphate in incubated brain slices. *J. Neurochem.* **16,** 1609–1619.
3. Takeda, T., Kuno, T., Shuntoh, H., and Tanaka, C. (1989) A rapid filtration assay for cAMP. *J. Biochem. (Tokyo)* **105,** 327–329.
4. Salomon, Y., Londos, C., and Rodbell, M. (1974) A highly sensitive adenylate cyclase assay. *Anal. Biochem.* **58,** 541–548.
5. Brown, B. L., Ekins, R. P., and Albano, J. D. M. (1972) Saturation assay of cyclic AMP to an endogenous binding protein. *Adv. Cyc. Nuc. Res.* **2,** 25–40.
6. Bruns, R. F., Lawson-Wendling, K., and Pugsley, T. A. (1983) A rapid filtration assay for soluble receptors using polyethyleneimine-treated filters. *Anal. Biochem.* **132,** 74–89.

CHAPTER 8

Mass Measurements
of Cyclic AMP Formation
by Radioimmunoassay,
Enzyme Immunoassay,
and Scintillation Proximity Assay

Jeffrey K. Horton and Peter M. Baxendale

1. Introduction

Adenosine 3',5'-cyclic monophosphate (cAMP) is involved in a myriad of normal and pathological processes. Indeed, this cyclic nucleotide serves as a second messenger for the action of endogenous and exogenous agents in organisms ranging from bacteria to humans *(1)*. The ubiquitous nature of cAMP has made its measurement essential to the study of numerous hormones, local mediators, neurotransmitters, drugs, and toxins. Methods for the estimation of cAMP include enzymatic radioisotopic displacement *(2)*, high-pressure liquid chromatography (HPLC) *(3)*, protein kinase activation *(4)*, luciferin-luciferase bioluminescence *(5)*, competitive protein binding *(6)*, and immunoassay techniques *(7–10)*. Here we describe rapid, selective, and highly sensitive methods for measuring cAMP by competitive immunoassay. Estimation involves the techniques of:

1. Radioimmunoassay (RIA), using a high specific activity adenosine 3',5'-cyclic phosphoric acid 2'-O-succinyl-3-[^{125}I]iodotyrosine methyl ester together with a second antibody that is bound to magnetizable polymer particles.

From: *Methods in Molecular Biology, Vol. 41: Signal Transduction Protocols*
Edited by: D. A. Kendall and S. J. Hill Copyright © 1995 Humana Press Inc., Totowa, NJ

2. Enzyme immunoassay (EIA), involving the linking of succinyl cAMP to horseradish peroxidase (HRP) and combining this with stable second-antibody-coated microtiter plates.

3. Homogeneous RIA (nonseparation), incorporating the novel method of scintillation proximity assay (SPA).

2. Materials

1. Human serum albumin (HSA) and 2'-*O*-monosuccinyl adenosine 3',5'-cyclic monophosphate are obtained from Sigma (Poole, Dorset, UK). Triethylamine, dimethylformamide, and ethylchloroformate are all obtained from Aldrich (Gillingham, Dorset, UK). Sephadex G-25 is obtained from Pharmacia-LKB Biotechnology (Uppsala, Sweden).

2. Freund's complete and incomplete adjuvants (Sigma).

3. Adenosine 3',5'-cyclic monophosphate (Sigma).

4. 2'-*O*-Monosuccinyl adenosine 3',5'-cyclic monophosphate tyrosine methyl ester, chloromine-T, and sodium metabisulfite are obtained from Sigma. Freshly prepare the sodium metabisulfite solution by dissolving 25 mg in 5 mL 0.05M sodium phosphate buffer, pH 7.4; 2 mCi sodium [^{125}I]iodide, carrier-free, are obtained from Amersham International (Amersham, Bucks, UK).

5. RIA, EIA, and SPA assay buffer: 0.05M sodium acetate pH 5.8, store at 2–8°C. Acetic anhydride is obtained from Aldrich. Adenosine 3',5'-cyclic phosphoric acid 2'-*O*-succinyl-3-[^{125}I]iodotyrosine methyl ester: 2000 Ci/mmol is obtained from Amersham. Store in *n*-propanol/water (1:1) below –20°C. Magnetic separator is obtained from Amersham. Amerlex-M anti-rabbit second-antibody reagent is also obtained from Amersham, code RPN510.

6. Horseradish peroxidase is obtained from Biozyme (Blaenavon, Gwent, UK). *N*-hydroxysuccinimide, dimethyl sulfoxide (DMSO) ethyl-3-(3 dimethylaminopropyl) carbodiimide hydrochloride, and guaiacol are all obtained from Sigma.

7. SPA anti-rabbit second-antibody reagent is obtained from Amersham, Code RPN140. Microtiter plates are obtained from Dynatech (Alexandria, VA).

8. Amprep minicolumns, Manifolds, and Super Separator-24 are all obtained from Amersham.

9. Tween-20, hydrogen peroxide, and tetramethylbenzidine hydrochloride are all obtained from Sigma.

10. Commercially available kits for estimating cAMP levels: The methods described here enable the development of immunoassay systems for cAMP. In addition, assay kits are available from a number of commercial sources:

Competitive protein binding assay
    Amersham International (Little Chalfont, Bucks, UK), code TRK 432
Radioimmunoassay
    Amersham International, code RPA 509
    Dupont NEN Research Products (Boston, MA)
    Advanced Magnetics Inc. (Cambridge, MA)
    Biomedical Technologies Inc. (Stoughton, MA)
    Incstar Corp. (Stillwater, MN)
    Immunotech (Marseille, France)
    Paesel and Lorei GmbH & Co. (Frankfurt, Germany)
Scintillation proximity assay
    Amersham International, code RPA 538
Enzyme immunoassay
    Amersham International, code RPN 225
    Cayman Chemicals (Ann Arbor, MI)
    Advanced Magnetics Inc. (Cambridge, MA)

3. Methods

3.1. Preparation of Immunogens

HSA cAMP conjugates are prepared essentially by the method of Cailla et al. *(11)* (*see* Note 1).

1. Add 100 µL triethylamine (TEA) to 1.1 mL dimethylformamide (DMF).
2. Dissolve 10 mg HSA in 1.5 mL redistilled water.
3. Add 40 µL of the TEA:DMF mixture to the HSA.
4. Dissolve 6 mg 2'-O-monosuccinyl adenosine 3',5'-cyclic monophosphate in 400 µL DMF.
5. Add 60 µL of the TEA:DMF solution (a 1:11 mixture) to the succinyl cAMP solution followed by 20 µL of an ethylchloroformate (ECF):DMF mixture (prepared by adding 100 µL ECF to 1.6 mL DMF).
6. Add the HSA solution to the mixture containing the succinyl cAMP. Incubate for 30 min at room temperature.
7. Remove low-mol-wt components from the reaction mixture by gel filtration on a Sephadex G-25 column (5 × 1 .5 cm) equilibrated and eluted with 0.01M sodium chloride.

3.2. Immunization

Antibodies are raised in rabbits by repeated immunizations.

1. Emulsify the succinyl cAMP-HSA conjugate (1 mL; *see* Section 3.1.) with an equal volume of complete Freund's adjuvant.

2. Inject intracutaneously 0.5 mg of immunogen (in 0.25 mL of emulsion) into multiple sites on the animal's back.
3. Boost once every 4 wk for 2 mo with 0.2 mg of immunogen in incomplete Freund's adjuvant. Test the bleeds. Immunize at 4-wk intervals for a maximum of 16 wk with the same dose of antigen.
4. Remove blood from the carotid artery 7–10 d after the last injection.
5. Separate serum, aliquot, and store at –70°C.

3.3. Preparation of Standards

1. Dilute adenosine 3',5'-cyclic monophosphate in 0.05M sodium acetate buffer, pH 5.8, to prepare a stock standard. Dilute the stock standard with the same buffer to give an appropriate standard curve range (0.25–16 pmol/mL).
2. Check the concentration of each of the stock standards by spectrophotometric measurement at 258 nm. The extinction coefficient for cAMP is 14,700 L/mol/cm.

3.4. cAMP Radioiodination

Adenosine 3'5'-cyclic phosphoric acid 2'-O-succinyl-3-[^{125}I]iodotyrosine methyl ester is prepared by a method based on that of Steiner et al. *(12)* whereby 2'-O-monosuccinyl adenosine 3',5'-cyclic monophosphoric acid tyrosine methyl ester is reacted with sodium [^{125}I]iodide and chloramine-T to give a specific activity of 2000 Ci/mmol. The product is purified by HPLC (*see* Note 2).

1. Transfer 20 µL Na ^{125}I solution containing 2 mCi, 74 MBq into a glass tube with a glass stopper.
2. Add 50 µL 2'-O-monosuccinyl adenosine 3',5'-cyclic monophosphoric acid solution, prepared by dissolving 1 mg in 10 mL 0.5M phosphate buffer, pH 7.4, and vortex mix for 10 s.
3. Add 50 µL chloramine-T solution, prepared freshly by dissolving 17 mg chloramine-T in 5 mL 0.05M phosphate buffer, pH 7.4.
4. Mix thoroughly, and allow to react for 90 s exactly.
5. Add 50 µL sodium metabisulfite solution, mix, and then cap the vial.
6. Purify the product by reverse-phase HPLC using a Whatman M9, ODS-2 column, and a linear gradient of acidified water to 30% acetonitrile over 45 min at a flow rate of 1 mL/min. The material can be detected by UV absorption at 254 nm using an LKB 2238 Uvicord II detector and radioactivity detected, using a Harwell ratemeter and associated detector.

3.5. Antisera Binding Assay

In order to obtain high-affinity antisuccinyl cAMP sera, each test bleed is assessed by the method of Scatchard *(13)* using adenosine 3',5'-cyclic monophosphoric acid 2'-*O*-succinyl-3-[^{125}I]iodotyrosine methyl ester.

1. Prepare standards of cAMP in 0.05*M* sodium acetate buffer, pH 5.8, to give seven solutions ranging from 40–2560 fmol/mL.
2. The acetylation reagent is made ready by mixing 1 mL of acetic anhydride with 2 mL of TEA.
3. Add 25 µL of acetylation reagent to each of the standards (0.5 mL), including 0.5 mL of sodium acetate buffer, pH 5.8, and thoroughly mix the solutions (*see* Note 3).
4. Duplicate assay tubes are prepared by the addition of aliquots from each of the standards (100 µL), tracer (100 µL; 20,000–30,000 cpm), and antisera (100 µL).
5. Mix each assay tube and incubate at 4°C for 18 h.
6. Add Amerlex-M antirabbit second-antibody reagent (500 µL), and incubate at room temperature for 10 min.
7. Place tubes on a magnetic separator, and after 15 min, decant the supernatant.
8. Count the antibody-bound fraction in a γ spectrophotometer.
9. Nonspecific binding is assessed in the absence of specific rabbit antibody.
10. Carry out several experiments using a range of antisera concentrations from 1/1000 to 1/500,000, each dilution prepared in 0.05*M* sodium acetate buffer, pH 5.8.

3.6. Preparation of Peroxidase-Labeled cAMP

cAMP is conjugated to HRP through an activated *N*-hydroxysuccinimide ester *(14)*. The ester is prepared by reaction of the carboxylic acid group of 2'-*O*-monosuccinyl adenosine 3',5'-cyclic monophosphate with *N*-hydroxysuccinimide in the presence of 1-ethyl-3-(3-dimethylaminopropyl) carbodiimide hydrochloride.

1. Dissolve 17.6 mg *N*-hydroxysuccinimide in 1 mL of DMSO, and add 4.5 mg 2'-*O*-succinyl cAMP.
2. Add 19.2 mg of ethyl-3-(3-dimethylaminopropyl) carbodiimide hydrochloride to 0.2 mL of DMSO, and mix with the solution containing the 2'-*O*-succinyl cAMP and the *N*-hydroxysuccinimide.
3. Incubate the mixture for 48 h at room temperature. When stored at –70°C, this intermediate is highly stable.
4. Dissolve 6 mg HRP (242 U/mg solid using guaiacol as substrate) in 0.5 mL of 0.1*M* sodium phosphate buffer, pH 8.0.

5. Add 180 µL of nucleotide-containing solution, prepared as in step 3, and incubate the mixture overnight at room temperature.
6. Remove low-mol-wt components from the reaction mixture by gel filtration on a Sephadex G-25 column (5 × 1.5 cm) equilibrated and eluted with 0.1M sodium phosphate buffer, pH 6.0, containing 5 mM EDTA.
7. The succinyl cAMP-HRP conjugate obtained is stable for at least 1 yr when aliquoted and stored at –70°C. The enzyme activity is not significantly reduced by conjugation of the enzyme with the hapten.

3.7. Immunoassay Separation Methods

Numerous methods have been used to separate antibody-bound cAMP from unbound cAMP in immunoassays. The efficiency with which this is done is crucial to the overall assay performance and to the simplicity and convenience of the protocol. Examples of the most widely used methods are described in the following sections and in detail in Section 3.9.

3.7.1. Adsorption

Separation is carried out by adding to the antigen/antibody reaction mixture a suspension of particles that can adsorb unbound cAMP onto their surface. Activated charcoal is commonly used in assays using [^{3}H]cAMP as the tracer. Following centrifugation, an aliquot of the supernatant is transferred to a scintillation vial to enable determination of the antibody-bound fraction.

3.7.2. Precipitation

These methods rely on the precipitation of immunoglobulin from the reaction mixture. This process can be selective for the primary antibody by use of a specific second antibody or protein A. Alternatively, less selective protein precipitation by ammonium sulfate or polyethylene glycol can be achieved. After incubation with the precipitating agent, the antibody-bound cAMP is separated from the unbound fraction by centrifugation. Supernatants are decanted away, and the remaining activity determined in the pellet.

3.7.3. Solid Phase

These methods, which are specific and convenient, greatly simplify cyclic nucleotide assays. There are two main types of solid-phase assays: those involving coated particles (such as Amerlex-M), and those involving coated wells or tubes.

In coated particle-based assays, antibody is coated onto the surface of small polymer or glass particles. The beads are added as a suspension, precoated with either primary or second antibody. After incubation, the beads, containing the antibody-bound cAMP, are separated, usually by centrifugation, and the activity remaining in the pellet determined. Separation can be further simplified with Amerlex-M, because these beads contain a magnetizable component, thus enabling magnetic separation. These types of assays are particularly suited to iodine-125-based methods.

Coated well techniques are becoming increasingly used, as the trend away from radioactive techniques gains momentum. A primary or secondary antibody is immobilized onto the walls of microtiter wells or tubes. Separation is achieved by pouring the contents of the wells away, leaving the bound fraction physically attached to the walls of the wells. This principle is fundamental to many enzyme immunoassay systems.

3.7.4. SPA

SPA is a homogeneous, nonseparation RIA system. It is based on the principle that relatively weak β emitters, such as [^{3}H]β particles and [^{125}I]Auger electrons, need to be close to scintillant molecules in order to produce light; otherwise, the energy is dissipated and lost to the solvent. This concept has been used to develop homogeneous RIAs by coupling second antibodies onto fluomicrospheres (beads containing scintillant). When a second-antibody-coupled fluomicrosphere is added to an RIA tube instead of a separation reagent, any radiolabeled ligand that is bound to the primary antibody will be immobilized on the fluomicrosphere. This process will bring into close proximity the radiolabel and the scintillant, activating the scintillant to emit light.

The level of light energy emitted, which is indicative of the extent to which the ligand is bound to the antibody, may be measured in a liquid scintillation counter. Any unbound radioligand remains too distant to activate the fluomicrosphere. Hence, the need for a physical separation process is eliminated. Examples of protocols used for solid-phase RIA and EIA, and an SPA assay are shown in Section 3.9.

3.8. Specimen Collection and Sample Preparation

1. Withdraw whole blood from individuals, and collect on ice into tubes containing 7.5 mM EDTA. The blood is immediately centrifuged at 1000g for 10 min at 4°C, and the plasma stored at –20°C prior to analysis.

2. Random, timed, or 24-h urine collections may be analyzed. Remove particulate matter by centrifuging at 500g for 10 min at 4°C, and store samples at −20°C before measurement.
3. Numerous procedures have been described for the extraction of cAMP from biological samples, such as tissues and cell cultures. These include acidic extraction methods using trichloroacetic acid, perchloric acid, diluted hydrochloric acid, and extraction with aqueous ethanol. Some investigators recommend the use of ion-exchange chromatography following one of these extraction techniques. For full details of these methods, the reader is referred to Goldberg and O'Toole *(15)*, Harper and Brooker *(7)*, Steiner *(16)*, and Rosenberg et al. *(17)*. Representative procedures are described in the following sections for the extraction of cAMP from tissues and cell cultures.

3.8.1. Liquid-Phase Extraction Method
1. Tissues:
   a. Homogenize frozen tissue in cold 6% (w/v) trichloroacetic acid at 2–8°C to give a 10% (w/v) homogenate.
   b. Centrifuge at 2000g for 15 min at 4°C.
   c. Recover the supernatant and discard the pellet.
   d. Wash the supernatant four times with 5 vol of water-saturated diethyl ether. The upper ether layer should be discarded after each wash.
   e. The aqueous extract remaining should be lyophilized or dried under a stream of nitrogen at 60°C.
   f. Dissolve the dried extract in a suitable volume of assay buffer prior to analysis.
2. Cell suspension:
   a. Add ice-cold ethanol to cell suspension to give a final suspension volume of 65% (v/v) ethanol. Allow to settle.
   b. Draw off the supernatant into test tubes.
   c. Wash the remaining precipitate with ice-cold 65% (v/v) ethanol, and add the washings to the appropriate tubes.
   d. Centrifuge the extracts at 2000g for 15 min at 4°C, and transfer the supernatant to fresh tubes.
   e. Dry the combined extracts under a stream of nitrogen at 60°C or in a vacuum oven.
   f. Dissolve the dried extracts in a suitable volume of assay buffer prior to analysis.

3.8.2. Solid-Phase Extraction Method
Amersham has developed a simple protocol for the extraction of cAMP from tissue extracts and cell-culture supernatants by ion-exchange

chromatography using disposable Amprep™ minicolumns. Maximum recovery of cAMP is obtained using columns containing anion-exchange silica sorbents, for example, Amprep SAX, code RPN 1918 (500 mg), which are available from Amersham. These columns provide a rapid sample clean-up and effectively reduce sample handling compared with solvent extraction methods.

The columns can be used with the Amprep Manifold-10 (RPN 1930) or the Super Separator-24 (RPN 1940) to facilitate the extraction of 10 or 24 samples simultaneously. Alternatively, a syringe can be used for a simple manual extraction. The syringe is easily connected to the column with the column adaptor in each pack of Amprep minicolumns. Representative procedures are described in the following section for the extraction of cAMP from biological samples using Amprep minicolumns.

3.8.3. Amprep Extraction of cAMP

1. Condition of the column:
   a. Rinse an Amprep SAX 500-mg minicolumn (code RPN 1918) with 2 mL methanol.
   b. Rinse the column with 2 mL distilled water (*see* Note 4).
2. Sample treatment:
   a. Homogenize 1 g (wet wt) tissue in 10 mL Hank's balanced salt solution (without calcium and magnesium) containing 5 mM EDTA.
   b. Centrifuge the homogenate for 10 min at $1000g$ at 4°C.
   c. Dilute homogenate supernatant 1:10 with Hank's, and apply 1 mL directly to the conditioned SAX column. Alternatively, mix 1 mL of supernatant with 1 mL undiluted acetonitrile. Vortex mix for 20 s, and centrifuge for 10 min at $1500g$ at 4°C. Apply 1 mL of supernatant to the column. Cell-culture supernatant may be applied directly to the column.
3. Remove interferences and wash column with 3 mL methanol.
4. Analyte elution:
   a. Pass 3 mL acidified methanol through the column, and collect the eluate. Prepare the acidified methanol by diluting concentrated HCl to $0.1M$ with absolute methanol.
   b. The eluate can be dried under nitrogen, reconstituted in assay buffer, and then assayed directly (*see* Note 5).

3.9. Assay Procedures

cAMP in test samples is analyzed by either acetylation or a nonacetylation assay procedure. Preliminary acetylation of samples and standards markedly increases the detection limit of the cAMP immunoassays (*see* Note 6).

3.9.1. RIA

3.9.1.1. NONACETYLATION METHOD

1. Prepare standards in assay buffer (0.05*M* acetate, pH 5.8) with cAMP diluted over the range of 0.25–16 pmol/mL. The zero standard tube consists of assay buffer without standard.

2. Diluted sample or standard (100 µL) is incubated with specific antisera (100 µL, diluted 1/11,000) and [^{125}I]cAMP (100 µL) for 3 h at 4°C. Nonspecific binding is determined in the absence of specific rabbit antisera. The total counts added are measured by the addition of 100 µL of [^{125}I]cAMP only (20,000–30,000 cpm).

3. Gently shake and swirl the bottle containing Amerlex-M second antibody reagent to ensure a homogenous suspension. Then add 500 µL to each tube except to total counts. Vortex mix all tubes thoroughly, and incubate for 10 min at room temperature. Separate the antibody-bound fraction by using magnetic separation or centrifugation as described in steps 4 and 5.

4. Magnetic separation: Attach the rack on the Amerlex-M Separator base, and ensure that all the tubes are in contact with the base plate. Leave for 15 min. After separation, do not remove the rack from the Separator base. Pour off and discard the supernatant liquids, keeping the Separator inverted. Place the tubes on a pad of absorbent tissue, and allow to drain for 5 min.

5. Centrifugation: Centrifuge all tubes for 10 min at 1500*g* or greater. After centrifugation, place the tubes into suitable decantation racks, and then pour off and discard the supernatant liquids. Keeping the tubes inverted, place them on a pad of absorbent tissues, and allow to drain for 5 min.

6. On completion of either magnetic or centrifugal separation, firmly blot the rims of the inverted tubes on the tissue pad to remove any adhering droplets of liquid. Do not reinvert the tubes once they have been turned upright.

7. Determine the radioactivity present in each tube by counting for at least 60 s in a γ scintillation counter.

3.9.1.2. ACETYLATION METHOD

cAMP in test samples is acetylated prior to assay as described by Harper and Brooker *(7)*.

1. Prepare standards in assay buffer with cAMP diluted over the range of 0.04–2.56 pmol/mL. The zero standard tube contains assay buffer without standard.

2. Standards and samples (0.5 mL) are acetylated by the addition of 25 μL of a 2:1 (v/v) mixture of TEA and acetic anhydride (*see* Note 3). For the assay, an aliquot (100 μL) of acetylated samples or standards is incubated with antisera (100 μL) and [^{125}I]cAMP (100 μL) overnight at 4°C. Nonspecific binding is determined in the absence of specific rabbit antisera.

3. The antibody-bound fraction is separated by using magnetic separation or centrifugation as described in Section 3.9.1.1., steps 3 and 4.

3.9.2. Enzyme Immunoassay

For full details of this method, the reader is referred to Horton et al. *(18)*. The acetylation step is carried out essentially as described for the RIA (*see* Note 3).

1. Sample or standard antisera (100 μL) is incubated with specific antisera (100 μL) in the donkey antisera-coated microtiter plates (Note 7) for 2 h at 4°C before addition of HRP-succinyl cAMP conjugate (100 μL; approx 0.004 U enzyme).

2. The microtiter plate is incubated for a further 1 h at 4°C and thoroughly washed with 0.01M sodium phosphate buffer, pH 7.5, containing 0.05% (v/v) Tween-20.

3. Activity of enzyme bound to the solid phase is determined by the addition of 0.03% (w/v) tetramethylbenzidine hydrochloride in 10 mM citric acid, pH 3.5, containing 0.01% (v/v) hydrogen peroxide.

4. The substrate is incubated for 1 h at room temperature, and the reaction terminated with 1.0M sulfuric acid (100 μL/well).

5. The yellow color produced in the presence of bound conjugate is read at 450 nm using a microtiter plate reader.

6. Standard curves are constructed in assay buffer with cAMP diluted over the range of 0.125–32 pmol/mL (nonacetylation) and 0.04–2.56 pmol/mL (acetylation assay).

3.9.3. SPA

See Section 3.7.4. for details of the assay principle.

1. Sample or standard antisera (100 μL) is added to specific antisera (100 μL) and [^{125}I]cAMP (100 μL) (*see* Note 8). Nonspecific binding is determined, as before, in the absence of specific rabbit antisera.

2. Place the SPA anti-rabbit reagent onto a magnetic stirrer, and adjust stirring speed to ensure a homogeneous suspension. Add 100 μL to all the tubes.

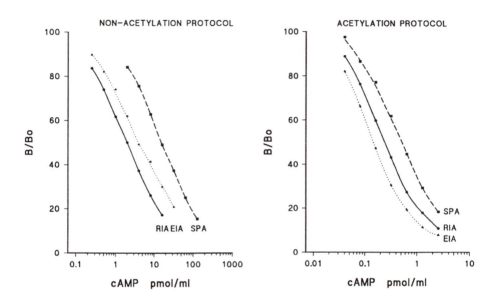

Fig. 1. Comparison of cAMP standard curves by RIA (●), EIA (▲), and SPA (■).

3. All vials should now contain a total volume of 400 μL.
4. Cap the vials, and mix on an orbital shaker for 15–20 h at room temperature. The shaker speed should be optimized for individual shakers; 200 rpm is sufficient for most orbital shakers.
5. Determine the amount of [^{125}I]cAMP bound to the fluomicrospheres by counting the vials in a β-scintillation counter for 2 min.
6. Standard curves are prepared in assay buffer with cAMP diluted over the range 2–128 pmol/mL (nonacetylation) and 0.04–2.56 pmol/mL (acetylation assay).
7. A comparison of standard curves prepared by RIA, EIA, and SPA is shown in Fig. 1.

4. Notes

1. In order to prepare immunogens with a high degree of potency, fresh reagents should be used. Dry dimethylformamide is essential.
2. Iodinated cAMP (adenosine 3'5'-cyclic phosphoric acid 2'-*O*-succinyl-3-[^{125}I]iodotyrosine methyl ester can be obtained commercially from Amersham International Code IM 106 or from Dupont NEN.
3. Carefully add 25 μL of the acetylation reagent to all acetylation tubes containing standards and unknowns. Optimum precision is attained by placing

the pipet tip in contact with the test tube wall above the aqueous layer and allowing the acetylation reagent to run down the test tube wall into the liquid. Each tube should be vortexed immediately following addition of the acetylating reagents.

4. Do not allow the sorbent in the column to dry. The flow rate should not exceed 5 mL/min.

5. If lyophilization is the preferred method of drying samples, 0.1M HCl in distilled water rather than methanol can be used to elute the analyte.

6. An important advance in increasing the sensitivity of cAMP RIA is substitution at the 2'-O position. In this case, acetylated cAMP has a higher affinity for antibody and displaces the labeled cyclic nucleotide derivatives employed in the assays far better than the unsubstituted cyclic nucleotides.

7. Antibodies for the preparation of second-antibody-coated microtiter plates are raised to rabbit IgG in the donkey using the following procedure. Rabbit IgG (1.5 mg in 0.5 mL of water) is emulsified with an equal volume of Freund's complete adjuvant, divided into several portions, and injected intramuscularly into the thighs of a donkey. The procedure is repeated after 12 wk, but with Freund's incomplete adjuvant. Blood is taken 7 d after the final injection. The serum is separated by centrifugation at 1000g for 10 min at 4°C, and aliquots are stored at –20°C until required.

   Microtiter plates are coated with the donkey antirabbit sera by the addition of 200 µL of a solution of 1 µg/mL of whole donkey serum in 35 mM sodium bicarbonate buffer, pH 9.5. The plates are incubated at 4°C overnight, and the antibody solution decanted and replaced with 200 µL bovine serum albumin (BSA; 0.5 [w/v]) in 50 mM phosphate-buffered saline (PBS), pH 7.4. After 4 h at room temperature, the BSA solution is removed and plates washed once with PBS containing Tween-20 (0.05% [v/v]) The coated microtiter plates are stable for at least 6 mo when stored desiccated at 4°C.

8. Assays should be carried out in the tubes, which are then used directly for counting. Assays that are to be counted in conventional counters should be performed in miniscintillation vials, which can be obtained from suppliers, such as Wallac (Turkur, Finland), Canberra Packard (Meriden, CT), and Beckman (Fullerton, CA) (e.g., Bio-vial™). These can be placed directly into the counting racks of the appropriate scintillation counter. Alternatively, stoppered glass, polypropylene, and polystyrene tubes that fit into standard scintillation vials can be used. Assays that are to be counted in microtiter plate format counters should be carried out in 96-well sample plates compatible with the instrument to be used. These plates are available from instrument suppliers.

References

1. Schram, M. and Selinger, Z. (1984) Message transmission: receptor controlled adenylate cyclase system. *Science* **225,** 1350–1356.
2. Butcher, R. W., Ho, R. J., Meng, H. C., and Sutherland, E. W. (1965) Adenosine 3',5'-monophosphate in biological materials, II. The measurement of adenosine 3',5'- mono-phosphate in tissues and the role of the cyclic nucleotides in the lipolytic response of fat to epinephrine. *J. Biol. Chem.* **240,** 4515–4523.
3. Brooker, G. (1972) High-pressure anion exchange chromatography and enzymatic isotope displacement assays for cyclic AMP and cyclic GMP, in *Advances in Cyclic Nucleotide Research,* vol. 2 (Greengard, P., Paoletti, R., and Robinson, G. A., eds.), Raven, New York, pp. 111–129.
4. Kuo, J. K. and Greengard, P. (1970) Cyclic nucleotide-dependent protein kinases. VIII. An assay method for the measurement of adenosine 3',5'-monophosphate in various tissues and a study of agents influencing its level in adipose cells. *J. Biol. Chem.* **245,** 4076–4083.
5. Ebadi, M. S., Weiss, B., and Costa, E. (1971) Microassay of adenosine 3',5'-monophosphate (cyclic AMP) in brain and other tissues by luciferin-luciferase system. *J. Neurochem.* **18,** 183–192.
6. Tovey, K. C., Oldham, K. G., and Whelan, J. A. M. (1974) A simple, direct assay for cyclic AMP in plasma and other biological samples using an improved competitive protein binding technique. *Clin. Chim. Acta* **56,** 221–234.
7. Harper, J. F. and Brooker, G. (1975) Fentomole sensitive radioimmunoassay for cyclic AMP and cyclic GMP after 2'-*O*-acetylation by acetic anhydride in aqueous solution. *J. Cyclic Nucleotide Res.* **1,** 207–218.
8. Steiner, A. L. (1974) Assay of cyclic nucleotides by radioimmunoassay methods, in *Methods in Enzymology,* vol. 38 (Langone, J. J. and Van Vunakis, H., eds.), Academic, London, pp. 96–105.
9. Cailla, H. L., Racine-Weisbuch, M. S., and Delaage, M. A. (1973) Adenosine 3',5'-cyclic monophosphate at 10^{-15} mole level. *Anal. Biochem.* **56,** 394–407.
10. Tsugawa, M., Fida, S., Fujii, H., Moriwaki, K., Tarui, S., Sugi, M., Yamane, M., Yamane, R., and Fujimoto, M. (1990) An enzyme-linked immunosorbent assay (ELISA) for adenosine 3',5'-cyclic monophosphate (cAMP) in human plasma and urine using monoclonal antibody. *J. Immunoassay* **11,** 49–61.
11. Cailla, H. L., Roux, D., Kurtziger, H., and Delaage, M. A. (1980) Antibodies against cyclic AMP, cyclic GMP and cyclic CMP. Their use in high performance radioimmunoassay. *Hormones Cell Reg.* **4,** 1–25.
12. Steiner, A. L., Parker, S. W., and Kipnis, D. M. (1972) Radioimmunoassay for cyclic nucleotides; 1. Preparation of antibodies and iodinated cyclic nucleotides. *J. Biol. Chem.* **247,** 1106–1113.
13. Scatchard, G. (1949) The attraction of proteins for small molecules and ions. *Ann. NY Acad. Sci.* **51,** 660–672.
14. Erlanger, G. F. (1973) Principles and methods for the preparation of drug-protein conjugates for immunochemical studies. *Pharmacol. Rev.* **25,** 271–280.

15. Goldberg, N. D. and O'Toole, A. G. (1971) Analysis of cyclic 3',5'-adenosine monophosphate and cyclic 3',5'-guanosine monophosphate, in *Methods of Biological Analysis,* 20 (Glick, D., ed.), Interscience Publishers, Wiley, London, pp. 1–39.
16. Steiner, A. L. (1979) Cyclic AMP and cyclic GMP, in *Methods of Hormone Radioimmunoassay* (Jaffe, B. M. and Behrman, H. R., eds.), Academic, New York, pp. 3–17.
17. Rosenberg, N., Pines, M., and Sela, I. (1982) Adenosine 3',5'-cyclic monophosphate—its release in a higher plant by an exogenous stimulus as detected by radioimmunoassay. *FEBS Lett.* **137,** 105–107.
18. Horton, J. K., Martin, R. C., Kalinka, S., Cushing, A., Kitcher, J. P., O'Sullivan, M. J., and Baxendale, P. M. (1992) Enzyme immunoassays for the estimation of adenosine 3',5'-cyclic monophosphate and guanosine 3',5'-cyclic monophosphate in biological fluids. *J. Immunol. Methods* **155,** 31–40.

CHAPTER 9

Measurement of Cyclic GMP Formation

Félix Hernández

1. Introduction

The recent advances in the understanding of the mechanism of action of molecules like nitric oxide and natriuretic peptides have increased the interest in cyclic GMP (cGMP) metabolism. Several techniques have been described to measure cGMP levels that include:

1. A binding assay using cGMP-dependent protein kinase purified partially;
2. Activation of cGMP-dependent protein kinase; and
3. Other methods based on enzymatic coupled assays (for a description of these methods, consult ref. *1*).

However, mainly because of its sensitivity, radioimmunoassay (RIA) is the method chosen by most investigators. This method, described first by Steiner et al. *(2)*, has been the object of multiple reviews (RIA methodology is well described in ref. *3*). Furthermore, kits from different commercial sources contain detailed information as well as experimental protocols. Acetylation of the cGMP is recommended for higher sensitivity in RIA. Perhaps the only problem is to choose the way to stop the reaction. Typically, cGMP is extracted from slices with dilute acid, for example, $0.3M$ perchloric acid, 5% trichloroacetic acid (TCA), or $0.1M$ HCl. Heating of the samples in a boiling water bath to coagulate protein can be used, too.

A prelabeling technique is an alternative approach to measure cGMP levels. This technique applied to cerebellar slices is described here. Slices are incubated with [^{3}H]guanine in order to label GTP pools. Then, they are treated with agents, and [^{3}H]cGMP is generated. [^{3}H]cGMP is separated

From: *Methods in Molecular Biology, Vol. 41: Signal Transduction Protocols*
Edited by: D. A. Kendall and S. J. Hill Copyright © 1995 Humana Press Inc., Totowa, NJ

from other [3H]metabolites by sequential Dowex-alumina chromatography *(4)*. In the first column (Dowex), GTP, GDP, and GMP are washed out, whereas cGMP adsorbs to the column. cGMP is then eluted onto alumina columns. Cyclic nucleotides are finally eluted with imidazole, whereas contaminants remain bound to these columns.

The method described here permits up to 60 samples to be processed in a single experiment. The assay procedure can easily be performed within a working day. Furthermore, consumable costs are substantially lower than for RIA The prelabeling technique detects increases in cerebellar cGMP in the presence of activators of either particulate or soluble guanylate cyclase.

2. Materials

1. Flat-bottomed plastic vials.
2. Plastic columns (Poly-Prep chromatography columns, 0.8 × 4 cm, Bio-Rad, Richmond, CA).
3. Racks for plastic columns: Ideally, the Dowex columns rack must allow that columns will drip onto alumina columns. On the other hand, alumina columns rack must allow that columns drip into scintillation vial rack.
4. [8-3H]guanine (324 GBq/mmol) from Rotem (Beer-Sheva, Israel).
5. [8-14C]cGMP (1.85–2.22 GBq/mmol) from Moravek (Brea, CA). Aliquots of 1 mL (1×10^5 dpm/mL) can be stored at –20°C. When required, add 4 mL of water (it is enough for 50 assays).
6. Plastic columns containing 1 mL (packed volume) of Dowex resin (Bio-Rad AG 50W-X4, 200–400 mesh, H$^+$ form) previously treated with 3 mL of 1N HCl and 20 mL of distilled water.
7. Plastic columns containing 0.6 g of neutral alumina for column chromatography (A-9003 Sigma, St. Louis, MO) previously washed with 20 mL of 0.1M imidazole.
8. Krebs-bicarbonate buffer, which contains 118 mM NaCl, 4.7 mM KCl, 1.2 mM NaSO$_4$, 1.2 mM KH$_2$PO$_4$, 11.7 mM glucose, 25 mM NaHCO$_3$, and 1.2 mM CaCl$_2$, equilibrated with 95% O$_2$:5% CO$_2$.
9. Dispensers containing: Distilled water, 1N HCl, 0.1M imidazole, and scintillation fluid for aqueous samples.

2.1. Preparation and Regeneration of Dowex Columns

Wash the resin by stirring in H$_2$O. Allow to settle. Decant the supernatant and the fines. Repeat several times. Finally, resuspend the resin in water (50% [v/v]) and pipet 2 mL of the stirred resin into plastic columns.

Variations in the elution profiles between different batches can be found. Thus, elution profile should be determined by loading [^{14}C]cGMP (100-µL stock) and collecting the eluate in 1-mL fractions.

Dowex columns can be used repeatedly (the same columns can be used daily for 6 mo). For storage, column luers are stored under water at room temperature. Prior to reuse, Dowex columns have to be washed with 3 mL 1M HCl and 20 mL of H$_2$O.

When a fall in the recovery of [^{14}C]cGMP is observed, columns can be washed with 1M NaOH (5 mL) followed by 10 mL water and one HCl/ H$_2$O cycle.

2.2. Preparation and Regeneration of Alumina Columns

Neutral alumina (0.6 g) is added to plastic columns. Prior to use, columns have to be washed with 20 mL of 0.1M imidazole. Elution profile should be determined by loading [^{14}C]cGMP and collecting the eluates. Columns can be stored dried at room temperature. Prior to reuse, alumina columns have to be washed with 20 mL of 0.1M imidazole.

3. Methods
3.1. Labeling of Slices with [^{3}H]Guanine

1. Incubate cross-chopped cerebellar slices (350 × 350 µm, prepared using a McIlwain tissue chopper) in a shaking water bath for 60 min at 37°C in Krebs-bicarbonate buffer. Change the buffer each 20 min (*see* Note 2).
2. Suspend slices in fresh Krebs-bicarbonate buffer, and add [^{3}H]guanine at a final concentration of 4 µM (0.18 MBq/mL). Incubate for 60 min.
3. Wash the slices three times. Aliquot gravity-packed slices (25 µL) into flat-bottomed plastic vials containing Krebs-bicarbonate buffer (275 µL). The tubes are sealed under 95% O$_2$:5% CO$_2$.
4. Allow to equilibrate for 15 min prior to addition of agents or appropriate vehicle controls in 10 µL. The tubes are resealed under 95% O$_2$:5% CO$_2$ after each addition.
5. Terminate the incubation after the appropriate length of time by addition of 200 µL of ice-cold 1N HCl.
6. Add a known amount of radiochemically pure [8-^{14}C]cGMP (100 µL of store, 2000 dpm/tube) to each experimental sample.
7. Add 0.75 mL of cold H$_2$O.
8. Allow to stand at 4°C for 30 min (it does not require homogenization or sonication of the samples).

3.2. Isolation of Labeled cGMP

1. Centrifuge samples for 10 min at 2000*g*.
2. Apply 0.9 mL of the supernatants to Dowex columns previously equilibrated with 3 mL of 1*M* HCl and 20 mL of distilled water.
3. Discard the eluates. Wash with 1 mL of H₂O and discard again (*see* Note 1).
4. Place Dowex columns rack over alumina columns rack. Alumina columns have been previously equilibrated with 0.1*M* imidazole.
5. Add 4 mL of H₂O to Dowex columns (*see* Note 3). The eluants from Dowex columns drip onto alumina columns. Discard alumina column eluants (*see* Note 1).
6. Remove Dowex columns rack. Add 1 mL of H₂O to alumina columns, and discard eluants.
7. Place alumina columns rack onto scintillation vial rack. Add 5 mL of 0.1*M* imidazole to alumina columns. Collect eluants into the vials.
8. Add 10 mL of scintillation fluid, cap scintillation vials, mix, and count in a liquid scintillation counter on a dual channel for ^{3}H and ^{14}C.

3.3. Expression of the Results

Values for [^{3}H]cGMP are corrected for the recovery of [^{14}C]cGMP (routinely 70–80%). Accumulation of [^{3}H]cGMP can be expressed as either radioactivity in dpm or as percent conversions (the percent of total radioactive guanine taken up that is converted to radioactive cyclic nucleotide). The basal levels of [^{3}H]cGMP found in adult guinea pig cerebellar slices are $0.27 \pm 0.13\%$ ($n = 22$) of the radioactivity taken up by the slices. This value is increased by the addition of different activators of either soluble or particulate guanylate cyclase. An example of results using this technique to measure cGMP accumulation in adult guinea pig cerebellar slices is shown in Fig. 1.

4. Notes

1. Volumes should be calculated by calibration of the columns with the appropriate standards. Volumes given here are merely indicatives and reflect the author's personal experience.
2. The principle of the method consists of labeling of intracellular GTP pools by incubation of the slices with [^{3}H]guanine. For this reason, an important control consists of testing the incorporation of [^{3}H]guanine. The author has observed that although in guinea pig cerebellar slices the incorporation of the [^{3}H]guanine (0.18 MBq/mL 60 min) is 4×10^5 dpm/25 µL of packed slices, in cortex and hippocam-

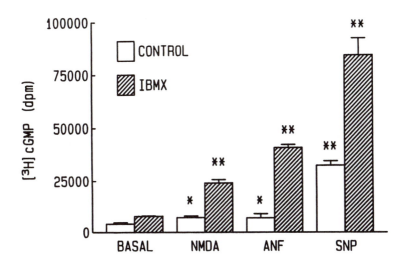

Fig. 1. The effects of *N*-methyl-D-aspartate (NMDA), sodium nitroprusside (SNP), and atrial natriuretic factor (ANF) on [³H]cGMP accumulation. [³H]guanine prelabeled slices were incubated for 10 min in absence or presence of 3-isobutyl-1-methylxanthine (1 m*M*, IBMX). After that, NMDA (1 m*M*), SNP (1 m*M*), or ANF (1 μ*M*) was added, and the incubations continued for 5 min. Data are means ± SEM of triplicate determinations from a single experiment repeated on three additional occasions with essentially similar results. *$p < 0.05$, **$p < 0.01$, vs the respective basal value.

pus, it is 1×10^5 and 5×10^4 dpm/25 μL of packed slices. Thus, before adopting this technique, it is necessary to test the incorporation of the radiolabeled purine.

3. It should be noted that specific activity of [³H]GTP can change during the experiment. Thus, it can be interesting to quantitate [³H]GTP and relate the [³H]cGMP dpm to the [³H]GTP dpm. If the effluent from the first column (which contains most of the [³H]GTP) is not discarded, but applied to a third column (Dowex-1 chloride form), [³H]GTP can be quantitated. This resin separates GTP from other contaminants by differential elution with increasing concentration of HCl.

Acknowledgments

The work described here was supported by NATO and the University of Nottingham. I thank Dr. David A. Kendall for his support and comments during this work.

References

1. Hardman, J. G. and O'Mallay, B. W., eds. (1974) *Methods in Enzymology*, vol. XXXVIII, section II, Academic, New York.
2. Steiner, A. L, Parker, C. W., and Kipnis, D. M. (1972) Radioimmunoassay for cyclic nucleotides. I. Preparation of antibodies and iodinated cyclic nucleotides. *J. Biol. Chem.* **247,** 1106–1113.
3. Brooker, G., Harper, J. F., Terasaki, W. L., and Moylan, R. D. (1979) Radioimmunoassay of cyclic AMP and cyclic GMP, in *Advances in Cyclic Nucleotide Research*, vol. 10, (Brooker, G., Greengard, P., and Robinson, G. A., eds.), Raven, New York, pp. 1–33.
4. Salomon, Y., Londos, C., and Rodbell, M. (1974) A highly sensitive adenylate cyclase assay *Anal. Biochem.* **58,** 541–548.

CHAPTER 10

Cyclic AMP-Dependent Protein Kinase Activity Ratio Assay

Kenneth J. Murray

1. Introduction

A multitude of natural and pharmacological agents, including hormones, neurotransmitters, and phosphodiesterase inhibitors, raise intracellular cAMP levels and, in the vast majority of instances, the physiological effects are the result of activation of cAMP-dependent protein kinase (cA-PrK) and subsequent protein phosphorylation. The involvement of cA-PrK in the response may be investigated by the use of the "cA-PrK activity ratio" assay, first introduced in 1973 *(1)*.

cA-PrK is an inactive holoenzyme composed of two classes of subunits: regulatory (R) subunits, which bind cAMP, and catalytic (C) subunits, which perform the phosphotransferase function. The activity ratio assay takes advantage of the observation that dissociation of the R and C subunits occurs on "activation" (i.e., binding of cAMP) of the holoenzyme:

$$R_2C_2 + 4cAMP \rightleftarrows R_2cAMP_4 + 2C \tag{1}$$
$$\text{inactive holoenzyme} \qquad \text{active holoenzyme}$$

Providing that this equilibrium is preserved during the homogenization and assay of the cell or tissue extract, then reliable information about the activity of cA-PrK in the intact system is obtained. However, it should be noted that a number of potential artifacts of this assay have been reported, and these have cast doubt on its usefulness (*see* Note 17). In contrast, a re-investigation *(2)* of the method using peptides rather than histone as

From: *Methods in Molecular Biology, Vol. 41: Signal Transduction Protocols*
Edited by: D. A. Kendall and S. J. Hill Copyright © 1995 Humana Press Inc., Totowa, NJ

the substrate has shown that meaningful results may be obtained using a modified assay (*see* Note 18). Therefore, in this chapter, the latter method for the cA-PrK activity ratio assay is described, and experimental approaches to investigate its validity are suggested.

2. Materials and Stock Solutions

1. Homogenization buffer (HB) *(1,2)*: 10 mM NaH$_2$PO$_4$, 10 mM EDTA, and 0.5 mM 3-isobutyl-1-methylxanthine, pH 6.8. This solution may be stored at 4°C for considerable periods of time. When required, NaCl or Triton X-100 may be added directly to this buffer.
2. Assay mixture *(2)*: Two assay mixtures, one of which contains cAMP, are required; they are composed of 70 mM NaH$_2$PO$_4$, pH 6.8, 14 mM MgCl$_2$, 1.4 mM EGTA, 0.014% Tween-20, 28 µM malantide (or 140 µM kemptide), and one mixture contains 14 µM cAMP. These assay mixtures may be stored frozen at –20°C, but repeated freezing and thawing should be avoided.
3. Radiolabeled ATP: [γ-^{32}P]- or [γ-^{33}P]-ATP may be used for the assay; it is diluted with stock unlabeled ATP to obtain the desired concentration (usually 2.1 or 0.7 mM) and specific radioactivity (typically 50–200 cpm/pmol).
4. Source of materials: Malantide is available from Ocean Biologicals (Edmonds, WA) and Peninsula Laboratories (Belmont, CA). Kemptide and PKI-derived peptide are available from a number of commercial sources including Peninsula Laboratories and Sigma (St. Louis, MO). Both [γ-^{32}P]- and [γ-^{33}P]-ATP can be obtained from Amersham (Arlington Heights, IL) and NEN. P81 (phosphocellulose) paper is from Whatman (Maidstone, Kent, UK).

3. Methods

3.1. Homogenization

The precise homogenization parameters vary according to the tissue or cell type being investigated, and the exact conditions can only be determined experimentally. Suggested steps are outlined.

1. Typically, the samples will be frozen at the end of the experimental protocol and stored at –70°C prior to the activity ratio assay. The intact tissue or cell pellet should be frozen; storage of extracts is not recommended. It is convenient to store the samples in the tubes in which they are to be homogenized, and this procedure also minimizes the risk of the samples thawing prior to homogenization. For the same reasons, tissue samples should be preweighed.

2. The first step is to determine the volume of HB required per unit weight, since this has impact on practical design (e.g., choice of tubes, homogenizer). The aim should be to use as large a volume of HB as possible (*see* Notes 5–9), such that no subsequent dilution of the extract is required prior to assay. In addition to being convenient practically, homogenization in a large volume minimizes the possibility of subsequent artifactual dissociation of cA-PrK. Once the appropriate volume of HB per unit weight has been established, this ratio should be adhered to, since altering the amount of cA-PrK could cause the assay to become nonlinear.

3. Mild homogenization conditions should be used because the activity of the free C subunit is particularly susceptible to vigorous conditions. If sufficient tissue is available, the preferred method is to produce a powder under liquid nitrogen in a pestle and mortar, and to homogenize this by use of a motor-driven glass/Teflon™ homogenizer. However, a mechanical homogenizer (e.g., Polytron) will be required for tough tissues or when only small amounts are available. In this case, care should be taken that the minimum time and speed are used in order to get a consistent homogenization.

4. It is usually necessary to subject the tissue homogenate to centrifugation prior to assay, although some direct extracts that contain little particulate matter may be assayed with care. The points to consider here are: depending on the tissue type, a significant proportion of cA-PrK may be naturally associated with the "particulate" fraction so that choice of centrifugation conditions must be made carefully; and unless NaCl is included in the HB, free catalytic subunit will artifactually bind to the particulate matter *(2,3)* (*see* Note 12).

5. The following procedure has been found to be suitable for both freshly isolated cells (e.g., platelets) and those in culture. Subject the cells to the desired treatment in microcentrifuge tubes, after which place the tubes in liquid nitrogen. Alternatively, the cells may be rapidly pelleted, the supernatant removed, and the cell pellet frozen. Add HB containing 50 mM NaCl and 0.2% Triton X-100 to the tube, which is then subjected to vigorous vortex mixing followed by a brief centrifugation to remove the ensuing foam. If the cells have been frozen as a suspension, then the volumes of the suspension and HB should be such that the suspension thaws rapidly (<10 s). If this does not occur, then use of a sonicator (probe or water bath) may increase the rate of thawing. Failing this, the cells should be pelleted prior to homogenization. Conditions for the assay of adipocytes are described in ref. *2*.

6. With both tissues and cells, it is essential that consistent conditions be used and that complete homogenization takes place. It is preferable that the samples are processed in small batches (four to eight), so that the time from the start of homogenization to completion of the assay is kept to a minimum.

3.2. Protein Kinase Assay

For further details of the assay, *see* Notes 1–4. Suggested preliminary experiments are given in Notes 5–9, whereas Notes 10–18 refer to experiments that may be used to validate the assay.

1. The assay is usually carried out in duplicate, i.e., four tubes (two with cAMP, two without cAMP) per tissue extract. Ten microliters of tissue extract are mixed with 50 µL assay mixture and pre-equilibrated at 30°C.
2. Following equilibration for 1 min, initiate the reaction by the addition of 10 µL radiolabeled ATP. After incubation for the desired period, 10 µL $1M$ HCl are added to the tubes to terminate the reaction.
3. When all the tubes have been terminated, an aliquot (usually 20–40 µL) of the reaction mixture is spotted onto numbered pieces of P-81 paper (approx 1.5×2.5 cm). The papers are given four washes of 5–10 min each in 75 mM H_3PO_4 in a mesh basket using at least 10 mL of H_3PO_4/paper. Following a further wash in ethanol, the papers are dried with a hair dryer, placed into individual vials, and subjected to scintillation counting.

3.3. Calculation of Results

Blanks are subtracted from all values. The activity ratio (also referred to as "fractional activity") is calculated by dividing the average of the duplicates assayed without cAMP by the average of the duplicates assayed in its presence:

$$(\text{Assay} - \text{cAMP})/(\text{Assay} + \text{cAMP}) = (C/(R_2C_2 + C) = \text{Active/Total activity} \qquad (2)$$

The activity ratio gives a value between 0 and 1, where 1 indicates full activity or complete dissociation of the holoenzyme. The results may also be expressed as "% active" by multiplying the activity ratio by 100.

3.4. Measurement of ATP Specific Activity and of ATP Depletion

The proportion of radioactivity that is associated with ATP may be determined by a quick and simple assay that utilizes the binding of ATP to activated charcoal.

1. A suspension is made by mixing activated charcoal (25 g/L) in $1M$ HCl, allowing the charcoal to settle, decanting the fines, and by repeating this process; prior to use, the charcoal is resuspended by use of a magnetic stirrer.

2. A sample of the ATP solution is diluted into a microfuge tube, and an aliquot (A) taken to determine the total radioactivity; subsequently, an equal volume of the activated charcoal is added to the sample in a microfuge tube and, following centrifugation (15,000g, 2 min), a second aliquot (B) of the sample is removed and subjected to scintillation counting.

3. To compensate for the dilution of the sample by the charcoal suspension, the second aliquot should be twice the volume of the first, or the appropriate calculation made. The counts in the second aliquot are those not associated with ATP. Therefore, the percentage of counts owing to ATP is obtained from: $-100 [(A-B)/A]$.

4. The specific activity of the ATP is determined by taking aliquots of the diluted stock solution prior to and after the addition of the activated charcoal; the radioactivity associated with ATP is obtained from the difference in counts of the two aliquots (as in step 3). The chemical concentration of the ATP is determined spectrophotometrically ($A_{259} = 15.4$), and thus, the specific radioactivity may be calculated. Knowledge of the ATP specific activity is not required to calculate the cA-PrK activity ratio; however, it is required if absolute activities are to be determined and also is needed to calculate conversion of the peptide substrate.

5. The amount of ATP hydrolyzed during the course of the activity ratio assay can be obtained by diluting an aliquot of the acid-stopped reaction mix. The radioactivity associated with ATP is determined by scintillation counting of an aliquot taken before and after addition of activated charcoal, as described. If it has been shown that incorporation of radioactivity into the peptide substrate is linear with time, then it is reasonable to assume that depletion of ATP is not affecting the activity ratio assay; however, there are possible benefits to be obtained from determining the concentration of ATP remaining at the end of the assay (*see* Note 8). Depletion of ATP is a factor that may well be responsible if the activity ratio assay is not linear, and whether this is so can be ascertained from this simple assay.

4. Notes
4.1. Protein Kinase Assay

1. The concentration and specific radioactivity of the ATP and the incubation time are determined by preliminary experiments (*see* Notes 5–9).

2. When terminated with HCl, the reaction mix may be spotted onto the P-81 paper at leisure. This allows more tubes to be assayed per incubation period, since HCl can be added to the tubes at 5-s intervals, whereas spotting requires a 10- or 15-s gap between tubes. It is convenient to use repeating pipets for the addition of the radiolabeled ATP and HCl. A sepa-

rate reason for the addition of HCl is that acidification of the reaction mixture may aid binding of the phosphopeptide to the P-81 paper *(4)*, although this is not required for the binding of malantide.

3. *See* ref. 5 for details of design of the mesh basket.

4. Suitable blank values are obtained by adding the HCl prior to the tissue extract; blanks are required for each basket, since they monitor the efficiency of the washing procedure. A minimum of four washes should be performed in order to obtain low blank values. The time of each wash is not critical; in general, increasing the number of washes (e.g., to six to eight, in which case the washing period may be reduced to 2–3 min) is more effective than increasing the washing period. Care must be taken in the disposal of the washing solution, which, particularly in the case of the first wash, may contain considerable amounts of radioactivity. Blank values are usually <5 pmol ATP. If higher numbers are obtained, the number of washes should be increased. However, certain batches of P81 paper give higher blank values, and this appears to be because of binding of a radiolabeled contaminant in the ATP. In this case, the blank may be decreased by washing in 0.5% (v/v) tetraphosphoric acid and 38 mM H_3PO_4; under these conditions, malantide, but not kemptide, binds to P-81 *(2)*, so it is probably more convenient to obtain a different batch of P81 paper.

4.2. Preliminary Experiments

5. Both the amount of cA-PrK and that of potentially interfering enzymes (e.g., ATPases) vary according to cell type. Therefore, experimental conditions for the activity ratio assay must be established for each tissue, and it cannot be assumed that conditions previously determined for another tissue are applicable. The major objective of these preliminary experiments is to demonstrate that the assay is linear both with time and amount of tissue extract. Although the experimental approaches to this are straightforward, the results have considerable implications for the practical design of subsequent experiments, since they will determine the volume of homogenization buffer required and the concentration of ATP employed in the assay.

6. The initial experiment is to prepare serial dilutions of the cell or tissue extract, and assay these, at a number of time-points, in the presence of cAMP using 300 µM radiolabeled ATP. The results should be calculated in picomoles of radioactivity transferred to the substrate, so that the amount of substrate utilized can be determined. As a rough rule of thumb, the maximum incorporation should be 10–20%, and the period of incubation should not exceed 15 min. For convenience, the reaction may be terminated by spotting directly onto the phosphocellulose paper, thus dispensing with the need to set up separate tubes for each time-point.

7. If the assay is linear in the presence of cAMP, then it will be in its absence. However, problems could be encountered if the activity in the absence of cAMP is too low to obtain valid numbers of cpm. Therefore, a few preliminary assays plus and minus cAMP on control samples (which should have the lowest activity ratio) should be performed.

8. The percentage of ATP hydrolyzed is determined using the assay described in Section 3.4. If this is low (<10%), then 100 μM radiolabeled ATP may be used in subsequent assays, although the linearity of the assay should be rechecked. The advantage of using a lower concentration of ATP is that the amount of radiolabel is decreased.

9. Within these constraints, the volume of homogenization buffer should be chosen such that small variations in the amount of tissue extracted will not affect the linearity of the assay, Practical considerations, e.g., amounts available, tube size, and so on, may also need to be taken into account.

4.3. Control Experiments

10. The results obtained with the cA-PrK activity ratio assay may be interpreted more vigorously if the following control experiments are performed to validate the assay. As stated previously, these experiments should be carried out for each new tissue or cell type investigated, and reliance cannot be placed on data from dissimilar tissues.

11. The two most suitable peptides for this assay are kemptide and malantide; histone should not be used for a number of reasons (*see* ref. 2 and Note 18). Although both these peptides are kinetically good substrates for cA-PrK, they are both phosphorylated by a number of other protein kinases. The proportion of the substrate that is being phosphorylated by cA-PrK can be determined simply by the inclusion of a peptide derived from the specific inhibitor protein of cA-PrK (PKI). In a number of rat tissues, the phosphorylation of malantide was inhibited (>90%) by 1 μM PKI-peptide, indicating that the vast majority of phosphorylation was catalyzed by cA-PrK *(2)*. Similar results have been reported for kemptide in various respiratory tissues *(6,7)*.

12. NaCl is included in the HB for two reasons: to prevent artifactual association of free C subunit with the particulate fraction *(2,3)* and to prevent reassociation of the R_2 and C subunits of the type II holoenzyme *(8)*. The reasons stated previously for not including NaCl in the HB were that it inhibited the phosphorylation of histone by cA-PrK, and it caused dissociation of the holoenzyme *(9)*. However, peptide substrates are less sensitive to inhibition by NaCl, and NaCl did not cause activation of cA-PrK in either rat (mainly type I cA-PrK) or guinea pig (mainly type II) heart *(2)*.

13. Tissue is homogenized in HB containing a range of NaCl concentrations (0–500 m*M*) to determine an appropriate concentration, and the activity ratio determined at timed intervals to ascertain if reassociation has taken place. Generally, HB containing 50 m*M* NaCl has been found suitable for most tissues and cells; adipose tissue is an exception to this, since 200 m*M* NaCl is required to prevent reassociation of the holoenzyme *(2)*. Also, other effects of high NaCl in the HB should be considered; for example, in muscle, 500 m*M* NaCl will cause solubilization of myofibrils, which contain bound cA-PrK This makes the extracts difficult to pipet, but also means a different "compartment" of cA-PrK is being assayed.

14. The ultimate test of validity (first suggested in ref. *9*) is to add exogenous cA-PrK to the tissue just prior to homogenization and observe whether dissociation of the added enzyme takes place. This approach has been used to demonstrate that the conditions used for activity ratio assays that employ peptide substrates do not result in artifactual dissociation *(2,7)*. It is particularly important to perform this experiment with samples that have raised cAMP levels, since it could be argued that the basal concentration of cAMP found in the control tissues is less likely to produce artifactual results.

4.4. General Comments

15. The cA-PrK activity ratio assay, as introduced by Corbin et al. *(1)* described the conditions required to conduct this assay in adipose tissue. Later, it was realized that cA-PrK exists as two major isoenzymes, termed type I and type II (reviewed in refs. *10,11*), and specific conditions for the activity ratio assay were established for tissues containing differing proportions of the two isoenzymes (reviewed in ref. *12*). A subsequent investigation of these assay conditions indicated a number of potential artifacts and concluded that dissociation of cA-PrK was not blocked throughout the extraction procedure *(9)*; other possible pitfalls of the assay have also been noted (reviewed in ref. *12*).

16. In all of these experiments, histone was used as substrate for cA-PrK; reinvestigation of the activity ratio assay using synthetic peptides has shown that many of the pitfalls described were owing to the use of histone, and are not apparent when either malantide *(2)* or kemptide *(6,7)* is employed. The major advantages of the synthetic peptides over histone are:
    a. They are more selective substrates;
    b. They do not cause dissociation of cA-PrK;
    c. Their phosphorylation by cA-PrK is not inhibited by NaCl; and
    d. The peptides are kinetically superior substrates.
    Both factors (a) and (b) result in a lower activity ratio being measured with

the peptide substrates. It should be reemphasized that the clear demonstration that histone dissociates cA-PrK during the assay *(2)* means that it is not a suitable substrate. Indeed, histone, rather than the extraction conditions, may have caused the activation of cA-PrK noted in ref. *9*. Use of peptide substrates also increases the sensitivity of the assay, allowing a larger volume of HB to be used, which again reduces the potential for artifacts.

17. Investigation of the NaCl concentration in the HB has also revealed the importance of this factor in the assay *(2)*. The complete absence of NaCl results in a dramatic decrease in the activity ratio owing to trapping of free catalytic subunit *(2,3)*; therefore, NaCl must be included, although the exact concentration required is tissue dependent. In conclusion, there is experimental evidence to indicate that the use of peptide substrates and inclusion of NaCl in the HB result in a cA-PrK activity ratio assay that reflects accurately the state of activation of cA-PrK in the intact cell. However, when a new tissue is being assayed, control experiments must always be performed to validate fully the conditions employed.

18. There are limitations to the information that may be gathered with the activity ratio assay as there are with any procedure that involves homogenization, since subcellular structures or compartments may not be preserved in their original state. In this respect, it should be noted that spatial localization of cAMP *(13,14)* and cA-PrK *(15)* within the cell has been described. Nevertheless, the cA-PrK activity ratio assay provides a convenient method for investigation of cellular responses involving the cAMP second-messenger system.

Acknowledgment

I am grateful for the assistance of David Mills and Marie Hancock in the preparation of this manuscript.

References

1. Corbin, J. D., Soderling, T. R., and Park, C. R. (1973) Regulation of adenosine 3',5'-monophosphate-dependent protein kinase. I. Preliminary characterization of the adipose tissue enzyme in crude extracts. *J. Biol. Chem.* **248,** 1813–1821.

2. Murray, K. J., England, P. J., Lynham, J. A., Mills, D., Schmitz-Peiffer, C., and Reeves, M. L. (1990) Use of a synthetic dodecapeptide (malantide) to measure the cyclic AMP-dependent protein kinase activity ratio in a variety of tissues. *Biochem. J.* **267,** 703–708.

3. Keely, S. D., Corbin, J. D., and Park, C. R. (1975) On the question of translocation of heart cAMP-dependent protein kinase. *Proc. Natl. Acad. Sci. USA* **72,** 1501–1504.

4. Toomik, R., Ekman, P., and Engström, L. (1992) A potential pitfall in protein kinase assay: phosphocellulose paper as an unreliable adsorbent of produced phosphopeptides. *Anal. Biochem.* **204,** 311–314.

5. Corbin, J. D. and Reimann, E. M. (1974) Assay of cAMP-dependent protein kinases. *Methods Enzymol.* **38**, 287–290.

6. Giembycz, M. A. and Diamond, J. (1990) Evaluation of kemptide, a synthetic serine-containing heptapeptide, as a phosphate acceptor for the estimation of cyclic AMP-dependent protein kinase activity in respiratory tissue. *Biochem. Pharmacol.* **39**, 271–283.

7. Langlands, J. M. and Rodger, I. W. (1990) Determination of soluble cAMP-dependent protein kinase activity in guinea-pig tracheal smooth muscle. *Biochem. Pharmacol.* **39**, 1365–1374.

8. Corbin, J. D., Keely, S. L., and Park, C. R. (1975) The distribution and dissociation of adenosine 3',5'-monophosphate-dependent protein kinase in adipose, cardiac and other tissues. *J. Biol. Chem.* **250**, 218–225.

9. Palmer, W. K., McPherson, J. M., and Walsh, D. A. (1980) Critical controls in the evaluation of cAMP-dependent protein kinase activity ratios as indices of hormonal action. *J. Biol. Chem.* **255**, 2663–2666.

10. Scott, J. D. (1991) Cyclic nucleotide-dependent protein kinases. *Pharmacol. Ther.* **50**, 123–145.

11. Taylor, S. S., Buechler, J. A., and Yonemoto, W. (1990) cAMP-dependent protein kinase: framework for a diverse family of regulatory enzymes. *Annu. Rev. Biochem.* **59**, 971–1005.

12. Corbin, J. D. (1983) Determination of the cAMP-dependent protein kinase activity ratio in intact tissues. *Methods Enzymol.* **99**, 227–232.

13. Barsony, J. and Marx, S. J. (1990) Immunocytology on microwave-fixed cells reveals rapid and agonist-specific changes in subcellular accumulation patterns for cAMP or cGMP. *Proc. Natl. Acad. Sci. USA* **87**, 1188–1192.

14. Bacskai, B. J., Hochner, B., Mahaut-Smith, M., Adams, S. R., Kaang, B.-K., Kandel, E. R., and Tsien, R. Y. (1993) Spatially resolved dynamics of cAMP and protein kinase A subunits in *Aplysia* sensory neurons. *Science* **260**, 222–226.

15. Scott, J. D. and Carr, D. W. (1992) Subcellular localization of the Type II cAMP-dependent protein kinase. *News Physiol. Sci.* **7**, 143–148.

CHAPTER 11

Purification and Activation of Cyclic GMP-Dependent Protein Kinase

Takayoshi Kuno and Hideyuki Mukai

1. Introduction

Cyclic GMP (cGMP)-dependent protein kinase (G kinase) can be purified from bovine lung with the combination of ion-exchange chromatography, ammonium sulfate precipitation, and cyclic nucleotide affinity chromatography *(1)*. In this chapter, we describe the purification procedure that is routinely used in our laboratory. As a source, the bovine lung seems to be the best material to obtain a large amount of purified sample. However, the procedure described here can be applicable to the purification of G kinase from various mammalian or nonmammalian tissues.

The G kinase assay has been performed using mixed histone or F2b histone as substrates *(1)*. Recently, a synthetic heptapeptide substrate has been shown to be a more specific substrate for the G kinase relative to other kinases *(2)*. Preparation of histone substrates is tedious, and the quality of the preparation may change from batch to batch. In contrast, the peptide substrate is easy to synthesize (also commercially available), and the results obtained with the peptide substrate seem to be more reproducible probably because of its purity and uniformity compared with the histone substrates. In this chapter, a G kinase assay using the peptide substrate is described. Inclusion of the peptide inhibitor of the cyclic AMP (cAMP)-dependent protein kinase in the assay mixture also makes the assay specific for G kinase.

From: *Methods in Molecular Biology, Vol. 41: Signal Transduction Protocols*
Edited by: D. A. Kendall and S. J. Hill Copyright © 1995 Humana Press Inc., Totowa, NJ

For the study of cyclic nucleotide binding protein, ammonium sulfate precipitation assay has been widely used *(3)*. However, inclusion of ammonium sulfate in the precipitation and washing steps results in a change in the properties of the cGMP binding sites of G kinase *(4)*. To avoid such artifacts produced by high concentration of ammonium sulfate used for G kinase precipitation, a rapid filtration assay using polyethylenemine-treated glass filter has been developed *(4,5)*. Free and bound [^{3}H]cGMP can be separated by using the filter, because bound cGMP is trapped by the filter through the binding of G kinase to the polycationic filter and free cGMP passes through the filter. Because the binding between G kinase and the filter is mainly the result of the ionic interaction, time required for filtration is shorter than that for the ammonium sulfate precipitation assay. Furthermore, use of glass-fiber filters instead of fragile nitrocellulose filters results in more reproducible and consistent binding data.

2. Materials

2.1. Purification of cGMP-Dependent Protein Kinase

1. Fresh bovine lungs were obtained from a local slaughterhouse.
2. PEM buffer: 20 mM sodium phosphate, pH 7.0, 2 mM EDTA, and 25 mM 2-mercaptoethanol.
3. DEAE-cellulose: DE-52 (Whatman, Maidstone, UK).
4. 8-6-Aminohexylamino-cAMP agarose (Sigma, St. Louis, MO).

2.2. Protein Kinase Assay

1. Peptide substrate for G kinase: Arg-Lys-Arg-Ser-Arg-Ala-Glu (Peninsula, Belmont, CA).
2. Peptide inhibitor of the cAMP-dependent protein kinase (Sigma).
3. Phosphocellulose paper: P-81 (Whatman, 2 × 2 cm).

2.3. cGMP Binding Assay
Using Polyethylenemine-Treated Glass Filter

1. Glass fiber filters: Whatman GF/B or GF/C presoaked (from 30 min to 2 h) in 1% polyethylenemine prepared from 10% stock solution.
2. Membrane filtration device coupled to a vacuum source (pump): Device for one filter can be used for assays of small number, but a cell harvester apparatus (e.g., M-24, Brandel) is convenient for large assay series (*see* Note 1).

3. Methods

3.1. Purification of cGMP-Dependent Protein Kinase

1. Grind freshly obtained bovine lung (2 kg) with a meat grinder into a fine mass and homogenize in 2 vol of PEM in a 4-L Waring™ blender (*see* Note 2).
2. Centrifuge the homogenate at 12,000g for 20 min. Dilute the supernatant in an equal volume of PEM and mix thoroughly.
3. Apply the extract to DE-52 column (10 × 60 cm), which should be previously equilibrated with PEM buffer.
4. After washing the column with 20 L of PEM containing 50 mM NaCl, elute the fraction containing G kinase with 12 L of PEM containing 150 mM NaCl.
5. Mix the eluate with 5 kg of solid ammonium sulfate for 1 h with stirring. Centrifuge the suspension at 12,000g for 20 min. Dissolve the pellet in 500 mL of PEM and dialyze against 25 L of PEM.
6. Centrifuge the dialysate at 12,000g for 20 min, and apply the supernatant to 3 mL of 8-6-aminohexylamino cAMP agarose, which should be equilibrated with PEM.
7. After washing the column successively with 100 mL of PEM, 100 mL of PEM containing 2M NaCl, 20 mL of PEM containing 10 mM 5'-AMP, and 100 mL of PEM, elute the G kinase with 20 mL of PEM containing 10 mM cGMP.
8. Apply the eluate to 2 mL of DE-52 column equilibrated with PEM. After washing the column with 500 mL of PEM, allow 10 mM cAMP in the same buffer to pass through the column, and then halt the flow for approx 5 h at 4°C.
9. Wash the column with 500 mL of PEM containing 20% (v/v) glycerol, and then elute G kinase with 0.25M NaCl in the same buffer. Dialyze this eluate overnight against PEM containing 50% (v/v) glycerol at 4°C, and use the dialysate as purified G kinase or store at –80°C. Usually approx 5000-fold purification and 10% recovery are obtained with this procedure (*see* Note 3).

3.2. Protein Kinase Assay

1. The standard assay should be conducted in a volume of 50 µL containing 20 mM Tris-HCl, pH 7.5, 20 mM magnesium acetate, 0.2 mM ATP, 0.1 mM 3-isobutyl-1-methylxanthine, 0.1 mM heptapeptide substrate for G kinase, [γ-^{32}P]ATP (3000 Ci/mmol, 5–20 x 10^5 cpm), 0.1 mM synthetic peptide inhibitor of the cAMP-dependent protein kinase, and 1 µM cGMP when present.
2. Initiate the reaction with the addition of the enzyme, and carry out incubations at 30°C for 5 min.
3. Terminate the reaction by applying 50-µL aliquots to phosphocellulose papers, and washing with several changes with 75 mM phosphoric acid and one ethanol change.
4. Dry the paper, and count the radioactivity with scintillation fluid (*see* Note 4).

3.3. cGMP Binding Assay
Using Polyethylenemine-Treated Glass Filter

1. The standard assay should be conducted in a volume of 0.1 mL containing 20 mM Tris-HCl, pH 7.5, 20 mM magnesium acetate, 0.1 mM 3-isobutyl-1-methylxanthine, 1 mg/mL bovine serum albumin (BSA), [^{3}H]cGMP (10–20 Ci/mmol, 10 nM–1 μM), and 0.1 mM nonradioactive cGMP when present (*see* Note 5).
2. Initiate the reaction with the addition of enzyme and carry out incubations at 30°C for 30 min.
3. Terminate the reaction by the addition of 5 mL of ice-cold 20 mM Tris/HCl, pH 7.5, containing 1 mM EDTA, and rapid filtration under vacuum through a glass-fiber filter presoaked in 1% polyethylenemine. Wash the filter twice with 5 mL of the same buffer.
4. Dry the filter at 80°C for 1 h in the oven, and count the radioactivity with scintillation fluid. Nonspecific binding is the binding in the presence of 0.1 mM cGMP. Calculate specific binding by subtracting the nonspecific binding from the total binding.

4. Notes

1. M-24 has a 24-well manifold in which strips of polyethylenemine-soaked glass filters can be automatically fitted and sealed into individual filtration compartments. Incubation mixtures are individually and automatically aspirated and filtered. The total time required for filtration and washing is on the order of approx 10 s/24 samples.
2. All procedures should be performed at 4°C, and long bursts in the blender (>1 min) should be avoided in order to obtain an intact enzyme preparation.
3. Two isoforms (Iα and Iβ,) of G kinase have been identified in bovine aorta *(6,7)*. Reportedly, these two isoforms were separated by chromatography on a DEAE-Sephacel (Pharmacia, Uppsala, Sweden) column *(6)*.
4. If the radioactivity is high, it can be measured by liquid scintillation spectrometry with distilled water using the Cherenkov effect. In this case, phosphocellulose papers are not necessarily to be treated with ethanol and dried.
5. Incubation conditions can be changed unless ionic strength and pH do not affect the binding of G kinase to the polyethylenemine-treated glass filter. BSA is included in the mixture because of its lowering effect on nonspecific binding.

References

1. Lincoln, T. M. (1983) cGMP-dependent protein kinase. *Methods Enzymol.* **99,** 62–71.
2. Glass, D. B. and Krebs, E. G. (1982) Phosphorylation by guanosine 3':5'-monophosphate-dependent protein kinase of synthetic peptide analogs of a site phosphorylated in histone H2B. *J. Biol. Chem.* **257,** 1196–1200.
3. Døskeland, S. O. and Øgreid, D. (1988) Ammonium sulfate precipitation assay for the study of cyclic nucleotide binding to proteins. *Methods Enzymol.* **159,** 147–159.
4. Hirai, M., Hashimoto, S., Kuno, T., and Tanaka, C. (1988) Characterization of cyclic GMP-binding sites of cyclic GMP-dependent protein kinase by rapid filtration assay. *Biochem. J.* **255,** 477–482.
5. Takeda, T., Kuno, T., Shuntoh, H., and Tanaka, C. (1989) A rapid filtration assay for cAMP. *J. Biochem.* **105,** 327–329.
6. Wolfe, L., Corbin, J. D., and Francis, S. H. (1989) Characterization of a novel isozyme of cGMP-dependent protein kinase from bovine aorta. *J. Biol. Chem.* **264,** 7734–7741.
7. Wernet, W., Flockerzi, V., and Hofmann, F. (1989) The cDNA of the two isoforms of bovine cGMP-dependent protein kinase. *FEBS Lett.* **251,** 191–196.

CHAPTER 12

The Analysis and Assay of Cyclic Nucleotide Phosphodiesterase Isoenzyme Activity

Mohammed Shahid and C. David Nicholson

1. Introduction

The ubiquitous intracellular second messengers 3',5'-cyclic adenosine monophosphate (cAMP) and 3',5'-cyclic guanosine monophosphate (cGMP) mediate the effects of a large variety of hormones and neurotransmitters. Both of these nucleotides are inactivated by a large group of enzymes collectively known as 3',5'-cyclic nucleotide phosphodiesterases (PDEs; EC 3.1.4.17), which catalyze the hydrolytic cleavage of the 3' phosphodiesterase bond to form the corresponding 5' nucleotide (i.e., 5'AMP or 5'GMP; Fig. 1).

Although PDE was discovered over 30 yr ago, only recent studies have confirmed the presence of multiple molecular forms of this enzyme. It is now clear from molecular cloning analysis that there are at least seven and perhaps as many as eight distinct classes of PDEs arising from distinct gene families (Table 1). These PDE isoenzymes show marked differences in substrate specificity, as well as kinetic, regulatory, physical, and immunological properties (1–3). Furthermore, the presence of different PDE isoenzymes varies between tissues both in terms of activity and intracellular location. Moreover, PDE isoenzymes also differ in their sensitivity to a variety of inhibitors. Indeed, the discovery of PDE isoenzyme-selective inhibitors and their ability to modulate tissue function in a specific manner have provoked much interest in elucidating the thera-

From: *Methods in Molecular Biology, Vol. 41: Signal Transduction Protocols*
Edited by: D. A. Kendall and S. J. Hill Copyright © 1995 Humana Press Inc., Totowa, NJ

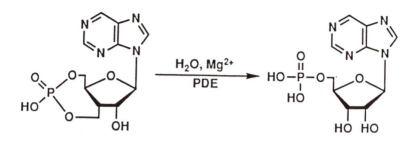

Fig. 1. The hydrolytic reaction catalyzed by 3',5'-cyclic nucleotide phosphodiesterases.

Table 1
Cyclic Nucleotide PDE Isoenzymes

Isoenzyme	Characteristic
I	Ca^{2+}/calmodulin stimulated
II	cGMP stimulated
III	cGMP inhibited
IV	cAMP specific (rolipram sensitive)
V	cGMP specific
VI	cGMP specific (light activated)
VII	cAMP specific (rolipram insensitive)
VIII	Others (e.g., IBMX insensitive)

peutic potential of these compounds for a variety of diseases *(4)*, two notable examples being the therapeutic areas of congestive heart failure *(5)* and asthma *(6)*. In parallel with and perhaps as a result of the enhanced interest in PDEs and the development of isoenzyme-selective inhibitors, the methods for isolating and assaying PDE isoenzymes have also been refined. The techniques described in this chapter have been successfully applied in studies aimed at investigating the role of PDE isoenzymes in regulating cell function in a variety of tissues *(7–10)*.

1.1. Separation of PDE Isoenzymes

Since a number of human PDE isoenzymes have now been cloned and expressed in eukaryotic *(11–16)* and prokaryotic *(14,17–19)* cells, for detailed molecular and pharmacological characterization of PDE isoenzymes, the emphasis has shifted from biochemical separation techniques to molecular biology methodology. However, stable cell lines transfected with PDE genes are not easily accessible. Consequently, purification from tissue extracts employing traditional biochemical methods is still an important route for preparing large quantities of PDE isoenzymes. This is particularly true for drug-screening projects aimed at discovering novel PDE isoenzyme-selective inhibitors. The method most commonly used to prepare PDE isoenzymes from tissue sources is ion-exchange chromatography alone or in combination with other procedures, such as gel-filtration and/or affinity chromatography. The latter techniques are usually secondary and are normally employed to resolve PDE isoenzymes that coelute during an ion-exchange separation. We have used both ion-exchange and affinity chromatography to separate PDE isoenzymes successfully from a variety of animal *(7,9,20)* and human tissues *(8,21,22)*. The method described in detail in Sections 2. and 3. was used to analyze the complement of PDE isoenzymes present in bovine tracheal smooth muscle. The same strategy could, however, be used for isolating PDEs from other tissues, such as platelets *(8)*, cardiac muscle *(7,10,21)*, and brain *(20)*.

2. Materials

All solutions should be prepared in distilled/deionized water, and filtered (0.22 μm) before use and storage.

2.1. Ion-Exchange Chromatography

Homogenization buffer, pH 6.5: 20 mM Bis-tris, 1 mM dithiothreitol (DTT), 2 mM benzamidine, 2 mM EDTA, 50 mM NaCl, and 0.1 mM phenylmethane sulfonyl fluoride (PMSF).

2.2. Affinity Chromatography

1. Buffer A, pH 6.5: 20 mM Bis-tris, 1 mM DTT, 2 mM benzamidine, 50 mM NaCl, 3 mM MgCl$_2$, 0.1 mM CaCl$_2$, and 0.1 mM PMSF.
2. Buffer B: buffer A plus 1M NaCl.
3. Buffer C: buffer B plus 1 mM EGTA.
4. Calmodulin agarose (Sigma, Poole, Dorset, UK).

2.3. Assay of PDE Activity

1. Assay buffer 50 mM Tris-HCl, pH 7.4.
2. 8-^{3}H cAMP/cGMP: The ammonium salt of these compounds with a specific activity of 15–30 Ci/mmol and a radioactive concentration of 1 μCi/μL (Amersham International, Amersham, UK) can be used. The radioisotopes are supplied in aqueous ethanol (1:1 [v/v]) solution, which should be

diluted fivefold with 50% ethanol and stored at –20°C. This stock solution can be further diluted 50-fold with assay buffer to give ≈0.1 μCi/25 μL for each assay tube (*see* Note 1).

3. Unlabeled cAMP/cGMP: The stock solutions (0.3–10 m*M*) of unlabeled cyclic nucleotide should be made in assay buffer. The exact concentration can be determined by spectral analysis using extinction coefficients of 1.465 and 1.370 L/mmol/mm at 259 nm for cAMP and cGMP, respectively. An independent crosscheck of the stock concentration should be performed using thin-layer chromatography or a commercially available cAMP assay kit. Stock solutions are stable for at least 1 yr when stored at –20°C. Repeated freezing/thawing has no adverse effects on cyclic nucleotide concentration. However at 4°C, both cyclic nucleotides deteriorate after 4–6 wk. Appropriate dilutions to prepare working solutions (e.g., 8 μ*M* for a final assay concentration of 1 μ*M*) should be made with assay buffer. For assays involving determination of kinetic parameters, make serial dilutions of stock solution in assay buffer to give final substrate concentrations ranging from 0.1–100 μ*M*.

4. 5'AMP/5'GMP: The 5' nucleotide stock solutions (8 m*M*) should be prepared in assay buffer and stored at –20°C.

5. Adenine-U-[14]C cAMP/cGMP: Use ^{14}C-labeled cyclic nucleotides as recovery markers to determine nonenzymatic loss of ^{3}H-cAMP/cGMP. The recovery markers are supplied as the ammonium salts in an aqueous solution containing 2% ethanol with specific activities of 260 mCi/mmol (cAMP) and 52 mCi/mmol (cGMP) and a radioactive concentration of 25 μCi/mL (Amersham). Dilute these 25-fold with a 2% ethanol/water (v/v) mixture, aliquot into 100-μL portions, and store at –20°C. Make each 100 μL up to 1 mL just prior to assay, and use 20 μL (≈10,000 dpm) in each assay tube. Surplus ^{14}C solution should be discarded and not refrozen.

6. Magnesium chloride: Magnesium chloride (24 m*M*) stock solution should be made in assay buffer and stored at 4°C.

7. Tris-HCl/BSA/DTT: Dissolve fat-free bovine serum albumin (BSA) and DTT (by gentle shaking) in assay buffer to make a 50 m*M* Tris-HCl, pH 7.4, 0.2% BSA, 4 m*M* DTT stock mixture for diluting enzyme samples.

8. Preparation of alumina columns: Pack chromatographic neutral alumina (1.5 g) into Pasteur pipets plugged with glass wool and link to a vacuum line, using a suitable holding perspex box, to expedite elution. Prior to use in assay, the elution profile of cyclic nucleotide on the alumina columns, pre-equilibrated with 10 mL of column buffer (0.4*M* Tris-HCl, pH 7.0), should be determined. This can be achieved by applying radioactive cyclic nucleotide (≈150,000 dpm) to the alumina and subsequently washing with 8 × 1 mL of column buffer. Each 1 mL of eluate should be col-

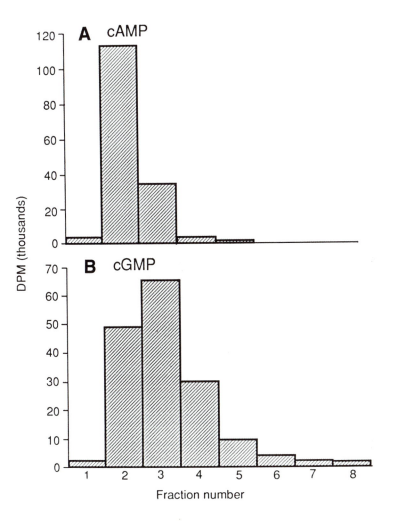

Fig. 2. Typical elution profile of cyclic nucleotides on alumina columns (*see* Section 2.3. for details).

lected in a scintillation vial and counted after addition of scintillant (3 mL of Picoflour 30). Typically most of the cAMP elutes in the second and third fractions, whereas cGMP comes off in the second, third, fourth, and fifth fractions (Fig. 2). 5' Nucleotide binds more tightly to the alumina and does not elute in these fractions.

9. Inhibitory compounds: Test compound stock solutions can be made in assay buffer or assay buffer/DMSO mixture, and diluted serially. Eight different concentrations of each compound should be examined.

3. Methods

3.1. Ion-Exchange Chromatography

1. Fresh bovine tracheae obtained from a local slaughterhouse should be transported to the laboratory in chilled Krebs-Henseleit solution (pH 7.4) pre-gassed with carbogen. All subsequent steps should be performed at 4°C. Tracheal smooth muscle must be carefully dissected, cut into smaller strips, and trimmed free of mucosal, connective, epithelial, and fat tissue in Krebs-Henseleit solution.

2. Chop the cleaned tissue into small cubes, and homogenize (Polytron: PT20 probe; 2 x 10 s bursts at speed setting 5 and 2 x 10 s bursts at speed setting 7) in 6 vol of homogenization buffer.

3. Pass the homogenate through two layers of cheesecloth, and centrifuge at 12,000*g* for 10 min.

4. Filter (0.22 μm) the supernatant fraction, and apply 10–27 mL (1 mL/min) to a 1-mL Mono Q column, pre-equilibrated in 10–20 bed vol of homogenization buffer, attached to a fast-protein liquid chromatography system (Pharmacia, St. Albans, UK).

5. Wash the column with 6 mL of homogenization buffer to remove unbound material. Bulk (70–80%) of the protein applied eluted in this fraction and is devoid of PDE activity.

6. Elute bound protein (1 mL/min) with a 32-mL nonlinear NaCl gradient ranging from 0.05–1.0M in seven steps: 0.05–0.16M (3 mL); 0.16–0.2M (3 mL); 0.2–0.2M (4 mL); 0.2–0.4M (10 mL); 0.4–0.4M (4 mL); 0.4–0.6M (4 mL); and 0.6–1.0M (5 mL).

7. Fractions (1 mL) should be assayed for PDE activity within 4 h of homogenization. Ethylene glycol (30%) can be added to active fractions, which may be frozen in liquid nitrogen and stored at –70°C. Under these conditions, the enzyme activity is normally stable for at least 3 mo.

3.2. Affinity Chromatography

Although chromatography with high-resolution ion exchangers can provide adequate separation of most PDEs, owing to the narrow range of isoelectric points for these proteins, there is invariably some crosscontamination of isoenzymes. In particular, coelution of PDEs I/V *(9,23,24)*, I/II *(10,24,25)*, II/III *(26)*, and III/IV *(9,24,27)* has been encountered. Thus, additional chromatographic steps are required to resolve these isoenzymes further. Although gel filtration is one useful option, the most convenient and effective solution to this problem is affinity chromatography. Affinity media with immobilized PDE isoenzyme-specific ligand (e.g., calmodulin, cGMP, cAMP, selective inhibitor) can be obtained commercially or made in the laboratory. We describe the method for resolving bovine tracheal smooth muscle PDE I and PDE V activities that coelute after ion-exchange chromatography:

1. Pack calmodulin-agarose into a column (10-mL bed vol; 1.6 x 4 cm), and equilibrate in at least 10 bed vol of buffer A. This and all subsequent steps should be performed at 4°C (*see* Note 4).
2. Prepare a combined sample (1.3 mL) of fractions representing PDE I/V activities, and increase the Ca^{2+} and Mg^{2+} concentrations to 2 mM (to counteract the EDTA present in the sample from the ion-exchange run).
3. This sample should then be applied (0.5 mL/min) to the calmodulin column, followed by 2–4 mL of buffer A to ensure penetration into the gel.
4. To allow complete adsorption, leave the sample on the column for 30 min.
5. Unbound and bound materials can then be eluted by serial stepwise treatment with 20 mL each for buffers A, B, and C.
6. Assay the fractions (1 mL) collected for PDE activity within 24 h.

This method can also be used to separate PDEs I and II *(10)*.

Affinity chromatography on media containing immobilized ligands, such as blue dextran *(26,28)*, heparin *(29,30)*, cyclic nucleotides *(29,31)*, PDE III *(27,32)*, or PDE IV *(33)* inhibitors has also been used to separate PDE isoenzymes. If absolute purity is difficult to achieve (e.g., in human tissue owing to limited supply), isoenzyme-selective inhibitors can be used to suppress to residual PDE III or IV in partially purified preparations *(22,24,34)*.

3.3. Assay of PDE Activity

A variety of radioactive *(35–39)* and nonradioactive *(40,41)* techniques for assaying PDE activity have been described. Because of higher sensitivity, however, the former methods are more commonly employed. PDE activity is assessed by quantitating substrate utilization or 5' nucleotide formation. In some instances, to ease substrate/product separation, the 5' nucleotide is converted to the corresponding nucleoside by a second reaction using 5' nucleotidase. The method used in our laboratory is a modification of the procedure described by Methven et al. *(36)* and involves determining the amount of substrate remaining at the end of reaction time. Batch chromatography on small alumina columns is used to recover nonhydrolyzed cyclic nucleotide, whereas an internal recovery marker allows correction for nonenzymatic loss of substrate.

Table 2
Composition of PDE Assays

Reagent	Volume of working solution	Final assay concentration
^{3}H cAMP or ^{3}H cyclic GMP (4 nCi/µL)	25 µL	≈150,000 dpm
Unlabeled cAMP or cGMP (8 µM)	25 µL	1 µM
MgCl$_2$ (24 mM)	25 µL	3 mM
5'AMP or 5'GMP (8 mM)	25 µL	1 mM
Test compound or solvent or buffer	50 µL[a]	Test compound (0.1–250 µM)
Enzyme sample	50 µL[b]	Mix and start time cycle
Total volume	200 µL	

[a] For PDE I and PDE II assays, test compound was added in a vol of 25 µL owing to inclusion of 25 µL each of Ca^{2+} and calmodulin or cGMP, respectively.

[b] For PDE I assays, 25 µL of enzyme is used owing to inclusion of 25 µL of Ca^{2+} in the assay.

3.4. Assay Procedure

1. Use small glass tubes (12 × 75 mm) or plastic microassay tubes as reaction vessels, with preassay additions made on ice. The order of addition of reagents and test compounds is shown in Table 2.
2. Start the reaction by addition of enzyme, and incubate the samples at 37°C for 15 min in a shaking water bath.
3. Stop the reaction by heat inactivation of enzyme (3 min in aluminum heating blocks set at 105°C), and add ^{14}C-labeled cyclic nucleotide recovery marker added to each assay tube.
4. Separate substrate in the reaction mixture from product by chromatography on alumina columns pre-equilibrated in 10 mL of column buffer. Apply the entire reaction mix to an alumina column, which should then be rinsed with sufficient column buffer (based on the elution profile) to elute cyclic nucleotide.
5. Discard the first 1 mL of the wash for both cAMP and cGMP. Typical recoveries for cAMP and cGMP are 70–80% and 60–70%, respectively.
6. Perform the assays in duplicate and in the linear range of the reaction, where <25% of the initial substrate is hydrolyzed. The assay blank should contain "excess" enzyme (usually 50 µL of undiluted 12,000 g supernatant fraction from a cardiac muscle homogenate) to ensure complete hydrolysis

of all the substrate (*see* Note 3). The ^{3}H counts recovered from these samples are between 3 and 6% of the total radioactivity used in each assay. Complete recovery (95–100%) of ^{3}H counts is achieved in blank samples lacking enzyme or containing boiled enzyme.

3.5. Calculation of Results

Determine the initial level of radioactivity (i.e., ^{3}H and ^{14}C) in each assay by preparing tubes ("totals") containing all assay ingredients except enzyme. Transfer these samples to scintillation vials, and make the volume up to 3 mL (for cAMP assay) or 4 mL (for cGMP assay) with 0.4*M* Tris-HCl buffer. Add suitable scintillation fluid such as Picofluor 30 (Canberra Packard, Berkshire, UK) and determine radioactivity in a liquid scintillation counter. A set of six totals should be included in each experiment. The enzyme activity (% hydrolysis of cAMP or cGMP) can be calculated using a computer program based on the following equations:

$$\text{Recovery of } {}^{14}\text{C marker } (R) = [({}^{14}\text{C dpm of sample})/(\text{totals } {}^{14}\text{C dpm})]$$
$$\text{Recovery corrected } {}^{3}\text{H dpm (Corr } {}^{3}\text{H dpm)} = [({}^{3}\text{H dpm of sample}) \times (1/R)]$$
$$\text{\% Hydrolysis} = [1-(\text{Corr } {}^{3}\text{H dpm/totals } {}^{3}\text{H dpm})] \times 100 \qquad (1)$$

In experiments examining the effects of inhibitory compounds inhibition percentage (%I) is calculated as:

$$\%I = [(A - B)/A] \times 100 \qquad (2)$$

where: A = % hydrolysis in the absence of inhibitor and B = % hydrolysis in the presence of inhibitor.

The concentration (IC_{50}) of test compound producing 50% inhibition of PDE activity can be determined using a nonlinear regression curve-fitting program (such as Inplot 4, Graphpad Software Inc., San Diego, CA).

3.6. Determination of Kinetic Constants

In these experiments, use seven to nine different substrate concentrations ranging from 0.1–1 m*M*. The concentration of radiolabeled cyclic nucleotide is kept constant, and the concentration of unlabeled nucleotide varies. Incubation time can also be varied (15–45 min) to adjust enzyme activity such that approx 25% of the substrate is hydrolyzed. The data should be analyzed with a commercial nonlinear curve-fitting program (e.g., Enzfitter, Elsevier-Biosoft) to determine K_m and $V_{\max}$ values. A suitable range of inhibitor concentrations bracketing the estimated inhibition constant should be used in experiments aimed at elucidating the nature of PDE isoenzyme inhibition by various compounds.

3.7. PDE Activity in Cell Homogenates

Analysis of overall PDE activity in homogenate samples can be a useful indicator of dominant PDE isoenzyme activity in a particular tissue. This usually involves examining the effects of established regulators (Ca/calmodulin, cGMP) and/or isoenzyme-selective inhibitors. It is important, however, in the latter case that appropriate maximal inhibitory, but isoenzyme-specific concentrations of inhibitors be used for clear interpretation of data. Comparison of the effects of low (<1 μM) and high (>100 μM) substrate concentration on homogenate PDE activity yields information on the presence of low- and high-affinity isoforms *(42)*. The results shown in Fig. 3 illustrate the usefulness of characterizing homogenate PDE activity. It is clear that the six tissues studied exhibit marked differences in the regulatory effects of cGMP and Ca/calmodulin on their cAMP hydrolytic activities. This is a reflection of the tissue-related variation in the distribution of cGMP and Ca/calmodulin-regulated PDE isoenzymes. Thus, for example, although cardiac and platelet cells contain substantial PDE III, as indicated by the inhibition of cAMP activity by 1 μM cGMP, the presence of this isoenzyme in the other tissues is not as obvious. It should be noted that lack of inhibition by cGMP does not necessarily suggest the complete absence of PDE III, since this effect may be canceled by the stimulatory action of cGMP on PDE II. The variable stimulatory effects of Ca/calmodulin also highlight the differing amounts of this isoenzyme in the tissues examined. Analysis of PDE activity in homogenates can also be used to double-check the PDE isoenzymes obtained by further purification. The degree of inhibition produced by isoenzyme-selective inhibitors can also be used to delineate the predominant PDE subtype in tissue homogenates. This is illustrated by the data in Fig. 4, which show the effects of the PDE IV-selective inhibitor denbufylline in a range of tissue homogenates. It is clear that denbufylline produced large inhibition of erythrocyte and skeletal muscle extracts, although only having a minor effect on adipocyte, aortic, and cardiac muscle homogenates. Thus, PDE IV is likely to be the principal isoform present in the former two tissues and a relatively minor component of the latter tissues. As expected, the nonselective PDE inhibitors theophylline and 3-isobutyl-1-methyl xanthine (IBMX) did not show any tissue dependence.

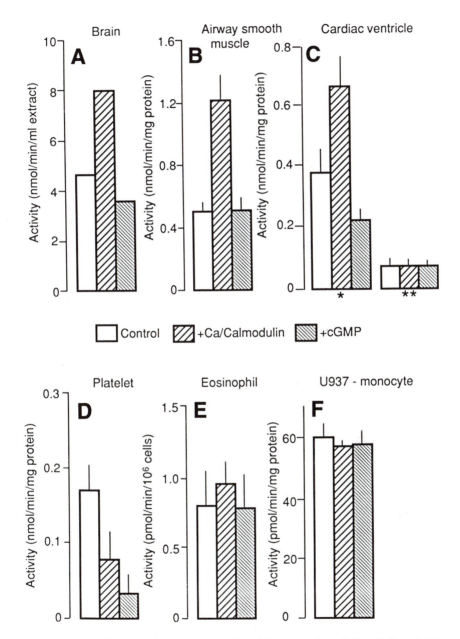

Fig. 3. The effects of the regulators Ca (20 μM)/calmodulin (1.5 μg/mL) or cGMP (1 μM) on the cAMP hydrolytic activity in homogenate fractions from a variety of tissues. Assays were performed at 1 μM cAMP. *Rabbit, **rat cardiac ventricle.

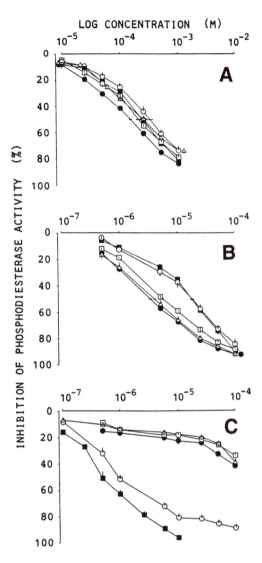

Fig. 4. Inhibition by theophylline (**A**), IBMX (**B**), and denbufylline (**C**) of the hydrolysis of cAMP (assay concentration 0.3 μ*M*) by cyclic nucleotide phosphodiesterase in homogenate fractions from rat erythrocytes (■), gastrocnemius muscle (○), adipocytes (□), abdominal aorta (●), and cardiac muscle (△).

3.8. PDE Isoenzymes in Bovine Tracheal Smooth Muscle

The PDE activities detected in bovine tracheal smooth muscle after ion-exchange chromatography on Mono-Q are shown in Fig. 5. Most

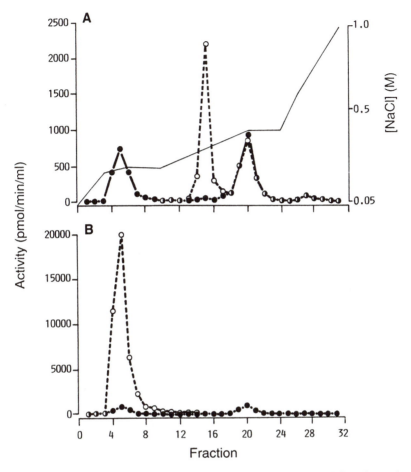

Fig. 5. Elution profiles of cyclic nucleotide PDE activities from bovine tracheal smooth muscle after separation on a Mono-Q ion-exchange column (*see* Section 3.1. for details). Fractions were assayed for the cAMP (1 μ*M*) PDE activity either in the absence (●) of any known regulator, or in the presence (○) of (**A**) cGMP (1 μM) or (**B**) Ca (20 μ*M*)/calmodulin (1.5 μg/mL).

(82–93%) of the applied PDE activity was recovered. Three major peaks of PDE activity eluting at 0.23 (PDE I), 0.34 (PDE II), and 0.44 (PDE IV) NaCl were identified and classified according to currently accepted nomenclature *(2)*. The cAMP activity of PDE I and PDE II was stimulated by Ca/calmodulin and cGMP, respectively. The hydrolytic activity of the third peak was not significantly modulated by either regulator.

The elution profile in the absence or in the presence of cGMP did not reveal PDE III activity. Although the elution profile did not show a distinct PDE V peak, it is possible that it coeluted with PDE I, as has been reported for dog tracheal smooth muscle *(23)*. To test this possibility, the Ca/calmodulin-stimulated peak shown in Fig. 5 was further purified on a calmodulin-agarose column. As shown in Fig. 6, calmodulin affinity chromatography resolved the Mono-Q peak into Ca/calmodulin-stimulated and independent PDEs. When Ca^{2+}-containing wash buffer was used, the PDE activity obtained in the eluate was not stimulated by Ca/calmodulin and was specific for cGMP. Hence, it could be classified as PDE V. The Ca/calmodulin-activated PDE could be released from the column by EGTA-containing buffer.

3.8.1. Inhibition of Bovine Tracheal PDE Isoenzymes

The data for the inhibitory effects of a wide range of compounds on bovine tracheal PDE isoenzymes are presented in Table 3. Examples of both nonselective and highly selective compounds are given. Org 9935, rolipram, and zaprinast were potent inhibitors of PDEs III, IV, and V, respectively, with weak activity against the other isoenzymes. In contrast, IBMX was completely nonspecific and inhibited all PDEs almost equipotently. Milrinone and Org 30029 show an interesting profile as dual inhibitors of PDEs III and IV.

3.8.2. Kinetic Parameters of PDE Isoenzymes

The kinetic characteristics (K_m and V_{max} values) of PDE isoenzymes from bovine tracheal smooth muscle are shown in Table 4. Although the PDE isoenzyme K_m values can be compared across tissues and laboratories, the same does not apply to V_{max} values, because variations in protein concentration have not been taken into account.

3.8.3. Kinetics of PDE IV Isoenzyme Inhibition

Although the functional and the PDE isoenzyme selectivities of PDE inhibitors have been well characterized, the exact nature of the enzyme inhibition produced by these compounds has been less well studied. Most investigators tend to assume that these drugs are acting through a competitive mechanism. However, it is clear from the Hanes plots shown in Fig. 7 for bovine tracheal PDE IV that this assumption is incorrect. The nonselective inhibitor IBMX produced concentration-dependent

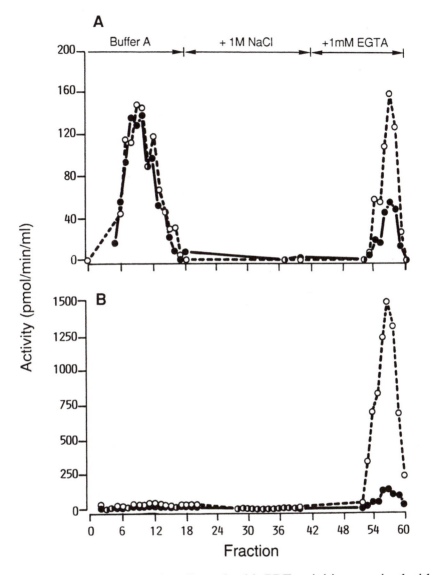

Fig. 6. Subfractionation of cyclic nucleotide PDE activities contained within the first activity peak obtained after ion-exchange chromatography of bovine tracheal smooth muscle. PDE activity was eluted from a calmodulin-agarose column with either Ca^{2+} or EGTA-containing buffers (*see* Section 3.2. for details). Fractions were assayed for cGMP (**A**) or cAMP (**B**) hydrolytic activities (1-μM substrate concentrations) in the absence (●) or presence (○) of Ca (1 mM)/calmodulin (1.5 µg/mL).

Table 3
Inhibition of Bovine Tracheal Smooth Muscle PDE Isoenzymes (I, II, IV, V) and Rabbit Heart PDE III

Inhibitor	$-\mathrm{Log\ IC}_{50}$[1a]				
	Ca+/calmodulin-dependent PDE I[b]	Cyclic GMP-stimulated PDE II[c]	Cyclic GMP-insensitive PDE IV[d]	Cyclic GMP-specific PDE V[d]	Cyclic GMP-inhibited PDE III[d,e]
IBMX (3-isobutyl-1-methylxanthine)	4.7 ± 0.1	4.3 ± 0.1	4.9 ± 0.1	4.5 ± 0.1	5.1 ± 0.1
Zaprinast	4.8 ± 0.1	4.5 ± 0.2	4.6 ± 0.1	6.6 ± 0.1	3.9 ± 0.1
Rolipram	<3.60 (24)	<3.60 (11)	5.9 ± 0.1	<3.60 (23)	4.3 ± 0.2
Milrinone	<3.60 (48)	<3.60 (22)	4.7 ± 0.1	<3.60 (34)	5.5 ± 0.1
Org 30029	<3.60 (32)	<3.60 (14)	4.6 ± 0.1	<3.6 (19)	4.5 ± 0.1
Org 9935	<3.60 (39)	<3.60 (6)	4.6 ± 0.1	5.01 ± 0.15	7.0 ± 0.2

[a]Mean ± SE mean of 4–12 assays performed with two to three separations; figures in parentheses indicate % inhibition at 250 µmol/L.
[b]cAMP (1 µmol/L) as substrate and in the presence of Ca^{2+} (1 mmol/L) and calmodulin (1.5 µg/mL).
[c]cAMP (1 µmol/L) as substrate and in the presence of cGMP (1 µmol/L).
[d]cAMP (1 µmol/L) as substrate.
[e]Inhibition of rabbit heart PDE III (7).

Table 4
Kinetic Characteristics of Bovine Tracheal PDE Isoenzymes[a]

PDE	Isoenzyme family	K_m (μmol/L) cAMP	K_m (μmol/L) cGMP	V_{max} (nmol/min/mL) cAMP	V_{max} (nmol/min/mL) cGMP
I	Ca^{2+} calmodulin dependent	2.9[c]	ND	4.0[b]	ND
II	cGMP stimulated	105[c]	85[d]	131[c]	72
IV	cGMP insensitive	2.9	ND	4.1	ND
V	cGMP specific	ND	2.6	ND	695

[a]ND = low activity at 1 μmol/L, precise value not determined. All values are means of two to three determinations performed in duplicate from at least two different separations.
[b]In the presence of Ca^{2+} (1 mmol/L) and calmodulin (1.5 μg/mL).
[c]In the presence of cGMP (1 μmol/L).
[d]Nonlinear kinetics (value given is the substrate concentration giving 50% of maximal activity).

increases in K_m without significantly altering V_{max}, indicating a direct competition with cAMP for binding to the catalytic site. In contrast, the selective compounds rolipram and denbufylline affected both K_m and V_{max} suggesting a mixed mechanism of inhibition. Interestingly, all three compounds produced competitive inhibition of rat brain PDE IV *(20)*, indicating the possible existence of a different PDE IV subtype in this tissue. Thus, it is important to perform these experiments not only for PDEs within one tissue, but also for the same isoenzyme across tissues and species.

4. Notes

1. The purity of radioactive substrates should be checked periodically by ion-exchange chromatography on QAE-sephadex *(39)*.
2. In crude enzyme preparations, nucleoside phosphates may interfere with the reaction by degrading the nascent, radiolabeled 5' nucleotide to the corresponding nucleotide (which being unchanged at pH 7 will not be retarded by aluminum oxide). This problem can be resolved by the inclusion of an unlabeled pool of the 5' nucleotide.
3. Excess enzyme, no enzyme, and boiled enzyme blanks should be performed routinely. Ideally, these should constitute <5% of the initiating radioactivity. Possible contamination with breakdown products (e.g., AMP and particularly adenosine) can be indirectly assessed by using excess enzyme (≈150 μg protein) to cause total hydrolysis of cyclic nucleotide. A more accurate and direct estimation of contamination can be achieved by thin-layer chromatography.

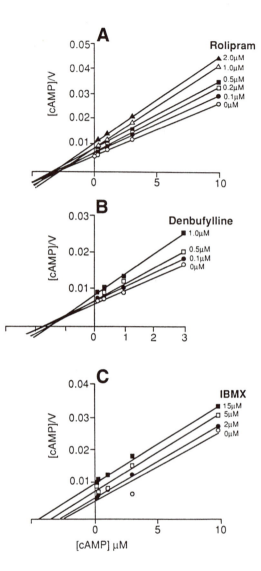

Fig. 7. Hanes plots highlighting the mechanism of enzyme inhibition produced by rolipram, denbufylline, and IBMX when tested on PDE IV from
bovine tracheal smooth muscle. Purified PDE IV was prepared by ion-exchange
chromatography as described in Section 3.1.

4. To avoid PDE activity at 4°C, it is useful to initiate the reaction by the
   addition of enzyme protein. With crude or partially purified preparations,
   it may be better to assay at 30°C rather than 37°C. This minimizes AMP-
   or GMP-deamination and myokinase-type reactions.

5. In the collection and treatment of samples, attention must be paid to conditions that might modify activity, especially in crude samples. Potential mechanisms for altering activity include proteolysis, covalent modification, sulfydryl oxidation, ionic strength, detergents and lipids, and sample dilution. Thus, chelating agents (EDTA and EGTA), protease inhibitors (benzamidine, PMSF, TLCK leupeptin), sulfydryl reagents (DDT, β-mercaptoethanol), BSA, and ethylene glycol (30%) have been used to enhance PDE stability. When using PMSF, it should be added, because of its short half-life, to the buffer just prior to homogenization.

6. PDEs show good long-term stability (at least 3 mo), as judged by kinetic, regulatory, and inhibitor sensitivity when stored in homogenization buffer containing 30% ethylene glycol. However, PDE separations from small amounts of tissue (e.g., human material) invariably yield fractions with low protein (<0.1 mg/mL) and should be supplemented with BSA (0.2 mg/mL). Otherwise enzyme activity declines rapidly within 2 wk.

7. The elution characteristics of cyclic nucleotides should be determined for each batch of new columns. There is significant variation in the chromatographic properties of neutral alumina between batches, as well as between suppliers.

5. Summary

Greater knowledge over the past decade on the biochemical properties, as well as the identification of specific pharmacological tools has led to a marked improvement in the methods employed for the analysis and assay of PDE isoenzymes. A major message has been the marked species and tissue-dependent variation in the distribution of the various isoenzymes and their subtypes. This has great implications not only in terms of extrapolating animal data to the human situation, but also from one tissue to another. Thus, it is critical, in particular for drug discovery efforts, to characterize human PDEs in the relevant tissue. Molecular cloning is probably the best and most direct route to achieve this objective since access to disease-free human tissue is heavily limited. Alternatively, well-characterized animal tissues showing PDE and pharmacological profiles similar to human tissues may be utilized. The methods described in this chapter have been successfully applied to study to the biochemistry and pharmacology of PDEs in both animal and human tissues, and in the discovery of novel selective inhibitors for these proteins.

References

1. Thompson, W. J. (1991) Cyclic nucleotide phosphodiesterases: pharmacology, biochemistry and function. *Pharmacol. Ther.* **51**, 13–33.
2. Beavo, J. A. and Reifsnyder, D. H. (1990) Primary sequence of cyclic nucleotide phosphodiesterase isoenzymes and the design of selective inhibitors. *Trends Pharmacol. Sci.* **11**, 150–155.
3. Michaeali, T., Bloom, T. J., Martins, T., Laughney, K., Ferguson, K., Riggs, M., et al. (1993) Isolation and characterisation of a previously undetected human cAMP phosphodiesterase by complementation of cAMP phosphodiesterase-deficient saccharomyces cerebisiae. *J. Biol. Chem.* **268**, 12,925–12,932.
4. Nicholson, C. D., Challiss., R. A. J., and Shahid, M. (1991) Differential modulation of tissue function and therapeutic potential of selective inhibitors of cyclic nucleotide phosphodiesterase isoenzymes. *Trends Pharmacol. Sci.* **12**, 19–27.
5. Cruickshank, J. M. (1993) Phosphodiesterase III inhibitors: long term risks and short term benefits. *Cardiovasc. Drugs and Ther.* **7**, 655–660.
6. Nicholson, C. D. and Shahid, M. (1993) Inhibitors of cyclic nucleotide phosphodiesterase isoenzymes—their potential utility in the therapy of asthma. *Pulmonary Pharmacol.* **6**, 101–117.
7. Shahid, M. and Nicholson, C. D. (1990) Comparison of cyclic nucleotide phosphodiesterase isoenzymes in rat and rabbit ventricular myocardium: positive inotropic and phosphodiesterase inhibitory effects of Org 30029, milrinone and rolipram. *Naunyn-Schmiedeberg's Arch. Pharmacol.* **342**, 698–705.
8. Shahid, M., Holbrook, M., Coker, S. J., and Nicholson, C. D. (1990) Characterisation of human platelet phosphodiesterase isoenzymes and their sensitivity to a variety of selective inhibitors. *Br. J. Pharmacol.* **100**, 443P.
9. Shahid, M., van Amsterdam, R. G. M., de Boer, J., ten Berge, R. E., Nicholson, C. D., and Zaagsma, J. (1991) The presence of five cyclic nucleotide phosphodiesterase isoenzyme activities in bovine tracheal smooth muscle and the functional effects of selective inhibitors. *Br. J. Pharmacol.* **104**, 471–477.
10. Shahid, M., Billington, C., and Nicholson, C. D. (1991) Evidence for two distinct Ca/calmodulin stimulated cyclic nucleotide phosphodiesterases in rabbit cardiac ventricle. *Br. J. Pharmacol.* **102(Suppl.)**, 30P.
11. Livi, G. P., Kmetz, P., McHale, M. M., Cieslinski, L. B., Sathe, G. M., Taylor, D. P., et al. (1990) Cloning and expression of cDNA for a human low-K_m, rolipram-sensitive cyclic AMP phosphodiesterase. *Mol. Cell. Biol.* **10**, 2678–2686.
12. Pittler, S. J., Baehr, W., Wasmuth, J. J., McConnell, D. G., Champagne, M. S., et al. (1990) Molecular characterization of human and bovine rod photoreceptor cGMP phosphodiesterase α-subunit and chromosomal localization of the human gene. *Genomics* **6**, 272–283.
13. Repaske, D. R., Swinnen, J. V., Jin, S. L. C., van Wyk, J. J., and Conti, M. (1992) A polymerase chain reaction strategy to identify and clone cyclic nucleotide phosphodiesterase cDNAs. *J. Biol. Chem.* **267**, 18,683–18,688.
14. Jin, S. L. C., Swinnen, J. V., and Conti, M. (1992) Characterization of the structure of a low K_m, rolipram-sensitive cAMP phosphodiesterase. *J. Biol. Chem.* **267**, 18,929–18,939.

15. McLaughlin, M. M., Cieslinski, L. B., Burman, M., Torphy, T. J., and Livi, G. P. (1993) A low-K_m, rolipram-sensitive, cAMP-specific phosphodiesterase from human brain. *J. Biol. Chem.* **268,** 6470–6476.

16. Sonnenburg, W. K., Seger, D., and Beavo, J. A. (1993) Molecular cloning of a cDNA encoding the "61-KDa" calmodulin-stimulated cyclic nucleotide phosphodiesterase. *J. Biol. Chem.* **268,** 645–652.

17. Henkel-Tigges, J. and Davis, R. L. (1990) Rat homologs of the drosophila dunce gene code for cyclic AMP phosphodiesterases sensitive to rolipram and RO-20-1724. *Mol. Pharmacol.* **37,** 7–10.

18. Torphy, T. J., Stadel., J. M., Burman, M., Cieslinski, L. B., McLaughlin, M. M., White, J. R., and Livi, G. P. (1992) Co-expression of human cAMP-specific phosphodiesterase activity and high affinity rolipram binding in yeast. *J. Biol. Chem.* **267,** 1798–1804.

19. Meacci, E., Taira, M., Moos, M., Smith, C. J., Movesian, M. A., Degerman, E., et al. (1992) Molecular cloning and expression of human myocardial cGMP-inhibited cAMP phosphodiesterase. *Proc. Natl. Acad. Sci.* USA **89,** 3721–3725.

20. Shahid, M., Chipperfield, K., and Nicholson, C. D. (1993) Tissue-dependent differences in the mechanism of cyclic nucleotide phosphodiesterase (PDE) IV isoenzyme inhibition by both rolipram and denbufylline. *Br. J. Pharmacol.* **110(Suppl.),** 127P.

21. Shahid, M., Bruin, J. C., Walker, G. B., Cottney, J. E., and Nicholson, C. D. (1991) Positive inotropic, Ca-sensitizing and cyclic nucleotide phosphodiesterase (PDE) isoenzyme inhibitory effects of Org 30029 in human atrial myocardium. *Br. J. Pharmacol.* **104(Suppl.),** 17P.

22. de Boer, J., Philpott, A. J., van Amsterdam, R. G. M., Shahid, M., Zaagsma, J., and Nicholson, C. D. (1992) Human bronchial cyclic nucleotide phosphodiesterase isoenzymes: biochemical and pharmacological analysis using selective inhibitors. *Br. J. Pharmacol.* **106,** 1028–1034.

23. Torphy, T. J. and Cieslinski, L. B. (1990) Characterisation and selective inhibition of cyclic nucleotide phosphodiesterase isoenzymes in canine trachealis smooth muscle. *Mol. Pharmacol.* **37,** 206–214.

24. Torphy, T. J., Undem, B. J., Cieslinski, L. B., Luttman, M. A., Reeves, M. L., and Hay D. W. P. (1993) Identification, characterization and functional role of phosphodiesterase isoenzymes in human airway smooth muscle. *J. Pharmacol. Exp. Ther.* **265,** 1213–1222.

25. Nicholson, C. D., Jackman, S. A., and Wilke, R. (1989) The ability of denbufylline to inhibit cyclic nucleotide phosphodiesterase and its affinity for adenosine receptors and the adenosine re-uptake site. *Br. J. Pharmacol.* **97,** 889–897.

26. Harrison, S. A., Reifsnyder, D. H., Gallis, B., Cadd, G. G., and Beavo, J. A. (1986) Isolation and characterisation of bovine cardiac muscle cGMP-inhibited phosphodiesterase: a receptor for new cardiotonic drugs. *Mol. Pharmacol.* **29,** 506–514.

27. Rascon, A., Lindgren, S., Stavenow, L., Belfrage, P., Anderson, K.-E., Manganiello, V. C., and Degerman, E. (1992) Purification and properties of the cGMP-inhibited cAMP phosphodiesterase from bovine aortic smooth muscle. *Biochim. Biophys. Acta* **1134,** 149–156.

28. Lavan, B. E., Lakey, T., and Houslay, M. D. (1989) Resolution of soluble cyclic nucleotide phosphodiesterase isoenzymes, from liver and hepatocytes, identified a novel IBMX-insensitive form. *Biochem. Pharmacol.* **38,** 4123–4136.

29. Murashima, J., Tanaka, T., Hockman, S., and Manganiello, V. C. (1990) Characterisation of particulate cyclic nucleotide phosphodiesterases from bovine brain: purification of a distinct cGMP-stimulated isoenzyme. *Biochemistry* **29,** 5285–5292.

30. Torphy, T. J., de Wolf, W. E., Green, D. W., and Livi, G. P. (1993) Biochemical characteristics and cellular regulation of phosphodiesterase IV. *Agents Actions* **43(Suppl.),** 51–71.

31. Martins, T. J., Mumby, M. C., and Beavo, J. A. (1982) Purification and characterization of a cyclic GMP-stimulated cyclic nucleotide phosphodiesterase from bovine tissues. *J. Biol. Chem.* **257,** 1973–1979.

32. Le Bon, T. R., Kasuya, J., Paxton, R. J., Belfrage, P., Hockman, S., Manganiello, V. C., and Fujitayamaguchi, Y. (1992) Purification and characterization of guanosine 3'5'-monophosphate-inhibited low K_m adenosine 3',5'-monophosphate phosphodiesterase from human placental cytosolic fractions. *Endocrinology* **130,** 3265–3274.

33. Fougier, S., Memoz, G., Prigent, A. F., Marivet, M., Bourguiigon, J. J., Wermuth, C., and Pacheco, H. (1986) Purification of cAMP-specific phosphodiesterase from rat heart by affinity chromatography on immobilised rolipram. *Biochem. Biophys. Res. Commun.* **138,** 205–214.

34. Harris, A. L., Connell, J., Ferguson, E. W., Wallace, A. M., Gordon, R. J., Pagani, E. D., and Silver, P. J. (1989) Role of low K_m cyclic AMP phosphodiesterase inhibition in tracheal relaxation and bronchodilation in the guinea-pig. *J. Pharmacol. Exp. Ther.* **251,** 199–206.

35. Davis, C. W. and Daly, J. W. (1979) A simple direct assay of 3',5'-cyclic nucleotide phosphodiesterase activity based on the use of polyacrylamide boronate affinity gel chromatography. *J. Cyclic Nucleotide Res.* **5,** 65–74.

36. Methven, P., Lemon, M., and Bhoola, K. (1980) The metabolism of cyclic nucleotides in the guinea pig pancreas. *Biochem. J.* **186,** 491–498.

37. Thompson, W. J. and Strada, S. J. (1985) Cyclic nucleotide phosphodiesterase (PDE), in *Methods of Enzymatic Analysis*, 3rd ed., vol. VIII (Bergmeyer, H. U., ed.), Verlag Chemie, Berlin, Germany, pp. 127–134.

38. Harrison, S. A., Beier, N., Martins, T. J., and Beavo, J. (1988) Isolation and comparison of bovine heart cGMP-inhibited and cGMP-stimulated phosphodiesterases. *Methods Enzymol.* **159,** 685–702.

39. Bauer, A. C. and Schwabe, U. (1980) An improved assay of cyclic 3'5'nucleotide phosphodiesterase with QAE-Sephadex columns. *Naunyn-Schmiedeberg's Arch. Pharmacol.* **311,** 193–198.

40. Weiss, B., Lehne, R., and Strada, S. (1972) A rapid microassay of adenosine 3'5'-monophosphate phosphodiesterase activity. *Anal. Biochem.* **45,** 222–235.

41. Kincaid, R. L. and Manganiello, V. C. (1988) Assay of cyclic nucleotide phosphodiesterase using radiolabeled and fluorescent substrates. *Methods Enzymol.* **159,** 457–470.

42. Wilke, R., Arch, J. R. S., and Nicholson, C. D. (1989) Tissue selective inhibition of cyclic nucleotide phosphodiesterase by denbufylline. *Arzneim.-Forsch. Drug Res.* **39,** 665–667.

CHAPTER 13

Separation of Labeled Inositol Phosphate Isomers by High-Pressure Liquid Chromatography (HPLC)

Stephen Jenkinson

1. Introduction

It has now been established that agonist stimulation of a large variety of cell-surface receptors promotes hydrolysis of inositol phospholipids through activation of a phosphoinositide-specific phospholipase C (1,2). Phosphodiesteric cleavage of phosphatidylinositol-4,5-bis-phosphate by phospholipase C produces Ins(1,4,5)P$_3$ and sn-1,2-diacylglycerol, which release Ca^{2+} from intracellular stores and activate protein kinase(s) C, respectively (1–4). Metabolism of Ins(1,4,5)P$_3$ is rapid and can follow two possible pathways (Fig. 1.; for review, see ref. 4). The simple dephosphorylation route proceeds via a 5-phosphatase to Ins(1,4)P$_2$, and Ins(1,3,4)P$_3$/Ins(1,4)P$_2$ 1-phosphatase to Ins(4)P$_1$. Alternatively, phosphorylation of Ins(1,4,5)P$_3$ by a 3-kinase generates the putative second messenger Ins(1,3,4,5)P$_4$, which then undergoes sequential dephosphorylation via a 5-phosphatase to Ins(1,3,4)P$_3$ and either a 4-phosphatase, an Ins(1,3,4)P$_3$/Ins(1,4)P$_2$ 1-phosphatase, or a 6-kinase route to produce Ins(1,3)P$_2$, Ins(3,4)P$_2$, or Ins(1,3,4,6)P$_4$, respectively. Subsequent dephosphorylation of Ins(1,3)P$_2$ and Ins(3,4)P$_2$ produces Ins(1)P$_1$ and Ins(3)P$_1$, which, along with Ins(4)P$_1$, are metabolized by the inositol monophosphatase to release myo-inositol (4). This complex pathway of Ins(1,4,5)P$_3$ serves to terminate second-messenger action and to conserve efficiently the cellular myo-inositol pool for the resynthesis of inositol phospholipids.

From: *Methods in Molecular Biology, Vol. 41: Signal Transduction Protocols*
Edited by: D. A. Kendall and S. J. Hill Copyright © 1995 Humana Press Inc., Totowa, NJ

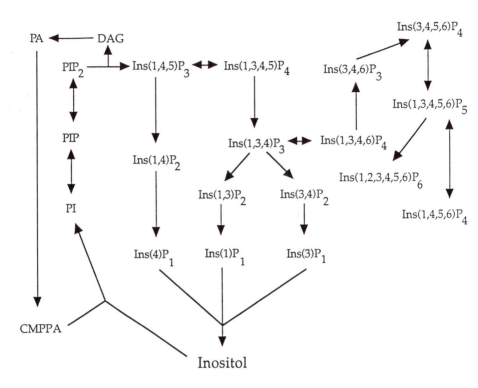

Fig. 1. The phosphoinositide cycle.

Assays have been developed that measure the mass of the unlabeled second messengers Ins(1,4,5)P$_3$ and Ins(1,3,4,5)P$_4$ in tissue extracts *(5,6)*. However, a general method that measured all the products of Ins(1,4,5)P$_3$ metabolism would clearly be of advantage, giving greater detail on the routes and kinetics of this complex pathway.

Head-group labeling of phosphoinositides is now the most favored approach for studies examining this signaling pathway. Incubation of cell lines with [³H]inositol for extended periods (2–5 d *[7,8]*) can allow radiolabeling that approaches isotopic equilibrium. Upon agonist challenge, the changes in radioactivity associated with the phospho-inositide cycle intermediates will closely approximate true concentration changes under these conditions, at least over initial time courses *(7)*. Unfortunately, many preparations, such as brain slices, are not sufficiently robust to allow such extended labeling periods. Under these circumstances a continuous labeling strategy must be adopted, where a prelabeling period, for example 60 min, is followed by agonist stimulation in the continued presence of the same concentration of radiolabel. Interpretation of data is complicated by the fact that rapid changes in the specific activity of both free and covalently bound inositol pools may occur. This is most noticeable with $Ins(1,4,5)P_3$ *(9)*. Therefore, changes in radioactivity associated with phosphoinositide cycle intermediates do not necessarily reflect an equivalent change in mass (for discussion, *see* ref. *10*).

A number of methods can be employed to separate the labeled inositol (poly)phosphate isomers according to their charge, including Dowex anion-exchange chromatography and high-pressure liquid chromatography (HPLC). Dowex separation has the advantage of being relatively quick and inexpensive with a large number of samples being able to be processed at one time. However, the separation is limited with only the total inositol phosphate fractions ($InsP_1$, $InsP_2$, and so forth) being resolved *(11)*. HPLC, as is discussed in this chapter, produces a much greater resolution of the individual inositol phosphate isomers *(12,13)* and may indeed highlight subtle changes in the profile of these isomers that could be missed by other less sensitive methods. The process, however, is expensive and time consuming. A modification of this HPLC methodology can also be used to determine the mass of individual unlabeled isomers *(14,15)*. However, this is not discussed in this chapter.

2. Materials

2.1. Instrumentation and Setup

The HPLC gradients discussed in this chapter require a two-pump HPLC system, as depicted in Fig. 2, and are routinely run as described in Section 3.2. Control of the system is via a 2152 controller (Pharmacia/LKB, Piscataway, NJ). However, any other controller would be appropriate if it can control the buffer switching valve (V1), the flow rate of both pumps (2150-type pumps [Pharmacia/LKB]), and an automatic fraction collector, such as a 2211 Super Rac (Pharmacia/LKB). The system is also equipped with a UV detector set to 254 nm that detects nucleotide markers spiked into the samples (*see* Section 3.3.). Fractions of the eluate are collected by a fraction collector (2211 Super Rac), and the radioactivity determined by liquid scintillation counting. Alternatively, a flowthrough radioactivity detector may be employed.

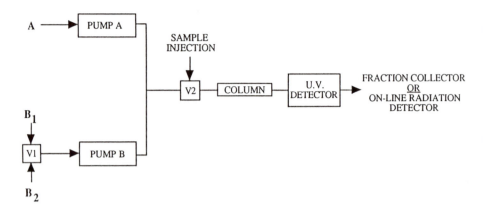

Fig. 2. Scheme of an HPLC system suitable for the separation of inositol phosphate isomers. A: H_2O; B_1 and B_2: Buffers (*see* Section 3.).

2.2. Chemicals

Analytical grade ammonium dihydrogen orthophosphate, ammonium acetate with orthophosphoric acid, acetic acid, and ammonia are used in the preparation of the elution buffers described in Section 2.3. (all obtained from Fisons). Prior to use, all buffers are filtered and degassed by vacuum filtration through inert 0.2-μm pore size membrane filters (Whatman, Maidstone, UK). The nucleotide markers used (adenosine and guanosine mono-, di-, tri-, and tetraphosphate) were purchased from Sigma (St. Louis, MO).

2.3. Buffers

1. Krebs/HEPES buffer: 118.6 mM NaCl, 4.7 mM KCl, 1.2 mM MgSO$_4 \cdot$ H$_2$O, 1.2 mM KH$_2$PO$_4$, 4.2 mM NaHCO$_3$, 1.3 mM CaCl$_2$, 10 mM HEPES (free acid), and 11.7 mM glucose, pH 7.4 at 37°C, made up in H$_2$O.
2. Perchloric acid solution: 10% (v/v) made up in H$_2$O.
3. 10 mM EDTA.
4. Freon: Octylamine: Make up fresh for each experiment—1:1 (v/v) mixture of freon (1,1,2-trichlorotriflouroethane) and tri-*n*-octylamine. Mix thoroughly before use.
5. 60 mM NaHCO$_3$.
6. Ammonium dihydrogen orthophosphate buffers: 0.5M (57.5 g/L), 0.8M (92 g/L), 1.4M (161 g/L), and 1.9M (218.5 g/L) ammonium dihydrogen orthophosphate solutions should be made up in H$_2$O. The pH of the solu-

tions should be corrected to 3.7 or 4.75 by addition of concentrated ortho-phosphoric acid. Solutions were vacuum-filtered (0.2-μm pore size) prior to use.

7. Ammonium acetate buffer: $0.2M$ (15.6 g/L) Ammonium acetate solution should be made up in H_2O, and the pH corrected to 3.75 by addition of acetic acid. Vacuum-filter the solution (0.2-μm pore size) prior to use.

2.4. Scintillation Cocktail

The scintillation cocktail of choice for HPLC is Floscint-IV from Packard (Downers Grove, IL). This cocktail is highly miscible with solutions of low pH and high-salt concentration. For example, 1 mL of $1.4M$ $NH_4H_2PO_4$ at a pH of 3.7 is completely miscible in 5 mL of scintillant.

2.5. Columns

The separations described in Section 3.2. are achieved using a Partisil/Partisphere (10 μm) strong anion-exchanger (SAX) analytical column (25 cm × 4.6 mm; Whatman) equipped with a precolumn packed with Whatman pellicular anion-exchange resin.

2.6. Labeled Inositol Phosphate Standards

Identification of sample [³H]inositol phosphates is based on the coelution with authentic [³H]labeled standards in separated runs or by spiking samples with the standard. [³H]-Labeled inositol monophosphates (Ins[1]P₁, Ins[4]P₁), bis-phosphate (Ins[1,4]P₂), trisphosphates (Ins[1,3,4]P₃, Ins[1,4,5]P₃), and tetrakisphosphate (Ins[1,3,4,5]P₄) are all commercially available from both Amersham (Arlington Heights, IL) and NEN DuPont, (Boston, MA). [³H]Ins(2)P₁ can be prepared in a mixture with [³H]inositol and [³H]Ins(1)P₁ by incubation of [³H]PtdIns with a crude preparation of brain phospholipase C as previously described *(16)*. [³H]Ins(4,5)P₂ can be prepared by mild alkaline phosphatase treatment of [³H]Ins(1,4,5)P₃ *(17)*. Ins(1,3)P₂ and Ins(3,4)P₂ are prepared by incubation of Ins(1,3,4)P₃ with rat brain supernatant *(12)*.

3. Methods

3.1. Preparation of Samples

Cells are incubated with [³H]inositol in order to achieve optimal labeling. The duration of this incubation period depends on how robust the tissue under examination is. The method described in this section is appropriate for cells in culture (for review, *see* ref. *10*).

1. Harvested cells are resuspended in media supplemented with [^{3}H]inositol (1–5 µCi/mL). Cultures are seeded into 24-well dishes (0.5 mL of media/ well) 48 h prior to the experiment and are allowed to reach equilibrium labeling at 37°C in 5% CO_2/95% O_2 (*see* Note 1).

2. After this labeling period the media is removed, and the cells are washed with Krebs/HEPES buffer (2 × 1 mL) to remove any exogenous inositol, both labeled and unlabeled. The buffer is replaced with 1 mL of fresh buffer, and the cells are allowed to stabilize for 15 min.

3. The buffer is removed from the cells, and 200 µL of buffer-containing drugs are added to the cells (*see* Note 2).

4. Incubations are terminated by addition of perchloric acid (PCA; 1 vol, 10%).

5. Samples are left for 5–15 min in order that the PCA can extract the labeled inositol phosphates from the cells. Note that this is not the only method of extraction that can be employed. However, it is the method of choice in our laboratory (*see* Note 3).

6. The supernatant is removed and neutralized by addition of EDTA (10 m*M*; 1–4 vol of supernatant) and freon:tri-*n*-octylamine (1:1 [v/v]; 1 vol). Samples are thoroughly mixed, centrifuged (3000*g* for 20 min), and 200 µL of the upper phase removed (*see* Note 4); the final pH is adjusted to 7 by addition of $NaHCO_3$ (60 m*M*; 1–4 vol of extract).

7. The samples are diluted to a final volume of 2 mL with water and can be stored at 4°C for up to 2 wk before HPLC analysis.

3.2. HPLC Elution Gradients

This section briefly describes a number of simple gradients that can be used for the separation of various labeled inositol phosphate isomers.

3.2.1. Gradient 1: Separation of InsP$_1$ to InsP$_4$ Isomers

A modification of the gradient developed by Dean and Moyer *(18)*, as described by Batty et al. *(12)*, allows for the extensive separation of InsP$_1$ to InsP$_4$ from sample extracts.

After sample injection (2 mL), free [^{3}H]inositol is eluted by washing the column for 5–15 min with water, depending on the extent of radio-labeling (*see* Note 5). This wash should also be carried out for all other gradients described in this section if the sample contains large quantities of free [^{3}H]inositol.

The gradient is for a three-solvent system comprised of:

A: H_2O
B$_1$: 0.5*M* ammonium dihydrogen orthophosphate
B$_2$: 1.4*M* ammonium dihydrogen orthophosphate.

The pHs of the solutions B_1 and B_2 are adjusted to 3.7 prior to use by the addition of orthophosphoric acid (*see* Note 6). The values are expressed as percentage contribution by B_1 and B_2 at each time point. The flow rate is 1 mL/min, 0.5-mL fractions are collected from T20–T95 and 1-mL factions from T95.1–T110.

T	%B$_1$	%B$_2$
0	0	0
30	12	0
30.1	38	0
45	38	0
60	60	0
60.1	100	0
95	100	0
95.1	0	100
110	0	100
110.1	0	0
120	0	0

An application of this gradient is illustrated in Fig 3. The approximate retention times for the isomers are as follows: Ins(1)P$_1$, ~24 min; Ins(4)P$_1$, ~27 min; Ins(1,3)P$_2$, ~49 min; Ins(1,4)P$_2$, ~53 min; Ins(3,4)P$_2$, ~56 min; Ins(4,5)P$_2$, ~60 min; Ins(1,3,4)P$_3$, ~81 min; Ins(1,4,5)P$_3$, ~84 min; Ins(1,3,4,5)P$_4$, 112 min.

Ins(3,4)P$_2$ is difficult to separate from Ins(1,4)P$_2$ and may appear as a peak on the tail of the elution of Ins(1,4)P$_2$. This is especially a problem, since the column deteriorates with age (*see* Note 7). Figure 2 also illustrates the elution time of the nucleotide markers (*see* Section 3.3.) with respect to the elution profile of the inositol phosphate isomers. This gradient is used routinely to examine phosphoinositide metabolism, the overall resolution being ideal for this purpose.

3.2.2. Gradient 2: Separation of InsP$_1$ Isomers

The gradient is for a two-solvent system comprised of:

A: H$_2$O
B$_1$: 0.2M ammonium acetate.

The pH of solution B$_1$ is adjusted to 3.75 prior to use by the addition of acetic acid. The values are expressed as percentage conribution by B$_1$ at each time point. The flow rate is 1 mL/min, and 0.5-mL fractions are collected throughout.

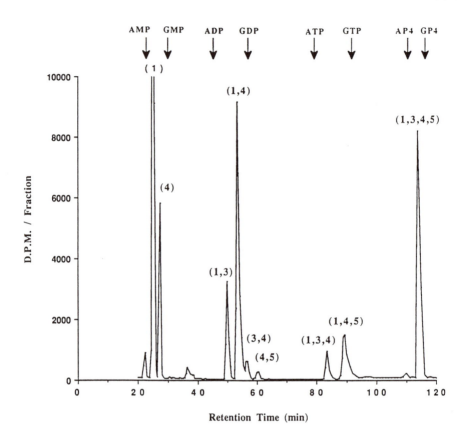

Fig. 3. HPLC separation, using Gradient 1, of the [³H]inositol phosphates present in an extract from carbachol-stimulated rat cerebral cortex slices labeled with [³H]inositol.

T	%B$_1$
0	0
5	0
15	35
100	35
105	100
110	100
110.1	0
120	0

An application of this gradient is illustrated in Fig. 4. Ins(1)P$_1$ and Ins(3)P$_1$ elute together at ~65 min followed successively by Ins(2)P$_1$ at

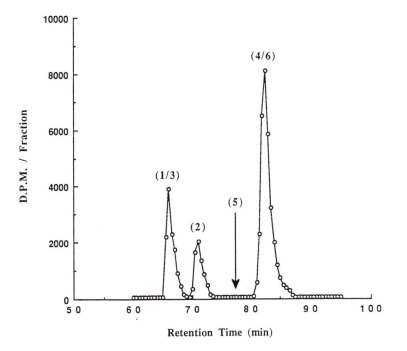

Fig. 4. HPLC separation of [^{3}H]Ins(1)P$_1$, [^{3}H]Ins(2)P$_1$, and [^{3}H]Ins(4)P$_1$ standards using Gradient 2. The elution Ins(5)P$_1$ would be intermediate to Ins(2)P$_1$ and Ins(4)P$_2$.

~20 min, Ins(5)P$_1$ at ~80–82 min, and Ins(4)P$_1$ and Ins(6)P$_1$ together at ~85 min. AMP runs prior to Ins(1/3)P$_1$ and GMP after Ins(4/6)P$_1$.

This gradient highlights the limitations of HPLC. Although most inositol phosphate isomers can be separated by this method, HPLC cannot resolve isomers that are enantiomers, such as Ins(1)P$_1$ and Ins(3)P$_1$. Other methods can, however, be employed for this purpose *(19)*. One advantage of this gradient is that the solvent can be removed from the samples simply by freeze drying, thus permitting easy desalting of the samples if required.

3.2.3. Gradient 3: Separation of Ins(1,3,4,5)P$_4$ and Ins(1,3,4,6)P$_4$

This gradient, developed by Batty et al. *(12)*, is for a two-solvent system comprised of:

A: H$_2$O
B$_1$: 0.8*M* ammonium dihydrogen orthophosphate.

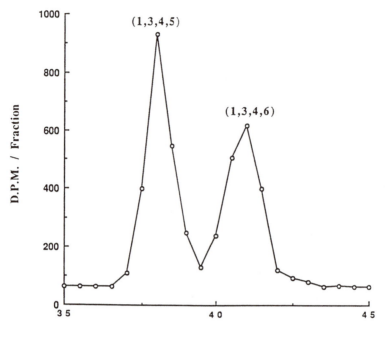

Fig. 5. HPLC separation of [³H]Ins(1,3,4,5)P$_4$ and [³H]Ins(1,3,4,6)P$_4$ standards using Gradient 3.

The pH of solution B$_1$ is adjusted to 4.75 prior to use by the addition of ammonia. The values are expressed as percentage contribution by B$_1$ at each time point. The flow rate is 1 mL/min, and 0.5-mL fractions are collected throughout.

T	%B$_1$
0	0
15	0
20	75
28	75
38	100
48	100
48.1	0
50	0

An application of this gradient is demonstrated in Fig. 5. Ins(1,3,4,5)P$_4$ elutes at ~40 min and Ins(1,3,4,6)P$_4$ at ~45 min, both running after the guanosine tetraphosphate nucleotide marker.

3.2.4. Gradient 4: Separation of $InsP_5$ and $InsP_6$

With the discovery of the higher inositol phosphates $InsP_5$ and $InsP_6$, a routine method was developed by Dean and Moyer *(20)* for their separation. Although the gradient separates $InsP_5$ and $InsP_6$, it does not resolve the individual isomers of $InsP_5$.

The gradient is for a two-solvent system comprised of:

A: H_2O
B_1: $1.9M$ ammonium dihydrogen orthophosphate.

The pH of solution B_1 is adjusted to 3.8 prior to use by addition of orthophosphoric acid. The values are expressed as percentage contribution by B_1 at each time-point. The flow rate is 1 mL/min, and 0.5-mL fractions are collected throughout.

T	%B
0	0
60	100
90	100
90.1	0
100	0

An application of this gradient is demonstrated in Fig. 6. $InsP_{1-4}$ elute at ~6, 12, 17, and 25 min, successively. $InsP_5$ and $InsP_6$ elute at ~30 and ~40 min, respectively.

These gradients give a general idea of the elutions that can be achieved using HPLC. It is possible, however, to customize any of them in order to obtain a more suitable gradient for a particular separation. Shallowing a gradient increases the length of the gradient and, therefore, the time required for each individual run. However, it can lead to a better resolution of the different isomeric species. The use of weak anion-exchange (WAX) columns may also be considered.

3.3. Standardization of HPLC Runs

To overcome problems associated with variations in retention times either between columns or owing to columns' aging, samples can be routinely spiked with 50–100 nmol each of adenosine and guanosine mono-, di-, tri-, and tetraphosphates before injection. Nucleotides are detected by continuous UV monitoring of the column eluate at 254 nm. Typically $InsP_1$s elute at retention times intermediate to AMP and GMP, $InsP_2$s between ADP and GDP (except [^{3}H]Ins[4,5]P$_2$, which

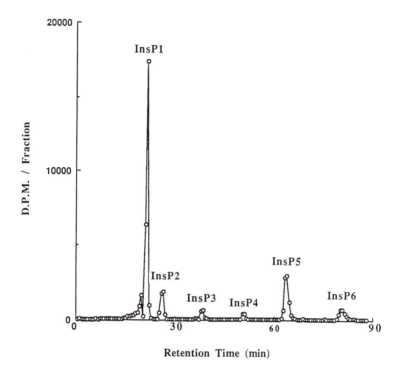

Fig. 6. HPLC separation, using Gradient 4, of the [³H]inositol phosphates present in an extract from carbachol-stimulated rat cerebellar granule cells labeled for 48 h with [³H]inositol.

runs 1–2 min after GDP), InsP$_3$s between ATP and GTP, and InsP$_4$s with the tetraphosphates. For a similar purpose, [³H]inositol phosphate standards can be spiked into samples in order to help confirm the identity of any unknown peak.

3.4. Contaminating Peaks

During preparation samples can become contaminated with small quantities of the deacylated phosphoinositide products glycerophosphoinositol, glycerophosphoinositol phosphate, and glycerophosphoinositol bis-phosphate (GroPIns, GroPInsP, and GroPInsP$_2$). These labeled water-soluble products are separated by HPLC and run prior to AMP, ADP, and ATP, respectively. Therefore, they should not be confused with unknown [³H]inositol phosphate peaks.

4. Notes

1. If the counts being obtained in the samples are low, it is possible to use larger multiwell dishes to increase the number of cells in the sample, i.e., 8-well plates rather than 12-well plates. Similarly, the use of inositol-free media may increase the labeling with [^{3}H]inositol.

2. A larger volume of buffer, containing the required drugs, may be required for long incubations, since evaporation from the multiwells can be a significant problem. This is especially true for larger plate sizes.

3. A number of other methods can be employed in the extraction of the inositol phosphates from the cells. These include: addition of trichloroacetic acid with subsequent removal of the acid by washing the samples with water-saturated diethyl ether, or alternatively, addition of an acidified chloroform methanol solution with subsequent addition of HCl and chloroform, resulting in the formation of aqueous and organic phases, with the neutralized inositol phosphates being found in the aqueous phase (for more detail on methods, *see* ref. *21*).

4. After neutralization with freon:octylamine, sample pH should be approx 5–6. If the pH is lower, then this step should be repeated until the pH reaches 5–6.

5. Always ensure that the pH of the sample to be injected is neutral and that the sample is particulate-free. The latter can be removed by attaching a 0.2-μm pore size acrodisk filter (Gelman, Ann Arbor, MI) to the injection syringe.

6. The pH and concentration of all buffers should be accurate. Small changes in concentration or pH can result in a change in the elution profile of the labeled inositol phosphates and the loss of some peaks, especially monophosphates.

7. Columns should be changed when the resolution of the separation starts to decrease. Generally, it is more economical to have the columns repacked by the supplier rather than having the column replaced.

8. Guard column inlet frits and resin should be changed regularly (once every 1–2 mo).

9. After use, wash HPLC system (especially the pumps) through thoroughly with water to prevent salt crystal formation in the pumps or pipes, which may lead to damage of the system.

References

1. Nahorski, S. R. (1988) Inositol polyphosphates and neuronal calcium homeostasis. *Trends Neurosci.* **11,** 444–448.
2. Fisher, S. K., Heacock, A. M., and Agranoff, B. W. (1992) Inositol lipids and signal transduction in the nervous system: an update. *J. Neurochem.* **58,** 18–38.

3. Downes, C. P. and MacPhee, C. H. (1990) *Myo*-inositol metabolites as cellular signals. *Eur. J. Biochem.* **193**, 1–18.

4. Shears, S. B. (1991) Regulation of the metabolism of 1,2-diacylglycerols and inositol phosphates that respond to receptor activation. *Pharmacol. Ther.* **49**, 79–104.

5. Bradford, P. G. and Rubin, R. P. (1986) Quantitative changes in inositol 1,4,5-trisphosphate in chemoattractant-stimulated neutrophils. *J. Biol. Chem.* **261**, 15,644–15,647.

6. Donie, F. and Reiser, G. (1989) A novel, specific binding assay for quantitation of intracellular inositol 1,3,4,5-trisphosphate (InsP$_4$) using a high-affinity InsP$_4$ receptor from cerebellum. *FEBS Lett.* **254**, 155–158.

7. Horstman, D. A., Takemura, H., and Putney, J. W. (1988) Formation and metabolism of [^{3}H]inositol phosphates in AR42J pancreatoma cells. *J. Biol. Chem.* **263**, 15,297–15,303.

8. Lambert, D. G., Challiss, R. A. J., and Nahorski, S. R. (1991) Accumulation and rate of Ins(1,4,5)P$_3$ and Ins(1,3,4,5)P$_4$ in muscarinic-receptor-stimulated SH-SY5Y neuroblastoma cells. *Biochem. J.* **273**, 791–794.

9. Challiss, R. A. J., Batty, I. H., and Nahorski, S. R. (1988) Mass measurements of inositol (1,4,5) trisphosphate in rat cerebral-cortex slices using a radioreceptor assay: effects of neurotransmitters and depolarisation. *Biochem. Biophys. Res. Commun.* **157**, 684–694.

10. Nahorski, S. R. and Challiss R. A. J. (1991) Modulation of receptor-mediated inositol phospholipid breakdown in the brain. *Neurochem. Int.* **19**, 207–212.

11. Kennedy, E. D., Challiss, R. A. J., and Nahorski, S. R. (1989) Lithium reduces the accumulation of inositol polyphosphate second messengers following cholinergic stimulation of cerebral cortex slices. *J. Neurochem.* **53**, 1652–1655.

12. Batty, I. H., Letcher, A. J., and Nahorski, S. R. (1989) Accumulation of inositol polyphosphate isomers in agonist-stimulated cerebral-cortex slices. *Biochem. J.* **258**, 23–32.

13. Jenkinson, S., Challiss, R. A. J., and Nahorski, S. R. (1992) Evidence for lithium-sensitive inositol 4,5-bisphosphate accumulation in muscarinic cholinoceptor-stimulated cerebral cortex slices. *Biochem. J.* **287**, 437–442.

14. Mayr, G. W. (1988) A novel metal-dye detection system permits picomolar-range h.p.l.c. analysis of inositol polyphosphates from non-radioactively labelled cell or tissue supensions. *Biochem. J.* **254**, 585–591.

15. Pestwick, S. A. and Bolton, T. B. (1991) Measurement of picomole amounts of any inositol phosphate isomer separable by H.P.L.C. by means of a bioluminescence assay. *Biochem. J.* **274**, 663–672.

16. Irvine, R. F., Letcher, A. J., and Dawson, R. M. C. (1979) Fatty acid stimulation of membrane phosphatidylinositol hydrolysis by brain phosphatidylinositol phosphodiesterase. *Biochem. J.* **178**, 497–500.

17. Hughes, P. J. and Drummond, A. H. (1987) Formation of inositol phosphate isomers in GH$_3$ pituitary tumour cells stimulated with thyrotropin-releasing factor. *Biochem. J.* **248**, 463–470.

18. Dean, N. M. and Moyer, J. D. (1987) Separation of multiple isomers inositol phosphates formed in GH$_3$ cells. *Biochem. J.* **242**, 361–366.

19. Ackerman, K. E., Gish, B. G., Honchar, M. P., and Sherman, W. R. (1987) Evidence that inositol 1-phosphate in brain of lithium-treated rats results mainly from phosphatidylinositol metabolism. *Biochem. J.* **242,** 517–524.

20. Dean, N. M. and Moyer, J. D. (1988) Metabolism of inositol bis-, tris-, tetrakis- and pentakisphosphates in GH$_3$ cells. *Biochem. J.* **254,** 585–591.

21. Challiss, R. A. J., Jenkinson, S., Mistry, R., Batty, I. H., and Nahorski, S. R. (1993) Assessment of neuronal phosphoinositide turnover and its disruption by lithium. *Neuroprotocols* **3,** 135–144.

CHAPTER 14

Mass Measurement of Key Phosphoinositide Cycle Intermediates

R. A. John Challiss

1. Introduction

The phosphoinositide cycle is now established as a major pathway by which a variety of hormones, neurotransmitters, and other signaling molecules can generate second messengers and thus affect the intracellular milieu *(1)*. Receptor-mediated activation of membrane-associated enzymic activities act on the minor membrane phospholipid phosphatidylinositol 4,5-bisphosphate [PtdIns(4,5)P_2] to generate three second messenger molecules: inositol 1,4,5-trisphosphate [Ins(1,4,5)P_3] and 1,2-diacylglycerol via inositol phospholipid-specific phospholipase C (PI-PLC) activity, and phosphatidylinositol 3,4,5-trisphosphate via phosphoinositide 3-kinase activity *(2)*.

Elucidation of the pathways by which the former two second messenger molecules are metabolized has been achieved almost exclusively by employing methodologies that involve the incorporation of radioisotopic labels into phosphoinositide cycle intermediates. Such studies have given rise to the concept that the complexity of the phosphoinositide cycle may be to allow the production of further biologically active molecules (e.g., inositol 1,3,4,5-tetrakisphosphate [Ins(1,3,4,5)P_4]), as well as allowing efficient second messenger signal termination and resynthesis of inositol phospholipids.

From: *Methods in Molecular Biology, Vol. 41: Signal Transduction Protocols*
Edited by: D. A. Kendall and S. J. Hill Copyright © 1995 Humana Press Inc., Totowa, NJ

Despite the undoubted worth of radiolabeling strategies, it is often desirable to measure directly the changes in concentrations of key pathway intermediates, such as Ins(1,4,5)P$_3$ and Ins(1,3,4,5)P$_4$, following receptor activation. Using radioisotopic labeling, this requires sophisticated separation methods to resolve the complex mixtures of inositol phosphate isomers generated by biological systems. Furthermore, it is possible that changes in intermediate concentrations are not truly reflected by radiolabeling patterns owing to changing specific activities of the precursor phosphoinositides on cycle activation *(3)*. Therefore, simple and sensitive methods that allow specific molecular recognition and direct mass assay of phosphoinositide cycle intermediates can provide important experimental advantages.

In this chapter, methods for the measurement of cellular levels of Ins(1,4,5)P$_3$, PtdIns(4,5)P$_2$, and Ins(1,3,4,5)P$_4$ are described. Space limitation prevents the presentation of data that fully validate such methods with respect to the inositol (poly)phosphate isomeric specificity of the binding proteins employed for each assay. Therefore, I recommend a number of publications that provide this information *(4–9)*.

2. Materials

1. D-Ins(1,4,5)P$_3$ (Research Biochemicals Inc., Natick, MA or Rhode Island University, Kingston, RI) can be stored in 1-mM aliquots at –20°C. For short periods (<3 mo), 40-µM aliquots of Ins(1,4,5)P$_3$ can also be stored for daily preparation of standard curves.
2. D-Ins(1,3,4,5)P$_4$ from some commercial sources often contains significant Ins(1,4,5)P$_3$ contamination. D-Ins(1,3,4,5)P$_4$ from Rhode Island University is one of the purest preparations available from a commercial source. Stocks of D-Ins(1,3,4,5)P$_4$ can be stored at millimolar concentrations at –20°C.
3. [^{3}H]Ins(1,4,5)P$_3$ (NEN DuPont [Boston, MA] NET-911, 17–20 Ci/mmol, or Amersham International [Amersham, UK] TRK.999, 30–50 Ci/mmol).
4. No commercial source of [^{32}P]Ins(1,3,4,5)P$_4$ is available at present. Therefore, preparation requires [^{32}P]Ins(1,4,5)P$_3$ (NEN DuPont NEG-066, 200–250 Ci/mmol) phosphorylation *(10)*. Alternatively, [^{3}H]Ins-(1,3,4,5)P$_4$ (NEN DuPont NET-941, 15–30 Ci/mmol, or Amersham International TRK 998, 20–60 Ci/mmol) can be used, although in my experience, this radioligand yields poorer data compared to the [^{32}P]-labeled ligand.

5. Homogenization buffer (buffer A): 20 mM NaHCO$_3$ and 1 mM dithiothreitol, pH 8.0.
6. Ins(1,4,5)P$_3$ assay buffer: 100 mM Tris-HCl and 4 mM EDTA, pH 8.0.
7. Ins(1,4,5)P$_3$ wash buffer for filtration assay: 25 mM Tris-HCl, 1 mM
   EDTA, and 5 mM NaHCO$_3$, pH 8.0.
8. Ins(1,3,4,5)P$_4$ assay buffer: 50 mM sodium acetate, 50 mM KH$_2$PO$_4$, 2 mM
   EDTA, and 0.25% bovine serum albumin, pH 5.0.
9. Ins(1,3,4,5)P$_4$ wash buffer for filtration assay: 25 mM sodium acetate,
   25 mM KH$_2$PO$_4$, 5 mM NaHCO$_3$, and 1 mM EDTA, pH 5.0.

3. Methods

3.1. Preparation of Ins(1,4,5)P$_3$ and Ins(1,3,4,5)P$_4$ Binding Proteins

Both binding proteins are prepared as crude "P$_2$" fractions from adrenal cortex (for Ins(1,4,5)P$_3$ assay, *see* Section 3.3.) and cerebellum (for
Ins(1,3,4,5)P$_4$ assay, *see* Section 3.5.) using identical methods:

1. Obtain bovine adrenal glands or porcine cerebella as fresh as possible (*see*
   Note 1). For the adrenal cortical preparation, cut each gland longitudinally,
   and remove the central medulla. Scrape the cortex from the outer capsule
   using a spatula, and maintain on ice until the required number of glands
   have been processed.
2. Roughly chop the cerebellum or adrenal cortex, and dispense into centrifuge tubes (approx 5 g tissue/50-mL tube). Homogenize (polytron setting
   5–6, 3 x 15 s) in ice-cold buffer A.
3. Centrifuge the homogenate (5000g, 10 min, 4°C). Pool the supernatants,
   and maintain on ice. Rehomogenize the pellet in buffer A and recentrifuge.
   Recover the supernatant, and discard the twice-extracted pellet.
4. Combine and dispense the supernatants from the two low-speed centrifugation steps into fresh tubes. Recover a P$_2$ fraction by centrifugation
   (40,000g, 20 min, 4°C). Discard the supernatant, and rehomogenize the
   pellet in buffer A. Recentrifuge (40,000g, 20 min, 4°C). Repeat this step to
   wash the P$_2$ fraction three times following its original isolation.
5. Rehomogenize the washed P$_2$ fraction in a known volume of buffer A, and
   determine the protein concentration. Adjust to 15–18 mg protein/mL for
   adrenal cortical preparation or 6–8 mg protein/mL for the cerebellar preparation. Dispense the P$_2$ fraction of known protein concentration as 1-mL
   aliquots into Eppendorf tubes, and store at –20°C until required. Both binding protein preparations can be stored for at least 6 mo without significant changes in binding properties.

3.2. Sample Preparation for Ins(1,4,5)P₃ and Ins(1,3,4,5)P₄ Mass Assay

It is impossible to give a single method for preparing samples for mass assay. For cell or tissue-slice preparations, ice-cold acid addition is sufficient to arrest metabolism (though care should be taken to avoid possible artifacts generated by this procedure; *see* Note 2). For experiments in vivo or organ-bath experiments, rapid arrest of metabolic activity (e.g., freezing in liquid nitrogen) should be achieved before tissue disruption into acid. In general, perchloric acid or trichloroacetic acid is used. An example of sample preparation used routinely in the author's laboratory is given here:

1. Add an equal volume of ice-cold $1M$ trichloroacetic acid (TCA). Transfer sample to ice bath and allow to extract for 20 min with intermittent vortex mixing.
2. Centrifuge ($3000g$, 15 min, 4°C) and remove the supernatant (the pellet can be washed in 0.9% NaCl and digested in $1M$ NaOH for determination of protein concentration, or processed for $PtdInsP_2$ mass measurement [*see* Section 3.4.]).
3. Wash the acid supernatant with 4 x 3 vol water-saturated diethylether (alternatively, samples can be neutralized using freon/tri-*n*-octylamine [*see* ref. *11*]).
4. Following acid extraction by diethylether, samples are allowed to stand for 30–60 min at 4°C. A known volume of sample is transferred to an Eppendorf tube, and $NaHCO_3$ (final conc. 10 mM) and EDTA (final conc. 5 mM) are added to bring the pH to 7. Under these conditions, samples may be stored at 4°C for up to 14 d prior to assay of $Ins(1,4,5)P_3$ and/or $Ins(1,3,4,5)P_4$.
5. A "buffer-blank" is also prepared to provide a suitable diluant for construction of the mass assay standard curve. This is achieved by mixing equal volumes of the original incubation medium and $1M$ TCA, extracting with 4 x 3 vol diethylether, and adding appropriate volumes of $NaHCO_3$ and EDTA (*see* Section 3.2., step 4).

3.3. Mass Determination of Ins(1,4,5)P₃

1. Construct standard curves by dilution of 40-µM stocks in the prepared buffer blank to give concentrations of 1.2, 4, 12, 40, 120, 400, and 1200 nM (for definition of nonspecific binding, aliquots of the 40-µM stock are used). These are four-times concentrates of the final assay concentration. Thus, if the final assay volume is 120 µL, pipet duplicate aliquots of 30 µL for each standard concentration corresponding to 0.036–36 pmol $Ins(1,4,5)P_3$/assay.

2. Add to 30 µL of standard Ins(1,4,5)P$_3$ or 30 µL of prepared sample in which concentration of Ins(1,4,5)P$_3$ is to be determined, and 30 µL [^{3}H]Ins (1,4,5)P$_3$ (appropriately diluted to give 6000–8000 dpm/assay). Take care to maintain assay tubes at 0–4°C at all times by performing this procedure in an ice bath.
3. Initiate the assay by addition of 30 µL of the bovine adrenal-cortical preparation (it may be necessary to rehomogenize the preparation on thawing). Incubate samples for 30 min on ice with intermittent vortex mixing.
4. Separation of bound and free radioligand can be achieved by either centrifugation or rapid vacuum filtration (*see* Note 3). For filtration, it is crucial that the Ins(1,4,5)P$_3$ wash buffer be ice-cold. Load Millipore vacuum manifolds with GF/B filters, and wet with wash buffer. Dilute assay samples with 3 mL wash buffer, and immediately filter; rapidly wash the sample tube with 2 x 3 mL wash buffer. This procedure should be completed within 5–10 s.
5. Following filtration, transfer GF/B filter disks to vials, and add scintillant. Samples should be allowed to extract for at least 6 h prior to scintillation counting. A typical standard curve is shown in Fig. 1. The method provides a high degree of intra- and interassay reproducibility *(4)* and can detect >0.1 pmol Ins(1,4,5)P$_3$ in a 30-µL sample.

3.4. Extraction/Hydrolysis
of PtdIns(4,5)P$_2$ for Mass Determination

1. Sequentially wash pellets from TCA-terminated incubations (*see* Section 3.2.) with 2 mL 5% (w/v) TCA/1 m*M* EDTA and 2 mL water. Following thorough aspiration of the supernatant, add 0.94 mL chloroform/methanol/conc HCl (40:80:1 by volume) to the pellet, and intermittently vortex mix for 20 min. Resolve phases by addition of 0.31 mL chloroform and 0.56 mL 0.1*M* HCl, and centrifugation (1000*g* for 15 min).
2. Remove solvent from a known volume of lower phase using a stream of N$_2$. Deacylate the lipid extract by addition of 0.25 mL 1*M* KOH. Tightly cap and heat the tubes in a boiling water bath for 15 min. Transfer samples to an ice bath, and neutralize by addition to columns (prepared by adding 0.5 mL of a 50% slurry of Dowex 50 (200–400 mesh; H$^+$ form). Elute from columns with 2.25 mL water.
3. Wash the total eluate (2.5 mL) with 2 x 2 mL butan-1-ol/light petroleum ether (5:1 [v/v]); if necessary, centrifuge the tubes (1000*g* for 15 min) to resolve phases, and take a 1-mL aliquot of the lower phase and lyophilize.
4. Redissolve the lyophilizate in the required volume of water, and take aliquots for Ins(1,4,5)P$_3$ mass measurement as detailed in Section 3.3.
5. It should be noted that the efficiency of alkaline hydrolysis of PtdIns(4,5)P$_2$ must be determined using [^{3}H]PtdIns(4,5)P$_2$ *(12)*. Hydrolysis yields

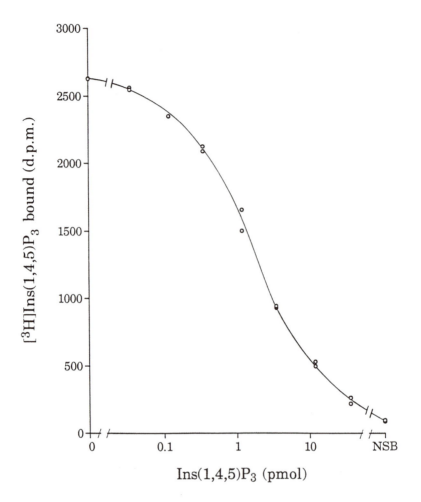

Fig. 1. A typical standard curve for the Ins(1,4,5)P$_3$ mass assay. In this experiment, each assay tube contained 7460 dpm [^{3}H]Ins(1,4,5)P$_3$ (17 Ci/mmol), the indicated concentration of Ins(1,4,5)P$_3$ and 460 µg of adrenal cortical binding protein in a final volume of 120 µL (assay pH 8.0). Nonspecific binding (NSB) was defined using 1.2 nmol Ins(1,4,5)P$_3$; a similar estimate of NSB was given using 1 mM InsP$_6$ or 100 µg/mL heparin (mol wt 4000–6000). Data are shown for duplicate determinations at each displacing concentration of Ins(1,4,5)P$_3$. The bound fraction was recovered by rapid vacuum filtration (*see* Section 3.3.).

Ins(2,4,5)P$_3$ and Ins(4,5)P$_2$, as well as Ins(1,4,5)P$_3$, and the molar ratio of these products must also be established; in ref. *12* a ratio of 20:14:66 was obtained. Note that Ins(2,4,5)P$_3$ and Ins(4,5)P$_2$ are 100–400 times weaker in their displacing activity in the Ins(1,4,5)P$_3$ mass assay, and therefore, do not interfere with the mass determination. There may also be a small loss of Ins(1,4,5)P$_3$ during the neutralization step; thus, it is imperative that the recovery of Ins(1,4,5)P$_3$ through this procedure is determined for each cell/tissue preparation to which it is applied.

3.5. Mass Determination of Ins(1,3,4,5)P$_4$

1. Construct standard curves to give 0.3–1000 nM final concentrations (i.e., 0.036–12 pmol Ins(1,3,4,5)P$_4$/assay for a 120-µL final assay volume).
2. Add to 30 µL of standard Ins(1,3,4,5)P$_4$ or 30 µL of prepared sample in which concentration of Ins(1,4,5)P$_3$ is to be determined, and 30 µL radiolabeled Ins(1,3,4,5)P$_4$ in assay buffer—if [^{32}P]Ins(1,3,4,5)P$_4$ is used, 10,000–12,000 dpm/assay is recommended. Take care to maintain assay tubes at 0–4°C at all times.
3. Initiate the assay by addition of 30 µL of the cerebellar preparation (it is important to rehomogenize the preparation prior to dispensing). Incubate samples for 30 min on ice with intermittent vortex mixing.
4. Separation of bound and free radioligand is best achieved by rapid vacuum filtration over GF/B filter disks (separation by centrifugation following introduction of a sucrose "cushion" has also been used *[7]*): It is crucial that the Ins(1,3,4,5)P$_4$ wash buffer be ice cold. Dilute the sample with 3 mL wash buffer, and immediately filter; rapidly wash the sample tube with 3 mL wash buffer twice. This procedure should be completed within 5–10 s.
5. Transfer filter disks to vials and add 4 mL of a suitable scintillant. Allow filters to extract in the scintillant for at least 6 h prior to counting. A typical example of an Ins(1,3,4,5)P$_4$ displacement curve is shown in Fig. 2. The biphasic displacement of [^{32}P]Ins(1,3,4,5)P$_4$ from preparations of rat and pig cerebellar membranes *(9)* does not compromise the mass assay, since displacement between 0.036 and 12 pmol Ins(1,3,4,5)P$_4$/assay can be adequately modeled using simple curve-fitting programs. The pig cerebellar preparation offers the advantage of a greater dynamic range for mass determination, owing to the greater proportion of high-affinity binding sites. The assay can reproducibly detect >0.05 pmol Ins(1,3,4,5)P$_4$ in a 30-µL sample *(8)*.

4. Notes

1. If obtaining cow or pig tissue is difficult, rat cerebellum (prepared as described in Section 3.1.) can be used for both Ins(1,4,5)P$_3$ and Ins(1,3,4,5)P$_4$ mass assays, since the affinities of the binding proteins present in the cerebellar "P$_2$" fraction are differentially pH-sensitive. The highest affinity of the Ins(1,4,5)P$_3$ binding protein has been reported to occur at pH 8.4 *(13)*. In contrast, optimal Ins(1,3,4,5)P$_4$ binding is observed

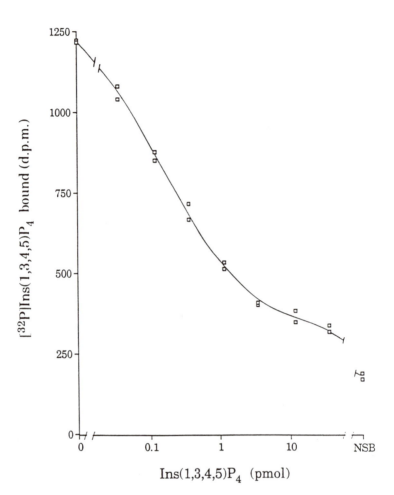

Fig. 2. A typical standard curve for the Ins(1,3,4,5)P$_4$ mass assay. In this experiment, each assay tube contained 10,595 dpm [^{32}P]Ins(1,3,4,5)P$_4$ (162 Ci/ mmol), the indicated concentration of Ins(1,3,4,5)P$_4$, and 168 µg of pig cerebellar binding protein in a final volume of 160 µL (assay pH 5.0). NSB was defined using 1 mM InsP$_6$; a similar estimate of NSB was given using 50 µM Ins(1,3,4,5)P$_4$ or 100 µg/mL heparin (mol wt 4000–6000). Data are shown for duplicate determinations at each displacing concentration of Ins(1,3,4,5)P$_4$. The bound fraction was recovered by rapid vacuum filtration (*see* Section 3.5.).

at acidic pH *(6,9)*. Therefore, manipulation of assay pH to 8.4 or 5.0 allows specific expression of Ins(1,4,5)P$_3$ or Ins(1,3,4,5)P$_4$ binding activities, respectively.

2. Conditions necessary for extraction of inositol (poly)phosphates should be optimized for the cell/tissue preparation under investigation. Some cells (e.g., the U937 human monocytic cell line) require homogenization of acid-arrested incubations to release these metabolites completely. Also, the concentration of acid used and the period of exposure to acid should be such that no chemical cleavage of inositol phospholipids occurs (e.g., by establishing that samples "spiked" with $[^3H]PtdIns(4,5)P_2$ are unaffected by the termination procedure). It should also be noted that some potential inositol phosphate metabolites will be chemically altered during the acid-termination procedure, most notably Ins(1:2-cyclic,4,5)P_3. Although this metabolite is a relatively poor displacer of specific $[^3H]Ins(1,4,5)P_3$ binding *(5)*, it is acid hydrolyzed to a mixture of Ins(2,4,5)P_3 and Ins(1,4,5)P_3 that will contribute to the Ins(1,4,5)P_3 concentration detected by the mass assay.

3. In my experience, the high protein concentration of the bovine adrenal cortex Ins(1,4,5)P_3 binding protein preparation can result in loose pellets following centrifugation. This causes problems with supernatant removal and may affect the quality of data obtained. Although this problem can be overcome (e.g., by increasing the salt concentration in the assay buffer), I routinely resolve bound and free ligand by filtration and recommend its use.

References

1. Berridge, M. J. (1993) Inositol trisphosphate and calcium signalling. *Nature* **361**, 315–325.
2. Downes, C. P. and Carter, A. N. (1991) Phosphoinositide 3-kinase: a new effector in signal transduction? *Cell. Signalling* **3**, 501–513.
3. Nahorski, S. R. and Challiss, R. A. J. (1991) Modulation of receptor-mediated inositol phospholipid breakdown in the brain. *Neurochem. Int.* **19**, 207–212.
4. Challiss, R. A. J., Batty, I. H., and Nahorski, S. R. (1988) Mass measurements of inositol 1,4,5-trisphosphate in rat cerebral cortex slices using a radioreceptor assay. *Biochem. Biophys. Res. Commun.* **157**, 684–691.
5. Palmer, S., Hughes, K. T., Lee, D. Y., and Wakelam, M. J. O. (1989) Development of a novel Ins(1,4,5)P_3-specific binding assay. *Cell. Signalling* **1**, 147–153.
6. Challiss, R. A. J., Chilvers, E. R., Willcocks, A. L., and Nahorski, S. R. (1990) Heterogeneity of $[^3H]$inositol 1,4,5-trisphosphate binding sites in adrenal-cortical membranes. *Biochem. J.* **265**, 421–427.
7. Donié, F. and Reiser, G. (1989) A novel, specific binding protein assay for quantitation of intracellular inositol 1,3,4,5-tetrakisphosphate using a high-affinity receptor from cerebellum. *FEBS Lett.* **254**, 155–158.
8. Challiss, R. A. J. and Nahorski, S. R. (1990) Neurotransmitter and depolarization-stimulated accumulation of inositol 1,3,4,5-tetrakisphosphate mass in rat cerebral cortex slices. *J. Neurochem.* **54**, 2138–2141.

9. Challiss, R. A. J., Willcocks, A. L., Mulloy, B., Potter, B. V. L., and Nahorski, S. R. (1991) Characterization of inositol 1,4,5-trisphosphate- and inositol 1,3,4,5-tetra-kisphosphate-binding sites in rat cerebellum. *Biochem. J.* **274,** 861–867.

10. Enyedi, P. and Williams, G. H. (1988) Heterogenous inositol tetrakisphosphate binding sites in the adrenal cortex. *J. Biol. Chem.* **263,** 7940–7942.

11. Sharps, E. S. and McCarl, R. L. (1982) A high-performance liquid chromatographic method to measure ^{32}P-incorporation into phosphorylated metabolites in cultured cells. *Anal. Biochem.* **124,** 421–424.

12. Chilvers, E. R., Batty, I. H., Challiss, R. A. J., Barnes, P. J., and Nahorski, S. R. (1991) Determination of mass changes in phosphatidylinositol 4,5-bisphosphate and evidence for agonist-stimulated metabolism of inositol 1,4,5-trisphosphate in airway smooth muscle. *Biochem. J.* **275,** 373–379.

13. Bredt, D. S., Mourey, R. J., and Snyder, S. H. (1989) A simple, sensitive and specific assay for inositol 1,4,5-trisphosphate in biological tissues. *Biochem. Biophys. Res. Commun.* **159,** 976–982.

CHAPTER 15

Measurement of Phospholipase C Activity in Brain Membranes

Enrique Claro, Elisabet Sarri, and Fernando Picatoste

1. Introduction

Phosphoinositide hydrolysis by phospholipase C (PLC) is a wide-spread transduction mechanism by which activation of many neuro-transmitter and hormone receptors triggers the formation of the second messengers inositol 1,4,5-trisphosphate [Ins(1,4,5)P_3] and 1,2-diacyl-glycerol *(1–3)*. Coupling between the receptor and PLC is often regulated by a guanine nucleotide binding regulatory protein from the G_q family *(4)*, similar to those involved in the regulation of adenylyl cyclase. In this chapter, we describe experimental protocols that allow the study of PLC activation by agonists of the M1/M3-muscarinic and 5HT$_2$-serotonergic receptors in washed membrane preparations from brain. These assays differ basically from the more extended experimental designs involving [^{3}H]inositol-labeled brain slices in the following aspects:

1. Besides the receptor/G_q mechanism, PLC is readily stimulated by increases in the cytosolic concentration of calcium *(5)*. Accordingly, a plethora of stimuli resulting in calcium entry (direct receptor gating, depolarization, and so forth) are expected to stimulate PLC in brain slices or other preparations, like synaptosomes or synaptoneurosomes. However, in membrane assays, there is no asymmetry in the calcium concentrations across the membrane, which are buffered down to resting intracellular levels. This features a clear difference between mechanisms of PLC activation involving the direct coupling receptor/G_q/PLC (as muscarinic or seroton-

From: *Methods in Molecular Biology, Vol. 41: Signal Transduction Protocols*
Edited by: D. A. Kendall and S. J. Hill Copyright © 1995 Humana Press Inc., Totowa, NJ

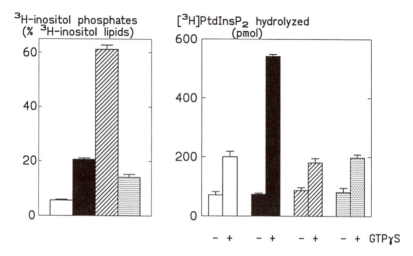

Fig. 1. Effect of PLC stimuli on the production of ^{3}H-inositol phosphates in brain cortical slices and in a membrane assay. Left panel: Rat brain cortical slices prelabeled with [^{3}H]inositol were incubated in Krebs-Henseleit buffer (2.5 mM CaCl$_2$) containing 10 mM LiCl and no further additions (☐), or with 1 mM carbachol (■), 30 µM norepinephrine (▨), or 20 mM KCl (▦). Right panel: Cortical membranes were assayed for the hydrolysis of exogenously added [^{3}H]PtdInsP$_2$ under basal conditions (☐), or with 1 mM carbachol (■), 30 µM norepinephrine (▨), or 20 mM KCl (▦), and in the presence or absence of 1 µM GTPγS as noted.

ergic stimulation), on the one hand, and those involving calcium activation of the enzyme (such as α$_1$-adrenergic, ionomycin, or KCl stimulation), on the other hand, which will not result in PLC stimulation (Fig. 1).

2. Because the membrane preparation is essentially free of endogenous GTP, receptor stimulation of PLC becomes strictly dependent on the inclusion of a hydrolysis-resistant analog of GTP to support activation of G$_q$. Guanosine-5'-O-(3-thiotriphosphate) (GTPγS) has been found to be four to five times more efficacious than 5'-guanylyl-imidodiphosphate (GppNHp) in terms of PLC stimulation.

3. The metabolism of inositol phosphates is simpler than in whole cells or tissue preparations. In brain membranes, Ins(1,4,5)P$_3$ is rapidly dephosphorylated to Ins(1,4)P$_2$, which is only marginally converted into Ins(4)P. However, Ins(1,3,4,5)P$_4$ and its dephosphorylation products Ins(1,3,4)P$_3$, Ins(1,3)P$_2$, and Ins(3,4)P$_2$ are not formed. Ins(1)P, which in cells appears to derive primarily from Ins(1,3)P$_2$, is produced in membranes by the direct breakdown of phosphatidylinositol (PtdIns) by PLC.

The following methods share many aspects in common, including the protocol for preparation of membranes, and the strict requirement of 1 mM sodium deoxycholate (0.04% [w/v]) in the assays to observe agonist effects synergistic to that of GTPγS. They only differ in the way the phosphoinositide substrate is presented, exogenous *(6)* or endogenous, and in this latter case, in the way the [³H]inositol label is incorporated into the lipids, *ex vivo* *(7,8)* or in vitro *(9)*. In this regard, three kinds of assays can be distinguished:

1. Assays with membranes derived from [³H]inositol-prelabeled slices;
2. Assays with exogenously added ³H-phosphoinositide substrates; and
3. Assays with membranes directly labeled with [³H]inositol.

2. Materials

2.1. Assay with Membranes Derived from [³H]Inositol-Prelabeled Tissue Slices

1. [³H]inositol (15–20 Ci/mmol): Store as indicated by the supplier.
2. Krebs-Henseleit buffer without added CaCl$_2$ (KH buffer): 116 mM NaCl, 25 mM NaHCO$_3$, 11 mM glucose, 4.7 mM KCl, 1.2 mM MgSO$_4$, and 1.2 mM KH$_2$PO$_4$, pH 7.4, equilibrated with O$_2$/CO$_2$ (95:5 [v/v]). Store at 4°C. Omission of CaCl$_2$ in this buffer enhances [³H]inositol incorporation into lipids.
3. Tris-EGTA buffer: 20 mM Tris-HCl, pH 7.0, with 1 mM EGTA. Store at 4°C.
4. 20 mM Sodium deoxycholate in water: Store at 4°C.
5. 25 mM Tris-maleate, 7.5 mM EGTA, pH 6.8, with KOH. Store at 4°C.
6. 1M CaCl$_2$ in water.
7. 25 mM Tris-maleate, 5 mM ATP, 15 mM MgCl$_2$, and 25 mM LiCl, pH 6.8, with KOH (resuspension buffer). Store 4-mL aliquots at –20°C.
8. 2 mM GTPγS in water: Store 40-µL aliquots at –20°C.
9. 25 mM Tris-maleate, pH 6.8, with KOH. Store at 4°C.
10. Chloroform/methanol (1:2 [v/v]).
11. Chloroform.
12. 0.25M HCl.

2.2. Assays with Exogenously Added ³H-Phosphoinositide Substrates

In addition to items 3–12 listed in Section 2.1., the following materials are required:

1. [³H]PtdInsP$_2$, [³H]PtdInsP, and [³H]PtdIns (all around 5 Ci/mmol): Store as indicated by the supplier.
2. PtdInsP$_2$ (2 mg/mL), PtdInsP (2 mg/mL), and PtdIns (10 mg/mL), all of them in chloroform/methanol (9:1 [v/v]): Store at –20°C.

2.3. Assay with Membranes Directly Labeled with [³H]Inositol

With the exception of Krebs-Henseleit buffer, in this protocol the same materials as in Section 2.1. are used. Also, 20 mM CMP in 25 mM Tris-maleate, pH 6.8, are needed. This can be either prepared fresh or stored at –20°C in 0.5-mL aliquots.

3. Methods

3.1. Assay with Membranes Derived from [³H]Inositol-Prelabeled Tissue Slices

3.1.1. [³H]Inositol Labeling

Brain cortices dissected free of meninges and white matter are cross-chopped (350 × 350 μm) with a McIlwain tissue chopper and dispersed in KH buffer. After a few washings, slices are incubated for 2 h at 37°C with 4 vol of KH buffer and [³H]inositol (2–50 μCi/mL) in a screw-capped Erlenmeyer flask, gassing with O_2/CO_2 (95:5 [v/v]) every 30 min. Avoid bubbling, since slices may stick all over the flask walls.

3.1.2. Preparation of Membranes

1. Aspirate KH buffer containing [³H]inositol, and wash slices five times with 20 vol of ice-cold Tris-EGTA buffer. All subsequent operations should be carried out at 4°C.
2. Transfer slices to a glass homogenizer, and homogenize them in 20 vol of Tris-EGTA buffer with 20 strokes of a motor-driven Teflon™ pestle at maximum setting. We do not find the use of protease inhibitors necessary.
3. Centrifuge the homogenate for 15 min at 40,000g. Discard the supernatant, and homogenize the pellet just as before in the same initial volume of Tris-EGTA buffer.
4. Centrifuge again for 15 min at 40,000g. Rehomogenize the pellet, and repeat step 3 twice more.
5. Resuspend the final pellet in Tris-EGTA buffer at 2.5 mg protein/mL, aliquot 1-mL samples in microfuge tubes, and centrifuge at maximum speed.
6. Aspirate supernatants, and store membranes at –80°C. We have not found loss of PLC responsiveness after 6 mo, provided the preparation has not been thawed and frozen again.

3.1.3. PLC A5mssay

The assay (250-μL final vol, although it can be scaled up or down for convenience) is done in 5-mL polypropylene tubes, and consists of 24 mM Tris-maleate, pH 6.8 (*see* Note 2), containing 1 mM sodium deoxycholate, 3 mM EGTA and CaCl$_2$, necessary to yield the desired free calcium concentration, 2 mM ATP, 6 mM MgCl$_2$, 10 mM LiCl, GTPγS, agonists, and membranes (250 μg protein, 20,000–500,000 dpm). Additions are made in the following order:

1. 12.5 μL of 20 mM sodium deoxycholate.
2. 100 μL of a freshly made solution consisting of 25 mM Tris-maleate/ 7.5 mM EGTA, pH 6.8, and CaCl$_2$ (*see* Notes 1 and 2). To obtain a free calcium concentration in the assay of 100 nM, mix 3.4 μL of 1M CaCl$_2$ with 5 mL of Tris-maleate/EGTA. The following offers a recipe with the volumes of 1M CaCl$_2$ to be mixed with 5 mL of 25 mM Tris-maleate/7.5 mM EGTA that will yield various free calcium concentrations in the assay:

Final free [Ca^{2+}]	Volume of 1M CaCl$_2$
50 nM	1.8 μL
100 nM	3.4 μL
200 nM	6.3 μL
300 nM	8.7 μL
1 μM	8.8 μL
3 μM	28.2 μL

3. 12.5 μL of 20X GTPγS in 25 mM Tris-maleate, pH 6.8: On a percentage basis, maximum effects of muscarinic agonists are found at 0.3 μM GTPγS, whereas serotonergic effects are best seen at 1-μM concentrations of the nucleotide.
4. 25 μL of 10X agonist in 25 mM Tris-maleate, pH 6.8. These are made fresh on the day of the assay.
5. Thaw membrane pellets in 1 mL each of cold resuspension buffer, and resuspend with the help of a 1-mL plastic syringe. Then start reactions by adding 100 μL of membranes. Vortex the tubes, and incubate 10–20 min at 37°C.

Reactions are stopped with 1.2 mL of cold chloroform/methanol (1:2 [v/v]). Then 0.5 mL each of chloroform and 0.25M HCl are added. Tubes are capped, shaken, and centrifuged to separate two phases. One milliliter samples of the upper (aqueous) phases containing ^{3}H-inositol phosphates are withdrawn and can be counted directly for radioactivity after evaporating traces of chloroform. However, signal is improved if [^{3}H]inositol that is carried over from the tissue labeling is separated from ^{3}H-inositol phosphates, which can also be separated from each other. This is achieved by anion-exchange chromatography in columns with Dowex 1 × 8 (100–200 mesh, formate form). Alternatively, ^{3}H-inositol phosphates can be separated by high-performance liquid chromatography (HPLC), which offers much better resolution (Fig. 2) and resolves

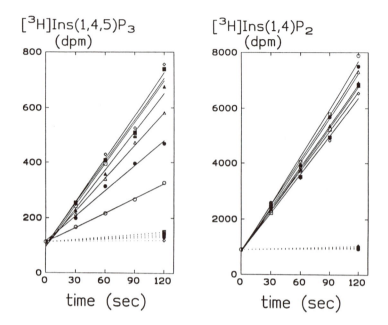

Fig. 2. $[^3H]Ins(1,4,5)P_3$ and $[^3H]Ins(1,4)P_2$ formed on stimulation of phospholipase C in $[^3H]$inositol-prelabeled rat brain cortical membranes. $[^3H]$inositol-prelabeled membranes (250 µg protein, 450,000 dpm) were stimulated for the times indicated with 1 µM GTPγS and 1 mM carbachol, and in the presence of different concentrations of unlabeled $Ins(1,4,5)P_3$ (in µM): 0 (○); 25 (●); 50 (△); 100 (▲); 150 (□); 200 (■); and 300 (◇). Controls run in the absence of GTPγS and carbachol were incubated for 2 min (dotted lines). Reproduced with permission from ref. *8*.

isomers, if necessary, although it is costly and time-consuming. To normalize results among different membrane preparations, ^3H-inositol phosphates produced are expressed as a percentage of total radioactivity in the lipid fraction, which is measured by counting 100-µL aliquots of the organic phases of zero time controls.

3.2. Assays with Exogenously Added ^3H-Phosphoinositide Substrates

These assays do not require $[^3H]$inositol prelabeling of the tissue, and therefore, the starting material can be either fresh or frozen. This allows work with tissue samples that are difficult to obtain metabolically active and even with human brain regions taken at the time of autopsy *(10)*.

Either exogenous [³H]PtdInsP$_2$ or [³H]PtdInsP can be used under the same experimental conditions. Unlike the situation with endogenous substrates, PLC is also able to hydrolyze exogenous PtdIns in a GTPγS and agonist-dependent manner *(6,8,11)*. Although this lipid does not appear to be a physiological substrate, its use may represent some cost advantage over that of polyphosphoinositides.

3.2.1. Preparation of Membranes

If necessary, the tissue is thawed in 20 m*M* Tris-HCl, 1 m*M* EGTA buffer, pH 7.0, then minced in small pieces, and processed exactly as in Section 3.1.2.

3.2.2. PLC Assays

In a final volume of 100 µL, the assays contain 30 µ*M* [³H]PtdInsP$_2$ or [³H]PtdInsP at a specific activity of 3–6 Ci/mol, representing 20,000–40,000 dpm/tube. If using PtdIns, its concentration is increased to 100 µ*M*, so that the specific activity is 1–2 Ci/mol. Additions are made in the same order as in Section 3.1.3.

1. Mix labeled and unlabeled phosphoinositide in a microfuge tube, and evaporate the solvent with a gentle stream of N$_2$. Then add 20 m*M* sodium deoxycholate necessary for a 20X phosphoinositide concentration (0.6 m*M* [³H]PtdInsP$_2$ or [³H]PtdInsP, 2 m*M* PtdIns). Resuspend the lipid by sonication with a probe dismembrator (two 15-s bursts, setting 35%), taking care not to sonicate with the tip near the surface, since this will cause foaming. Then place 5 µL of the resuspended substrate in 5-mL polypropylene tubes.
2. 40 µL of 25 m*M* Tris-maleate/7.5 m*M* EGTA and CaCl$_2$ (*see* Note 1). Basal PLC activity with exogenous substrates is more sensitive to activation by calcium. For optimum GTPγS and agonist effects, the free calcium concentration is dropped to 50 n*M* if using either ³H-polyphosphoinositide. However, breakdown of exogenous [³H]PtdIns requires higher calcium concentrations, and agonist effects are best observed at 300 n*M* free calcium.
3. 5 µL of 20X GTPγS in 25 m*M* Tris-maleate, pH 6.8.
4. 10 µL of 10X agonist in 25 m*M* Tris-maleate, pH 6.8.
5. Start reaction with 40 µL of membranes at 2.5 mg protein/mL in resuspension buffer. Vortex and incubate at 37°C. The resuspension buffer is modified when using PtdIns substrate, because ATP is no longer required to fuel polyphosphoinositide synthesis, and consequently, the optimum final concentration of MgCl$_2$ is reduced to 3 m*M*.

Reactions with [³H]PtdInsP₂ or [³H]PtdInsP are linear only for 3–5 min presumably because of dephosphorylation of the substrates to [³H]PtdIns by the active phosphoinositide phosphomonoesterases present in the membrane preparation. However, 10-min incubations give better resolution in terms of agonist effects. On the other hand, [³H]PtdIns breakdown is fairly linear for at least 30 min. Stop reactions and separate phases as in Section 3.1.3. One-milliliter aliquots of the aqueous phases are directly counted, since all the radioactivity present corresponds to ³H-inositol phosphates.

A modification of the assay with exogenous [³H]PtdIns takes advantage of the activities of PtdIns 4-kinase and PtdInsP 5-kinase in the membrane preparation, and reproduces better the situation with endogenous substrates. It consists of lowering the free calcium concentration to 30–50 n*M* to prevent direct breakdown of [³H]PtdIns, and supplying ATP and MgCl₂ for [³H]PtdIns conversion into ³H-polyphosphoinositides prior to the action of PLC.

3.3. Assay with Membranes Directly Labeled with [³H]Inositol

In this assay, another activity present in membranes, that of PtdIns synthase, is used to label directly endogenous PtdIns with [³H]inositol. Physiologically, this enzyme incorporates inositol into CDP-diacylglycerol, yielding PtdIns and CMP. Because of the reversibility of the reaction, the enzyme can also exchange free inositol by the inositol moiety of PtdIns in a process that requires CMP *(12)*:

$$\text{PtdIns} + \text{CMP} \quad \rightleftarrows \quad \text{Ins} + \text{CDP-diacylglycerol}$$

This reaction can be coupled in membranes to those of phosphoinositide kinases and PLC, so that a flux of tritium label is established from [³H]inositol to ³H-inositol phosphates that is dependent on the presence of GTPγS and agonists *(9)*.

3.3.1. Preparation of Membranes

Membranes are prepared by direct homogenization of the starting material as in Section 3.2.1., since the method does not require [³H]inositol prelabeling of brain slices.

3.3.2. PLC Assay

In a final volume of 100 μL, the assay contains the same components as in Section 3.1.3. (except that membranes are unlabeled) plus [³H]inositol (0.5 μCi) and 1 mM CMP. Additions are as follows:

1. 5 μL of 20 mM Sodium deoxycholate.
2. 40 μL of 25 mM Tris-maleate/7.5 mM EGTA and $CaCl_2$. The optimum free calcium concentration is 100 nM, a compromise between the extremely high sensitivity of PtdIns synthase to inhibition by calcium (IC_{50} around 200 nM) and the facilitating effects of the ion on PLC activation by agonists *(9)*.
3. 5 μL of 20 mM CMP in 25 mM Tris-maleate. PtdIns could also be potentially labeled by [³H]inositol through an enzymatic exchange mechanism distinct from PtdIns synthase and therefore independent of CMP. However, deoxycholate at 1 mM inhibits this reaction by 80%, so that labeling of phosphoinositides becomes essentially dependent on the presence of the nucleotide.
4. 5 μL of [³H]inositol (0.1 mCi/mL), diluted in 25 mM Tris-maleate.
5. 2.5 μL each of 40X concentrations of GTPγS and agonist in 25 mM Tris-maleate.
6. Start reactions by adding 40 μL of membranes (2.5 mg protein/mL in resuspension buffer). Vortex and incubate at 37°C.

At least, three concerted reactions have to take place for the generation of ³H-inositol phosphates, and therefore, these appear after a lag of 10 min. The reaction is linear thereafter for at least 50 min (Fig. 3). Reactions are stopped as before, and 1-mL samples of the aqueous phases are processed either for Dowex chromatography or HPLC to separate ³H-inositol phosphates from excess [³H]inositol, which may represent 99% of the total radioactivity after a 30-min incubation. This relatively low efficiency of [³H]inositol conversion to ³H-inositol phosphates can be improved if lipid labeling and PLC stimulation are conducted separately. Membranes (2.5 mg protein/mL) are incubated 30 min at 37°C in 20 mM Tris-HCl, pH 7.0, with 1 mM EGTA, 2 mM ATP, 6 mM $MgCl_2$, and 1 mM CMP (this buffer can be stored at –20°C), containing [³H]inositol (5 μCi/mL), then centrifuged and assayed as in Section 3.1.3.

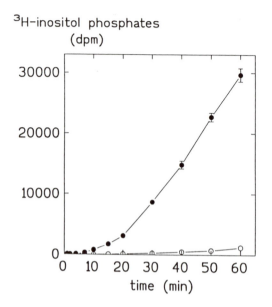

Fig. 3. Time-course of the coupled PtdIns synthase and PLC reactions. Rat brain cortical membranes (100 µg protein) were incubated with 1.5 µCi [³H]inositol and 1 m*M* CMP, and in the presence (●) or absence (○) of 1 µ*M* GTPγS and 1 m*M* carbachol. Reproduced with permission from ref. *9*.

4. Notes

1. Optimum responses to GTPγS alone or in combination with agonists in terms of signal-to-noise ratio are obtained at free calcium concentrations of 100–200 n*M*. Total calcium in the assay is calculated from the expression

$$Ca_T = Ca_F (EGTA\ K_a + Ca_F\ K_a + 1)\ /\ (1 + Ca_F\ K_a)$$

   where Ca_F is the desired free calcium molar concentration, EGTA is $3 \times 10^{-3} M$ and K_a is the association constant of EGTA and calcium ($10^6/M$ at pH 6.8).

2. The assay is very sensitive to pH changes, since acidifications will release calcium from the EGTA buffer, and this in turn will stimulate PLC. Therefore, agonist solutions should be checked for pH, particularly when they are supplied in the hydrochloride form. To avoid false agonist stimulations of PLC, controls with agonist, but in the absence of GTPγS, should be run. Of course, a true agonist effect is demonstrated by the inhibition with specific receptor antagonists. Antagonists are preincubated with the membranes

before the assay, and to avoid dilution, they are added as well in 12.5 µL Tris-maleate, pH 6.8 (agonists should then be added 20X in 12.5 µL).

3. On stimulation of PLC, [^{3}H]Ins(1,4)P$_2$ and ^{3}H]Ins(1,4,5)P$_3$ are the main products, which appear with no apparent lag. As the product of a marginal hydrolysis of PtdIns, some [^{3}H]Ins(1)P is formed as well. When membranes are prepared with less energetic procedures than tissue homogenization, [^{3}H]Ins(4)P is also formed as a product of [^{3}H]Ins(1,4)P$_2$ dephosphorylation. Conversion of [^{3}H]Ins(1,4,5)P$_3$ to [^{3}H]Ins(1,4)P$_2$ can be prevented by adding unlabeled Ins(1,4,5)P$_3$. Under these conditions, the generation of [^{3}H]Ins(1,4)P$_2$ still predominates over that of [^{3}H]Ins(1,4,5)P$_3$, revealing the balanced contribution of PtdInsP and PtdInsP$_2$ breakdown to the response (Fig. 2). The use of inhibitors of Ins(1,4,5)P$_3$ 5-phosphatase, like 2,3-bisphosphoglycerate, is not recommended, since they may inhibit PLC at the high concentrations required for their action.

Acknowledgment

The assays described in this work were developed with the financial support of CAICYT Grant PR85-177 to F. Picatoste, E. Claro, and Dr. Agustina García, DGICYT Grant PB91-503 to F. Picatoste, and E. Claro, and USPHS Grant DK 37004 to Professor John N. Fain. E. Claro is recipient of a postdoctoral fellowship RE90 from the Spanish Ministry of Education and Science.

References

1. Berridge, M. J. (1993) Inositol trisphosphate and calcium signalling. *Nature (Lond.)* **361,** 315–325.
2. Fisher, S. K. and Agranoff, B. W. (1987) Receptor activation of inositol lipid hydrolysis in neural tissues. *J. Neurochem.* **48,** 999–1017.
3. Fisher, S. K., Heacock, A. M., and Agranoff, B. W. (1992) Inositol lipids and signal transduction in nervous systems: an update. *J. Neurochem.* **58,** 18–38.
4. Sternweis, P. C. and Smrcka, A. V. (1992) Regulation of phospholipase C by G proteins. *Trends Biochem. Sci.* **17,** 502–506.
5. Eberhard, D. A. and Holz, R. A. (1988) Intracellular Ca^{2+} activates phospholipase C. *Trends Neurosci.* **11,** 517–520.
6. Claro, E., Wallace, M. A., Lee, H. M., and Fain, J. N. (1989) Carbachol in the presence of guanosine 5'-*O*-(3-thiotriphosphate) stimulates the breakdown of phosphatidylinositol 4,5-bisphosphate, phosphatidylinositol 4-phosphate, and phosphatidylinositol by rat brain membranes. *J. Biol. Chem.* **264,** 18,288–18,295.
7. Claro, E., García, A., and Picatoste, F. (1989) Carbachol and histamine stimulation of guanine nucleotide-dependent phosphoinositide hydrolysis in rat brain cortical membranes. *Biochem. J.* **261,** 29–35.

8. Claro, E., Sarri, E., and Picatoste, F. (1993) Endogenous phosphoinositide precursors of inositol phosphates in rat brain cortical membranes. *Biochem. Biophys. Res. Commun.* **193,** 1061–1067.

9. Claro, E., Wallace, M. A., and Fain, J. N. (1992) Concerted CMP-dependent [^{3}H]inositol labeling of phosphoinositides and agonist activation of phospholipase C in rat brain cortical membranes. *J. Neurochem.* **58,** 2155–2161.

10. Wallace, M. A. and Claro, E. (1993) Transmembrane signaling through phospholipase C in human brain membranes. *Neurochem. Res.* **18,** 139–145.

11. Carter, H. R. and Fain, J. N. (1991) Muscarinic cholinergic stimulation of exogenous phosphatidylinositol hydrolysis is regulated by guanine nucleotides in rabbit cortical membranes. *J. Neurochem.* **56,** 1616-1622.

12. McPhee, F., Lowe, G., Vaziri, C., and Downes, C. P. (1991) Phosphatidylinositol synthase and phosphatidylinositol/inositol exchange reactions in turkey erythrocyte membranes. *Biochem. J.* **275,** 187–192.

CHAPTER 16

Measurement of Diacylglycerol

Origins of Receptor-Stimulated Diacylglycerol Accumulation

Michael R. Boarder and John R. Purkiss

1. Introduction

Diacylglycerol (DAG) found within cells is derived from two broad sources. First, in the *de novo* biosynthetic pathway, glycerol 3-phosphate is converted by two acylation steps to phosphatidic acid (PA), which is converted to DAG by phosphatidate phosphohydrolase (PPH). This is part of the pathway for synthesis of triacylglycerol and phospholipids from the carbon backbone of glycerol. Second, DAG is formed from phospholipids by the actions of phospholipases. It is this second process that is directly regulated by cell-surface receptors, and to which this chapter is addressed.

The receptor regulation by phospholipases can itself be divided into two pathways, the single-step formation of DAG by phospholipase C (PLC) and the two-step process of sequential phospholipase D (PLD) and PPH activities. DAG formation by PLC is most commonly discussed in the context of phosphatidylinositol 4,5-bisphosphate hydrolysis. Here PLC will generate inositol 1,4,5-trisphosphate and DAG as its primary water-soluble and lipid products, respectively. By contrast, the action of PLD is most commonly reported to be on phosphatidylcholine, generating PA as its primary lipid product. The question of the nature of the lipid substrate is of significance to the methods discussed in Section 3. because of the differential acyl substitutions. For example, the inositol phospholipids are principally 1-steroyl, 2-arachidonyl, whereas palmitoyl

From: *Methods in Molecular Biology, Vol. 41: Signal Transduction Protocols*
Edited by: D. A. Kendall and S. J. Hill Copyright © 1995 Humana Press Inc., Totowa, NJ

and myristoyl are preferentially incorporated into phosphatidylcholine. This difference is reflected in the design of two different types of studies: those labeling the phospholipids with [³H] or [¹⁴C] fatty acids and those analyzing the fatty acid content of agonist-stimulated DAG. Useful information on patterns of acyl substitutions on different parent lipids can be found in recent publications *(1–4)*. The following points should be noted. Although there is a predominance of these acyl substitutions in the phospholipid mentioned, this is not absolute (e.g., there are some nonarachidonate substitutions in the sn-2 position of inositol phospholipids). There is a difference between cell types with respect to acyl substitutions on the phospholipids (e.g., compare refs. *3* and *4*). Also, it is highly likely that receptor-stimulated events mainly take their phospholipid substrates from specific subcellular pools that may have differing lipid structures. These points indicate limitations on using acyl labeling or acyl analysis to indicate the parent phospholipid. Furthermore, there are widespread reports of PLC and PLD acting on lipids other than those indicated. For example, in some cases, PLD may act on inositol phospholipids *(5)*, and in others, PLC may act on phosphatidylcholine *(6)*. This means that even if the parent lipid structures are known, the nature of acyl substitutions on DAG would not serve as a reliable indicator of the phospholipase responsible.

As a route for DAG synthesis, PLD needs to be linked to PPH activity, which will convert the PA initially generated into DAG. PPH is a regulated, and sometimes rate-limiting enzyme *(7,8)*, which has been relatively little studied despite being the final step in one of the two routes for receptor-regulated DAG synthesis.

DAG is a subset of a class of lipids called diradylglycerols, which includes ether-linked (alkyl and alkenyl) substitutions on the glycerol backbone (abbreviated to AAG), as well as the ester-linked acyl substitutions of DAG. These AAG species are known to be naturally occurring and involved in the receptor-regulated responses. Some reports describe the separate analysis of DAG and AAG.

What follows is an account of the measurement of DAG by radiolabeling and mass assay procedures, with some comments on other analytical procedures. Finally, the problem of distinguishing between agonist-stimulated DAG by the PLC and the PLD/PPH pathways is addressed.

2. Materials

2.1. DAG Analysis by Labeling Cells

1. Balanced salt solution (BSS): 3.64 g NaCl, 0.2 g KCl, 0.68 g NaHCO$_3$, 0.1 g MgSO$_4$, 0.5 g glucose, 3.6 g HEPES, and 0.08 g NaH$_2$PO$_4$. Adjust pH to 7.4 with 5M NaOH, make up to 500 mL, and then add 0.14 g CaCl$_2$. Sterilize by filtration through 0.2-μm filter.
2. [5,6,8,9,11,14,15-^{3}H]Arachidonic acid (150–230 Ci/mmol), [2-^{3}H]glycerol (0.5–1 Ci/mmol), or [9,10(n)-^{3}H]palmitic acid 40–60 Ci/mmol).
3. Medium 199 (M199) with supplements (2 mM glutamine, 25 IU/mL penicillin, 25 μg/mL streptomycin) or Dulbecco's Modified Eagle's Medium (DMEM): Ham's F12 nutrient mixture (1:1 [v/v]) with the same supplements.
4. Methanol, ice cold.
5. Chloroform.
6. Chloroform:methanol (95:5 [v/v]).
7. Whatman LK6D silica gel 60 thin-layer plates.
8. Thin-layer chromatography (TLC) solvent systems:
   a. Diethyl ether:hexane:acetic acid (70:30:1 [v/v/v]).
   b. Diethyl ether:hexane:acetic acid (50:50:1 [v/v/v]).
   c. A two solvent system (*see* Section 3.1., step 8c).
      i. Ethylacetate:acetic acid:trimethylpentane (9:2:5 [v/v/v]).
      ii. Hexane:diethylether:methanol:acetic acid (90:20:3:2 [v/v/v/v]).
9. TLC standards: DAG, monoacylglycerol, triacylglycerol (1 mg/mL in chloroform:methanol, 95:5 [v/v]).
10. TLC plate oxalating solution: 5 g potassium oxalate dissolved in 250 mL H$_2$O, made up to 500 mL with methanol.
11. Elemental iodine.

2.2. DAG Analysis by Enzymic Mass Assay

1. BSS (*see* Section 2.1., item 1).
2. Methanol, ice cold.
3. Chloroform.
4. Chloroform:methanol (95:5 [v/v]).
5. DAG, 50–1000 pmol/10 μL of chloroform.
6. Cardiolipin, 100 pmol/10 μL of chloroform.
7. 0.75 g Octylgluconide in 10 mL 1 mM of diethylenetriamine penta-acetic acid (DETAPAC: 0.393 g DETAPAC in 1L H$_2$O).
8. Reaction buffer: 6.8 g imidazole, 5.8 g NaCl, 2.375 g MgCl$_2$, and 0.76 g EGTA, pH to 6.6 with HCl. Make up to 1L with H$_2$O.

9. 2 m*M* Dithiothreitol: 0.308 g in 100 mL H$_2$O.
10. *E. coli* DAG kinase (Lipidex USA): 1 mg in 1 mL H$_2$O.
11. 10 m*M* Imidazole: 0.068 g imidazole in 100 mL of 1 m*M* DETAPAC solution (*see* step 7).
12. 11.1 m*M* ATP: 0.0565 g in 10 mL H$_2$O.
13. [^{32}P]ATP (2.5 mCi/mL).
14. Methanol:chloroform (2:1 [v/v]).
15. 1*M* NaCl, 5.8 g NaCl in 100 mL H$_2$O.
16. 1% (w/v) Perchloroacetic acid.
17. TLC solvent system. Chloroform:acetone:methanol:acetic acid:H$_2$O (10:40:3:2:1 [v/v/v/v/v]).
18. Thin-layer plates (as in Section 2.1., item 10).
19. TLC standard: PA, 1 mg/mL of chloroform:methanol (95:5 [v/v]).
20. Elemental iodine.

2.3. Distinguishing Between PLD- and PLC-Derived DAG

1. BSS (*see* Section 2.1., item 1).
2. [9,10(*n*)-^{3}H]Palmitic acid (40–60 Ci/mmol).
3. 10X final concentration of: 0–300 m*M* ethanol, 0–1.38 g in 10 mL H$_2$O; 0–100 m*M* butanol, 0–0.74 g in 10 mL H$_2$O.
4. 100 n*M* Phorbol 12-myristate 13-acetate (PMA). Make up stock solution (1 m*M*) of 1 mg PMA dissolved in 1.62 mL dimethyl sulfoxide. Use 5 µL of this in 5 mL BSS to give 1 µ*M* working solution. Use this at 100 µL/mL of incubating solution to give 100 n*M*.
5. Methanol, ice cold.
6. Chloroform.
7. Chloroform:methanol (95:5 [v/v]).
8. Whatman LK6D thin-layer plates.
9. TLC solvent system. Ethylacetate:acetic acid:trimethylpentane (9:2:5 [v/v/v]).
10. TLC standards: PA, phosphatidylethanol (PBut), phosphatidylbutanol (PBut, each at 1 mg/mL of chloroform:methanol (95:5 [v/v]).
11. TLC plate oxalating solution (*see* Section 2.1., item 10).

3. Methods

Here we assume that the requirement is for the assay of DAG in cultured cells. Typically, the experiments would be done with cells on 24-well plates, and the procedures are suitable for handling the samples generated from experiments requiring more than one (but not exceeding four) such plates. The procedures have been used on a variety of cell types, but in our hands have been mainly developed for adrenal chromaffin cells and aortic endothelial cells, both of bovine origin. The procedures described in some detail are DAG analysis by labeling of cells (with [^{3}H]glycerol, [^{3}H]arachidonate, or [^{3}H]palmitate), and the mass assay of DAG by an enzymic procedure

3.1. DAG Analysis by Labeling of Cells

Experimental procedures are essentially the same for labeling the glycerol backbone of the phospholipids with [^{3}H]glycerol or the acyl substitutions with [^{3}H]arachidonate or [^{3}H]palmitate. The DAG spot on the TLC plate at the end of the analysis has more interfering labeled species with glycerol than with the fatty acids. Glycerol will label phospholipids independent of their molecular form, whereas arachidonate and palmitate would be expected to label preferentially inositol phospholipids and phosphatidylcholine, respectively.

For a change in radioactivity in DAG to reflect a change in mass of the lipid and not a change in turnover, it is necessary to label to equilibrium (*see* Note 2). Time of labeling necessary to achieve apparent equilibrium should be determined for each cell type studied and for each label used, although in our experience, the equilibrium time for a given label is similar across cell types. To establish the apparent equilibrium labeling time, the cells in 24-well multiwells are incubated in the presence of the labeled precursor and extracted into methanol/H$_2$O/chloroform as described in the following steps 6 and 7. The organic layer is analyzed by one of two methods. An aliquot of the chloroform layer is counted directly to determine radioactivity incorporated into total lipid. Although this gives an indication of equilibrium of the pooled lipid, it is preferable to estimate equilibrium of separated lipid species. To achieve this, the extract is dried under a stream of N$_2$ and taken into 100 µL chloroform/methanol (95:5). This is then applied to a silica gel G TLC plate with lipid standards and developed in chloroform/methanol/acetic acid/H$_2$O (50:25:8:4). The position of lipid standards are determined by iodine staining, and the spots scraped and counted. The significance of this separated equilibrium determination can often be seen in results that are very different for each lipid species. This preliminary work will enable the selection of a suitable labeling period for the investigations into stimulation by different agonists using the following protocol.

Each of the steps described in this protocol should be at 37°C. Where this is not in a CO_2 incubator, temperature is maintained in a water bath.

1. Wash cells in 24-well multiwells twice with 1 mL of sterile BSS.
2. Remove last wash and add 0.5 mL of either 1 µCi/mL [5,6,8,9,11,14,15-^{3}H]arachidonic acid, 10 µCi [2-^{3}H]glycerol, or 10 µCi/mL [^{3}H]palmitic acid made up in M199 medium supplemented with glutamine (2 mM), penicillin (25 IU/mL), and streptomycin (25 µg/mL). M199 can be replaced with DMEM:F12 (50:50 [v/v]).
3. Place in CO_2 incubator for the time required for labeling to apparent equilibrium, determined separately for each label as just described. Typical times have been 8 h for [^{3}H]arachidonic acid, 24–48 h for [^{3}H]palmitic acid, and 48 h for [^{3}H]glycerol.
4. At end of the labeling period, wash cells with 1 mL M199 and return to the incubator for a "chase" period. The optimum chase period should be determined separately for each label; the chase period will enable further incorporation into lipids and reduce the amount of unincorporated label that is applied to the TLC plates. It is typically 1–2 h.
5. At the end of the chase period, wash cells with 1 mL BSS followed by incubation with the required stimulating drugs in 0.5 mL of BSS for the stimulation period required. In a typical experiment investigating agonist stimulations of DAG, this is likely to be between 5 s and 10 min. In any experimental program, a time-course is likely to be the first requirement. Preincubations, when necessary, can be substituted for the last wash.
6. Stop the reaction with 0.5 mL ice-cold methanol. Transfer each multiwell to ice on conclusion of the plate. Still on ice, scrape each well; the plunger of a 1-mL disposable syringe has been found to be useful for the wells of 24-well plates. Transfer the content of each well to an Eppendorf tube on ice; wash each well with 0.45 mL H_2O ensuring that all cellular debris is transferred in the process.
7. Add 0.5 mL chloroform to each Eppendorf. Vortex vigorously to form an emulsion, and spin for 1 min in a microfuge. Remove all the upper aqueous phase and cell debris, and dry the organic phase under a stream of N_2.
8. Prepare silica gel 60 TLC plates (Whatman LK6D preabsorbant prescored with 19 channels) and solvent systems. The following solvent systems are useful for the labeled DAG experiments:
   a. Diethyl ether:hexane:acetic acid (70:30:1) used with arachidonic acid labeling. Typical Rf values were 0.64 for DAG, 0.79 for arachidonic acid, and 0.17 for monoacylglycerol, whereas triacylglycerol ran with the solvent front and phospholipids remained at the origin. This system

can also be used with [³H]gycerol, since this remains at the origin. [³H]Palmitate is less satisfactory in this system, since it runs ahead of, but close to, DAG.

  b. Diethyl ether:hexane:acetic acid (50:50:1) gave a greater separation of arachidonic acid (*Rf* 0.45) from DAG (*Rf* 0.32).
  c. Ethylacetate:acetic acid:trimethylpentane (9:2:5) was run half way up oxalate-coated silica gel plates. The plates were then run to the top of the plate in a second system (hexane:diethyl ether:methanol:acetic acid [90:20:3:2]). This two-solvent system has been used with [³H]glycerol and [³H]palmitate. Oxalate coating was achieved by spraying the plates with oxalating solution until saturated and then oven drying.

9. To each dried extract add 80 µL 95:5 chloroform/methanol ([v/v]). Standards can be added at this stage to some samples (10 µL of 1 µg/µL in chloroform:methanol). Vortex and apply to preabsorbent loading strip of each TLC lane in two or more aliquots. When dry, place in TLC tanks pre-equilibrated with about 100 mL of the solvent system. Whatman No. 1 filter papers placed against the sides and dipping in the solvent system help to maintain a saturated atmosphere. Run each plate to the top.

10. Dry the plates in a fume hood, and place in a TLC tank containing crystalline iodine for about 30 min to visualize standards. When stained, mark their position with a pencil.

11. Scrape spots for DAG into a small scintillation vial, add 5 mL scintillant, vortex well, and count after leaving for at least 1 h (*see* Note 2).

3.2. DAG Analysis by Enzymic Mass Assay

The following procedure is an adaptation of those described in Kennerly et al. *(9)*, Preiss et al. *(10)*, Walsh and Bell *(11)*, and Wright et al. *(12)*. The procedure involves the phosphorylation of DAG by DAG kinase in the presence of [³²P]ATP to form [³²P]PA, which is separated by extraction and subsequent TLC.

1. Wash cells twice with 1 mL BSS, preincubate when required, and incubate with and without agonists. Stop the reaction by the addition of 0.5 mL cold methanol, and extract cells into chloroform as indicated in Section 3.1., steps 7 and 8. The organic phase is dried under nitrogen and taken into 95:5% chloroform/methanol. Various dilutions of this cellular extract may be made up in adapting the assay to a particular use; this is important not only to get within the limits of the standard curve, but also to achieve the right lipid levels for micelle formation. The lipid constitution varies between cells, so this must be undertaken for each cell type.

2. Place aliquots of cell extracts, or known amounts of DAG for the standard curve (*see* Note 3), into Eppendorf tubes with 0.1 μmol cardiolipin (Avanti Lipids, Birmingham, AL); dry down under a stream of N_2. Cardiolipin acts as a cofactor for the DAG kinase.

3. Add 20 μL of 7.5% octylglucoside in 1 mM DETAPAC, sonicate in a bath sonicator for 5 min, and shake at 25°C for 30 min. This process initially solubilizes the diacylglycerol in the detergent solution and then allows mixed micelle formation, so that the water-soluble enzyme can reach the diacylglycerol in the micelles.

4. Make up the reaction mixture as follows:
   a. 1 mL Reaction buffer of 100 mM imidazole, 100 mM NaCl, 25 mM $MgCl_2$, and 2 mM EGTA, pH to 6.6.
   b. 200 μL of 20 mM Dithiothreitol.
   c. 100 μL of 1 mg/mL *E. coli* DAG kinase.
   d. 100 μL of 10 mM Imidazole in 1 mM DETAPAC.
   e. 180 μL of 11.1 mM ATP.
   f. 20 μL of [^{32}P]ATP (2.5 mCi/mL).

5. Add 80 μL of this extract to each tube, vortex well, and incubate for 60 min at 25°C. Stop reaction with 0.5 mL methanol:chloroform (2:1), and leave on ice for 15 min. Add 167 μL chloroform and 267 μL of 1M NaCl, vortex, and spin in a microfuge for 5 min.

6. Discard the top layer, and wash the bottom organic layer twice with 1 mL 1% perchloroacetic acid to remove residual [^{32}P]ATP. Dry under N_2.

7. Take into 100 μL of 95:5 chloroform:methanol and apply with a PA standard 10 μL of 1 mg/mL) to the preabsorbent strip of Whatman silica gel G plates (*see* Section 3.1., step 8). Develop plates in chloroform:acetone:methanol:acetic acid:water (10:40:3:2:1).

8. Visualize standards with iodine vapor (see Section 3.1., step 10), and scrape and count the PA spots (*see* Notes 4–6).

3.3. DAG Analysis by Other Methods

Of other approaches that can be applied to the analysis of agonist-stimulated diradylglycerol levels, the most significant is the quantitative analysis of molecular species, defined by their radyl substitutions. These studies have been based on a combination of TLC separations and high-performance liquid chromatography (HPLC) of derivatized acyl groups, with gas chromatography of column eluates to identify the molecular species in each group *(1,3,4)*. These procedures provide an important advance in the information that is available about agonist-stimulated diradylglycerol levels. Their requirements for instrumentation and analysis time beyond that available for many projects is likely to result in limited application of these methods. However, their importance lies not only in the understanding of the biochemical origins of agonist-stimulated DAG or AAG, but also in the significance of these changes. For example, it has been suggested that the phospholipid of origin determines the ability of stimulated DAG levels to result in activation of particular isoforms of protein kinase C *(13)*. This must refer to the radyl substitutions, resulting in hypotheses that can only be addressed by direct analysis of molecular species of diradylglycerols.

3.4. Distinguishing Between PLD- and PLC-Derived DAG

Although PLC is the sole phospholipase that will form DAG directly from phospholipids, PLD is now known to be widely encountered as an enzyme that is stimulated in response to agonists. In conjunction with PPH, this can lead to agonist-stimulated DAG formation. PPH is probably a ubiquitous enzyme because of its endoplasmic reticulum-located role in lipid metabolism indicated at the beginning of this chapter. However, there is evidence that a different plasmalemma enzyme also exists, and that this enzyme is subject to regulation *(7,14)*. It is, therefore, the case that when agonist-stimulated DAG formation is investigated, it is important to establish whether it comes from PLC or PLD/PPH. Initially, the stimulation of PLD by the agonist should be investigated, by methods recently reviewed *(15)*. (The occurrence of plasmalemma PPH may be verified as indicated in Jamal et al. *[14]*, although in most studies, this has not been undertaken.) Given that PLD stimulation can be observed, then the possibility that PLD contributes to the elevated levels of DAG must be considered.

The selective labeling of acyl substitutions thought to occur predominantly in one parent lipid that is most commonly the substrate of either PLC or PLD may provide suspicions as to which pathway is involved, but it cannot provide unequivocal evidence. Analysis of molecular species of radyl substitutions in parent lipids and diradyglycerols can provide a clearer indication of the origin of the DAG, but not of the pathway involved.

No selective inhibitor of PLD is available. However, PLD does catalyze a unique reaction in the presence of primary alcohols in which the transphosphatidylation activity forms a phosphatidylalcohol instead of free phosphatidate; this, therefore, prevents the formation of DAG by the action of PPH. Widely used as the basis of assay for PLD activity, it can also be used to indicate the origin of agonist-stimulated DAG. To prevent formation of PLD fully, it is likely that high concentrations, such as 250 mM ethanol or 50 mM butanol, are necessary to divert essentially all the product of the PLD reaction away from possible formation of DAG. In the first instance, [³H]palmitate is generally applicable since phosphatidylcholine is the most frequently described PLD substrate. We have also used ³²Pi labeling as described in ref. *15;* this may be used to assess the diversion to the phosphatidylation reaction by a given concentration of alcohol, but of course cannot be used in the subsequent studies on DAG. To assess the extent of diversion of the PLD reaction, two approaches may be considered:

1. The use of increasing concentrations of alcohol, to determine when a maximal formation of phosphatidylalcohol is achieved. It may be assumed that there is no PLD activity resistant to the influence of alcohols and, therefore, that at the stage of maximal effect, all the PLD activity has been diverted away from possible DAG formation.
2. In many systems, protein kinase C-stimulating phorbol esters, such as PMA, inhibit agonist-stimulated PLC responses and activate PLD. If this has been established, then increasing concentrations of alcohols can be assessed for their ability to inhibit PMA-stimulated phosphatidic acid formation; when this has been prevented, then this concentration of alcohol has been established as suitable for effectively inhibiting PLD-derived DAG. Phosphatidylalcohol production can also be measured and should inversely reflect the decrease in PA production.

The following protocol can be used in [³H]palmitate-labeled cells to establish the optimal concentration of alcohol for diversion of PLD activity using the aforementioned approaches.

1. Label cells to equilibrium with [³H]palmitic acid as described in Section 3.1., step 2. Wash cells twice with 1 mL BSS and then incubate at 37°C in BSS containing the appropriate concentration of alcohol by addition of an aliquot of 10X final concentration of alcohol to the final wash. Suitable final concentration ranges are 0–300 mM ethanol and 0–100 mM butanol (*see* Note 7).
2. Remove this preincubate, and stimulate for 5 min with BSS containing the appropriate concentration of alcohol with and without agonist or PMA (100 nM of the latter should be maximal).

3. Terminate reaction with 0.5 mL ice-cold methanol; extract lipids into chloroform, and dry down as described in Section 3.1., steps 6 and 7.

4. Take dried extracts into 80 μL of chloroform:methanol (95:5), and apply to loading strips of oxalate-treated silica gel 60 TLC plates, prepared as described in Section 3.1., step 9.

5. Apply standards with the samples: PA is commercially available, but phosphatidylalcohols must be prepared by the method described in ref. *15*. Develop to the top of the TLC plates with ethylacetate:acetic acid:trimethylpentane (9:2:5) and visualize standards with I_2 vapor. PA has an *Rf* value of 0.34, and that for PBut is 0.58. PEt migrates more slowly than PBut, which combined with the fivefold lower potency of ethanol as an acceptor in the transphosphatidylation reaction makes butanol the alcohol of choice for these experiments.

Once this concentration of alcohol has been established, then the assay of agonist-stimulated DAG formation, by one of the procedures discussed in Sections 3.1., 3.2., and 3.3., can be undertaken in the presence and absence of alcohol. In our experience, 50 mM butanol has proven effective in a wide variety of cell types. The only complication for the DAG assays based on labeling of phospholipids applied in this way is that it is necessary that phosphatidylalcohols as well as PA do not interfere in the separation system used. In diethyl ether:hexane:acetic acid (70:30:1) described in Section 3.1., step 8a, both PA and PBut remain at the origin; this is therefore a suitable system for this application.

4. Notes

4.1. DAG Analysis by Labeling of Cells

1. In most cases, it will be necessary in the first instance to scrape and count entire lanes of TLC plates (e.g., in 0.5-cm segments) in order to determine the distribution of radioactivity. Once it is assessed that using a particular combination of cell type, radiolabel, and TLC system the radioactivity in the DAG spot forms a clean peak, then counting of entire lanes should be unnecessary.

2. The labeling procedures, such as that described in Section 3.1., enable the determination of changes in radioactive DAG that should correlate with changes in the amount of DAG present in the cells, assuming that the pool from which the DAG is formed is in true equilibrium, and that this is not disturbed during the procedure. The labeling procedures cannot, however, tell us how much DAG is present. To achieve this, we must use mass assays, such as the enzymic procedure described.

4.2. DAG Analysis by Enzymatic Mass Assay

3. Standard curves should be linear between 0 and 1000 pmol.
4. When DAG standards were used in the absence of cell extracts, then only one band of radioactivity, corresponding to the PA standard, could be seen. For this condition, then, it was unnecessary to run the TLC; the washed organic phase could be counted directly to generate a standard curve.
5. When cell extracts were used, three spots could be seen, corresponding to the *Rf* values for PA (0.92), ceramide phosphate (0.48), and *lyso*-PA (0.06), (ceramide and monoacylglycerol are known substrates of *E. coli* DAG kinase).
6. To confirm that micelle formation in the presence of cell extracts is permissive of the same degree of activity as the standard curve, known amounts of DAG should be added to aliquots of cell extracts for each application of the assay. In fact, since the reaction is intended to go to completion during the long (60 min) incubation, there is room for some decrease in reaction efficiency without compromising the results.

4.3. Distinguishing Between PLD- and PLC-Derived DAG

7. It is necessary to consider whether these concentrations of alcohol are selective with respect to effects on PLD. Although difficult to ascertain in many circumstances, we have investigated one cell type, the adrenal chromaffin cell, in which we find no evidence of agonist-stimulated PLD *(16)*. In this case, the presence of 50 m*M* butanol has no effect on the receptor-regulated PLC and protein kinase C events, which we have investigated.
8. However, as with any assay dependent on inhibitors, the procedure may modulate those processes that it is intended to investigate. We have reported in one cell type that there is a large protein kinase C-dependent feedback loop that attenuates agonist-stimulated PLC, and that this feedback is partly dependent on PLD *(17)*. Interference with the PLD reaction by butanol, therefore, enhances the PLC reaction *(17)*, so that in this cell type, the use of butanol to assess the relative contribution of PLC/PLD would result in an overestimate of the contribution that PLC makes.
9. The outcome of these points is that there is no simple way to assess the relative contribution of PLD to agonist-enhanced DAG levels. However, the use of carefully chosen [^{3}H] fatty acid to label cells, in conjunction with other procedures described, should enable an estimate of the significance of the two pathways.

Acknowledgment

We thank The Wellcome Trust for financial support.

References

1. Lee, C. and Hajra, A. K. (1991) Molecular species of diacylglycerols and phosphoglycerides and the postmortem changes in the molecular species of diacylglycerols in rat brains. *J. Neurochem.* **56,** 370–379.
2. Holbrook, P. G., Pannell, L. K., Murata, Y., and Daly, J. W. (1992) Molecular species analysis of a product of phospholipase D activity. Phosphatidylethanol is formed from phosphatidylcholine in phorbol ester and bradykinin stimulated PC12 cells. *J. Biol. Chem.* **267,** 16,834–16,840.
3. Lee, C., Fisher, S. K., Agranoff, B. W., and Hajra A. K. (1991) Quantitative analysis of molecular species of diacylglycerol and phospatidate formed upon muscarinic receptor activation on human SK-N-SH neuroblastoma cells. *J. Biol. Chem.* **266,** 22,837–22,846.
4. Pettitt, T. R. and Wakelam, M. J. O. (1993) Bombesin stimulates distinct time dependent changes in the sn-1,2-diradylglycerol molecular species profile from Swiss 3T3 fibroblasts as analysed by 3,5-dinitrobenzoyl derivatisation and h.p.l.c. separation. *Biochem. J.* **289,** 487–495.
5. Huang, C., Wykle, R. L., Daniel, L. W., and Cabot, M. C. (1992) Identification of phosphatidylcholine-selective and phosphatidylinositol-selective phospholipases D in Madin-Darby canine kidney cells. *J. Biol. Chem.* **267,** 16,859–16,865.
6. Loffelholz, K. (1989) Receptor regulation of choline phospholipid hydrolysis. *Biochem. Pharmcol.* **38,** 1543–1549.
7. Gumoz-Munoz, A., Hamza, E. H., and Brindley D. N. (1992) Effects of sphingosine, albumin and unsaturated fatty acids on the activation and translocation of phosphatidate phosphohydrolase in rat hepatocytes. *Biochim. Biophys. Acta* **1127,** 49–56.
8. Martinson, E. A., Trilivas, I., and Brown, J. H. (1990) Rapid protein kinase C dependent activation of phospholipase D leads to delayed 1,2-diglyceride accumulation. *J. Biol. Chem.* **265,** 22,282–22,287.
9. Kennerly, D. A. (1989) Diacylglycerol metabolism in mast cells. *J. Biol. Chem.* **264,** 16,305–16,313.
10. Preiss, J., Loomis, C. R., Bishop, W. R., Stein, R., Niedel, J. E., and Bell, R. M. (1986) Quantitative measurement of sn-1,2-diacylglycerols present in platelets, hepatocytes, and *ras* and *sis* transformed normal rat kidney cells. *J. Biol. Chem.* **261,** 8597–8600.
11. Walsh, J. P. and Bell, R. M. (1986) sn-1,2-Diacylglycerol kinase of *Eschericia coli.* Mixed micellar analysis of the phospholipid cofactor requirement and divalent cation dependence. *J. Biol. Chem.* **261,** 6239–6247.
12. Wright, T. M., Ranagan, L. A., Shin, H. S., and Raben, D. M. (1988) Kinetic analysis of 1,2-diacylglycerol mass levels in cultured fibroblasts. *J. Biol. Chem.* **263,** 9374–9380.
13. Leach, K. L., Ruff, V. A., Wright, T. M., Pessin, M. S., and Raben D. M. (1991) Dissociation of protein kinase C activation and sn-1,2-diacylglycerol formation: comparison of phosphatidylinositol and phosphatidylcholine derived diglycerides in alpha thrombin stimulated fibroblasts. *J. Biol. Chem.* **266,** 3215–3221.

14. Jamal, Z., Martin, A., Gomez-Munoz, A., and Brindley D. N. (1991) Plasma membrane fractions from rat liver contain a pbospatidate phosphohydrolase distinct from that in endoplasmic reticulum and cytosol. *J. Biol. Chem.* **266,** 2988–2996.

15. Boarder, M. R. and Purkiss, J. R. (1993) Assay of phospholipase D as a neuronal receptor effector mechanism, in *Neuroprotocols, vol. 3* (Bleasedale, J. E. and Fisher, S. K., eds.), Academic, New York, pp. 157–164.

16. Purkiss, J. R., Murrin, R. A. J., Owen, P. J., and Boarder M. R. (1991) Lack of phospholipase D activity in chromaffin cells: bradykinin stimulated phosphatidic acid formation involves phospholipase C in chromaffin cells but phospholipase D in PC12 cells. *J. Neurochem.* **57,** 1084–1087.

17. Challiss, R. J. A., Wilkes, L. C., Patel, V., Purkiss, J. R., and Boarder M. R. (1993) Phospholipase D activation regulates endothelin-1 stimulation of phosphoinositide-specific phospholipase C in SK-N-MC cells. *FEBS Lett.* **327,** 157–160.

CHAPTER 17

Measurement of Intracellular Free Calcium Ion Concentration in Cell Populations Using Fura-2

Philip A. Iredale and John M. Dickenson

1. Introduction

The early methods for calcium measurement involved microinjection of calcium-sensitive proteins, such as aequorin or obelin, into large cells *(1,2)* or the use of microelectrodes *(3)*. Both techniques are still employed, however, with much improved sensitivity allowing investigation of a greater range of cell types. In the early 1980s, the "Null Point method" was introduced *(4)*, which involved addition of a metallo-chromic calcium indicator (Arsenazo III) to cells permeabilized with digitonin. Using this technique, the accumulation and release of calcium from intracellular stores could be recorded. A major advance in calcium measurement was made when Tsien and his colleagues *(5,6)* introduced fluorescent calcium indicators. The first to be used was quin-2: its structure was based on the novel calcium chelator 1,2-bis-(O-aminophenoxyl-ethane-N,N,N',N'-tetraacetic acid (BAPTA) *(7,8)*, a double aromatic analog of EGTA. The major problem of inducing a hydrophilic polycarboxylate anion to cross the plasma membrane was overcome by the addition of an acetoxymethyl ester group (AM), thus producing a lipophilic, membrane-permeant molecule (quin-2 AM) that, once within the cytoplasm, was subject to attack by intracellular enzymes, which cleaved the ester bond and left the calcium-sensitive free acid trapped within the cell *(5)*. A number of improved calcium indicators have since

From: *Methods in Molecular Biology, Vol. 41: Signal Transduction Protocols*
Edited by: D. A. Kendall and S. J. Hil Copyright © 1995 Humana Press Inc., Totowa, NJ

been developed (e.g., fura-2, indo-1, and fluo-3), but the basic principles of dye loading and continuous calcium reporting remain the same.

Fura-2 was introduced in 1985 *(9,10)*, and its much improved absorption coefficient and quantum yield significantly reduced the intracellular concentrations required. Furthermore, the marked shift in excitation wavelength, following calcium binding, and the relatively high *Kd* (224 nm *[10]*), allowed more accurate, dual-wavelength recording of intracellular calcium ($[Ca^{2+}]_i$) over an improved range, from basal levels to as high as 10 μ*M*.

1.1. Dual-Wavelength Recording with Fura-2

The unbound form of fura-2 exhibits peak fluorescence following excitation at 380 nm (with an emission wavelength of approx 500 nm). Following calcium chelation, there is a shift in the excitation spectrum to a peak of 340 nm with no apparent change in the emission spectrum *(9)*. This shift is accompanied by significant changes in fluorescence intensity of both forms of the dye (i.e., there is a large increase in fluorescence at 340 nm and a large decrease in fluorescence at 380 nm). Thus, recording following excitation at both wavelengths can be used to produce a fluorescence ratio (340/380 nm) representative of $[Ca^{2+}]_i$. (There is a small shift in the excitation spectrum when quin-2 binds calcium, but unlike fura-2, the fluorescence of the unbound dye is very weak and susceptible to background interference.)

Dual-wavelength recording using fura-2 has a number of advantages *(10)*. The signal-to-noise ratio is much larger, therefore increasing the sensitivity of the indicator. Furthermore, the absolute concentration of the dye within the cells is not as important when a ratio is being recorded. For example, instrument fluctuations or loss of dye owing to bleaching or leakage, at single wavelength, might be misinterpreted as a change in $[Ca^{2+}]_i$. The 340/380 nm ratio, however, would not record such a change, since each excitation peak would be equally affected *(10)*. This lack of dependency on the intracellular dye concentration, theoretically, negates the necessity for routine calibration measurements after every experiment, although other affecting parameters would appear to necessitate at least day-to-day calibration for the production of reproducible results.

We have routinely used fura-2 to measure $[Ca^{2+}]_i$ in both suspensions and on coverslips *(11–19)* and this chapter details both methods. The assays described have been developed for specific cell types, but can be easily modified for other lines.

2. Materials

1. Fura-2/AM (Calbiochem, Nottingham, UK): 1 mM stock solution in dimethyl sulfoxide (DMSO).
2. 2M CaCl$_2$, 250 mM EGTA.
3. HEPES-buffered physiological saline solution (HBS): 145 mM NaCl, 10 mM glucose, 5 mM KCl, 1 mM MgSO$_4$, 10 mM HEPES, pH 7.4.
4. Phosphate-buffered saline (PBS) with EDTA (PBS/EDTA) consisting of 138 mM NaCl, 2.7 mM KCl, 12.9 mM Na$_2$HPO$_4$ · 2H$_2$O, and 1.5 mM KH$_2$PO$_4$, 0.53 mM EDTA, pH 7.4.
5. 5% (w/v) Stock solution of bovine serum albumin (BSA) (Fraction V, Sigma, Poole, Dorset, UK).
6. 10 × 24 mm Glass coverslips (No. 1; Richardsons, Leicester, UK).
7. Fluorimeter polymethacrylate cuvets (Sigma).
8. Fetal calf serum (FCS, Northumbria Biologicals, Cramlington, UK).
9. Ionomycin from *Streptomyces conglobatus:* free acid (Calbiochem) prepared as a 2-mM stock solution in DMSO.
10. 0.5 and 1M NaOH.
11. 0.2M MnCl$_2$.
12. Fluorimeter, e.g., Perkin-Elmer LS-50 spectrometer.
13. Straight watchmaker forceps.
14. Thermostatically controlled water bath.
15. Bench centrifuge and microcentrifuge.
16. IBM-compatible computer.

3. Methods

3.1. Measurement of [Ca²⁺]$_i$ Using Cell Suspensions

3.1.1. [Ca²⁺]$_i$ Measurements

1. The cells (N1E-115 neuroblastomas) are grown in 75-cm^2 flasks until they reach confluency (*see* Note 2).
2. Detach the monolayers from two flasks gently using PBS/EDTA, and centrifuge at approx 700g for 5 min. Resuspend the resultant pellet in 7.5 mL of HBS containing 2 mM Ca^{2+} and 10% (v/v) FCS. Add 37.5 µL of the fura-2/AM stock solution (giving a final concentration of 5 µM), and incubate in a water bath at 37°C for 20 min (*see* Note 1). During this period, the cells need to be kept in suspension by gently agitating the suspension every 5 min. After the initial loading period, dilute the cell suspension threefold using prewarmed HBS containing 10% FCS (v/v) and leave for another 5 min at 37°C. This last stage helps to ensure complete hydrolysis by intracellular esterases of fura-2/AM to the calcium-sensitive free acid form.

3. Centrifuge the fura-2-loaded cell suspension at approx 700*g* for 5 min, resuspend in 10–12 mL fresh HBS containing 2 m*M* Ca^{2+} without serum, and leave at room temperature (this reduces dye leakage). We generally perform six to eight calcium time-course experiments on the cells from two confluent 75-cm^2 flasks.
4. Take a 2-mL aliquot of the cellular suspension in an Eppendorf, and rapidly pellet the cells using a microcentrifuge (a 5-s pulse at 1000*g* is generally sufficient; this step again helps to reduce the extracellular dye concentration and is essential for "leaky" cells). Resuspend in 2 mL of fresh HBS (no serum) within a polymethacrylate cuvet containing a stirrer bar. The cuvet is then placed within a thermostatically controlled cell holder.
5. Measure the fluorescence at 500 nm following excitation at both 340 and 380 nm, and ratio the values (340/380 nm). Depending on the type of fluorimeter used, this can be done by manually switching between wavelengths or using a computer-controlled system. We routinely use a Perkin-Elmer LS 50, which has a motor-driven monochromator that switches between 340 and 380 nm approx every 1.6 s and a continuous display of the 340/380 nm ratio as a function of time.
6. Once a steady baseline ratio (generally between 1.5 and 2) has been achieved, drugs can be added directly into the cuvet, and the changes in the 340/380 nm ratio observed.

3.1.2. Calibration

$[Ca^{2+}]_i$ can be determined from the 340/380 nm ratio using the equation of Grynkiewicz et al. *(10)* (*see* Note 4).

$$[Ca^{2+}]_i = K \times [(R - R_{min}) / (R_{max} - R)]$$
$$K = (K_d \times F_0)/F_s$$

where K_d = the fura-2 dissociation constant (224 n*M* at 37°C), F_0 = the 380-nm fluorescence in the absence of calcium, F_s = the 380-nm fluorescence with saturating calcium, R = the 340/380 mn fluorescence ratio, R_{max} = the 340/380 nm ratio with saturating calcium, and R_{min} = the 340/380 nm ratio in the absence of calcium.

3.1.3. Calculation of R_{max} and R_{min}

These values can be calculated at the end of any calcium time-course using the following method:

1. Permeabilize the cells to calcium by addition of the calcium ionophore, ionomycin (20 μ*M*; 10 μL of a 2-m*M* stock solution). This allows sufficient

entry of calcium (provided the extracellular calcium concentration is at least 2 mM) to saturate fully the intracellular fura-2. The resulting 340/380 nm ratio is R_{max}.

2. After a steady value for R_{max} has been obtained, a value for R_{min} can be determined by chelating all of the calcium using EGTA (6.25 mM; 200 µL of a 250-mM stock solution) and increasing the pH to >8.5 using approx 20 µL of 0.5M NaOH.

3.1.4. Autofluorescence

Before using these values in the equation, it is necessary to correct all of the data for autofluorescence. There are a number of fluorescent species that occur naturally within the cytosols of living cells (e.g., NADPH and NADP$^+$), and their contribution to the overall fluorescence signals needs to be removed.

This can be assessed by taking a separate cuvet of cells from that used to calculate R_{max} and R_{min} and again adding ionomycin (20 µM), this time followed by manganese ions (Mn^{2+}, 5 mM; 50 µL of a 0.2M solution of MnCl$_2$). Mn^{2+} enters the cells, is preferentially bound to fura-2, and quenches its fluorescence. Thus, the remaining fluorescence is the contribution from the autofluorescence of the cells. The individual 340 and 380 nm values need to be subtracted from all data, thus generating corrected values for R, R_{max} (around 30 after correction), and R_{min} (around 1 after correction), which can now be entered into the equation. The values for F_0 and F_s can be extracted from the values for R_{max} and R_{min} (Fig. 1).

3.2. Measurement of [Ca^{2+}]$_i$ Using Cell Monolayers

3.2.1. Preparation of Buffers

1. Cells are loaded with fura-2/AM (the cell-permeant ester derivative of fura-2) using HEPES-buffered HBS containing 2 mM CaCl$_2$ (10 µL of 2M CaCl$_2$/10 mL of HBS), 10% (v/v) FCS, and 3 µM fura-2/AM (30 µL of the fura-2/AM stock solution/10 mL of HBS). Each coverslip requires 1 mL of "loading" buffer.
2. [Ca^{2+}]$_i$ measurements are performed using HEPES-buffered HBS containing 0.1% BSA and 2 mM CaCl$_2$ (BSA-HBS). If nominally Ca^{2+}-free conditions are required, use HEPES-buffered HBS containing 0.1% BSA and 0.1 mM EGTA.

3.2.2. [Ca^{2+}]$_i$ Measurements

1. For these experiments, cell monolayers are grown on 10 × 24 mm glass coverslips in 90-mm Petri dishes (10–12 coverslips/dish). Cells are ready to use when the glass coverslip is covered with a confluent cell monolayer.

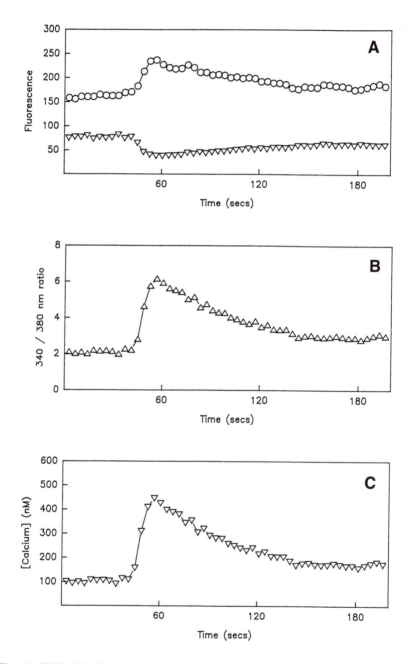

Fig. 1. CHO-NAX2 cells loaded with fura-2 and stimulated with N[6]-cyclo-pentyladenosine (100 n*M*). (**A**) The individual 340- and 380-nm fluorescence values, (**B**) the 340/380 nm ratio, and (**C**) the final calibrated trace.

2. Using straight watchmaker forceps, place individual coverslips in 35-mm Petri dishes. Add 1 mL of "loading" buffer/Petri dish, and incubate for 30 min at 37°C.

3. Remove the "loading" buffer, and add 1 mL of BSA-HBS/Petri dish. Leave at 37°C for a further 15 min. This 15-min "standing" period (in the absence of fura-2/AM) is to allow the hydrolysis of fura-2/AM into the calcium-sensitive free acid form (fura-2) by cell esterases.

4. Add 2.9 mL of BSA-HBS to a polymethacrylate cuvet (containing a magnetic stirrer bar), and place into the cell holder of the spectrometer. The cell holder can be maintained at 37°C by circulating thermostatically controlled water through the chamber.

5. Using watchmaker forceps, mount the fura-2 "loaded" coverslip into a coverslip holder (*see* Fig. 2). The holder enables the coverslip to be positioned across the diagonal of the cuvet. Note: The coverslip holders are not commercially available and will need to be specially made.

6. The basal ratio (R) of 340/380 fluorescence should be between 1.5 and 2. Ca^{2+} mobilizing agents are added to the cuvet through the hole in the top of the coverslip holder using a Finn pipet or equivalent (*see* Fig. 2). We routinely use 100-µL aliquots (to enable rapid mixing) from 30-fold stock solutions. The increase in fluorescence ratio varies greatly between cell types. For example, in DDT_1MF-2 cells, a typical response would be an increase in 340/380 from 2 to 3. In contrast, human airway and guinea pig aortic smooth muscle cells produce increases in fluorescence ratio of up to five (J. M. Dickenson, unpublished work).

3.2.3. Calibration of R_{max}, R_{min}, and Autofluorescence

1. Increases in $[Ca^{2+}]_i$ can be calculated from the ratio (R) of the 340 nm/380 nm fluorescent values using the equation of Grynkiewicz et al. *(10)* as described previously. R_{max} and R_{min} values are determined using separate coverslips under saturating and calcium-free conditions, respectively.

2. To determine R_{max} (this can be done at the end of an experimental run or on a separate coverslip), initially increase extracellular $[Ca^{2+}]_i$ to 20 mM (add 30 µL of 2M $CaCl_2$ stock solution) and then add the calcium ionophore, ionomycin (10 µM; 15 µL of a 2-mM stock solution; we normally add this in 85 µL of BSA-HBS to allow rapid mixing). Following the addition of ionomycin, there should be a large increase in the 340/380 ratio (up to 4 in DDT_1MF-2 cells).

3. Because of the high $[Ca^{2+}]_i$ used to calculate R_{max} (20 mM), R_{min} is determined on a separate coverslip using calcium-free conditions (again this can be done at the end of an experimental run or on a separate coverslip). Add 15 µL of 2 mM ionomycin (10 µM final; again in 85 µL of BSA-HBS),

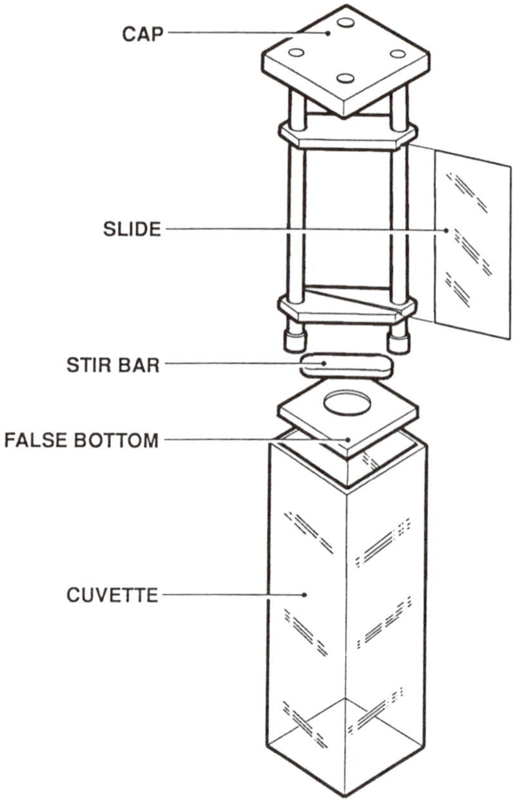

Fig. 2. Illustration of the glass coverslip holder used for measuring increases in $[Ca^{2+}]_i$ in cell monolayers. The base of a polymethacrylate cuvet has a small plastic "lump" that, unfortunately, hinders the movement of the magnetic stirrer bar. We use perspex "false bottoms," which enable the stirrer bar to move freely.

followed by 100 µL of 250 m*M* EGTA (8 m*M* final). After the addition of EGTA, immediately add 25 µL of 1.0*M* NaOH to compensate for the decrease in pH associated with the chelation of calcium by EGTA. Following the addition of EGTA, there should be a decrease in the 340/380 ratio.

4. Before applying the R, R_{max}, and R_{min} values to the Grynkiewicz et al. equation, they need to be corrected for autofluorescence. Autofluorescence is determined by simply measuring the fluorescence produced by coverslips that have not been loaded with fura-2. Values for R, R_{max}, and R_{min} are subsequently corrected by subtracting the 340- and 380-nm autofluorescence values.

5. In DDT_1MF-2 cells, corrected values for R_{max} and R_{min} were 12 and 0.7, respectively. Values for F_0 and F_s can be extracted from the corrected R_{max} and R_{min} (in DDT_1MF-2 cells, F_0/F_s was typically around 3.6).

6. Intracellular $[Ca^{2+}]_i$ can be now calculated using the software supplied with the Perkin-Elmer LS-50 spectrometer. Alternatively, the raw fluorescence data can be imported into a spreadsheet, e.g., AsEasyAs (TRIUS Inc., North Andover, MA).

4. Notes

1. Dye loading conditions: It is necessary to optimize the times and conditions of dye loading depending on the cell type under investigation. The final concentration of fura-2 can be critical, since too much dye can be toxic to the cells and too little can result in a poor signal-to-noise ratio. Likewise, the density of cells is important when using both suspensions and coverslips. Too few cells, again, reduces the signal-to-noise ratio. However, using a large number of cells per experiment can be very costly, especially when measuring $[Ca^{2+}]_i$ in suspensions.

2. Suspensions vs coverslips: Generally, measuring $[Ca^{2+}]_i$ in cell suspensions results in signals that are proportionally bigger and more reproducible than using coverslips. However, it is much cheaper to use coverslips, and most cells grow as monolayers. Therefore, using coverslips is less "stressful" to the cells. Indeed, certain cell lines can only be measured on coverslips, since removing the monolayer results in very high basal $[Ca^{2+}]_i$.

3. Dye leakage: The rate of loss of fura-2 from the cytoplasm is very much a feature of the cell type under investigation. Certain lines are no problem at all and "leak" dye into the extracellular medium only very slowly. In other cell lines (e.g., DDT_1MF-2 cells), the intracellular concentration of fura-2 decreases quite rapidly. This leads to the contribution to the overall signal from extracellular dye becoming more and more significant. This is easily overcome when measuring monolayers by staggering the loading times of individual coverslips. It can, however, be more of a problem when measuring $[Ca^{2+}]_i$ in suspensions, and can be partially overcome by spinning down aliquots of cells and replacing the extracellular buffer prior to beginning recording.

4. Calibration: Calibration of the signals can be quite problematical and is best approached with caution. When comparing responses, it is advisable to compare the 340/380 nm ratios (corrected for autofluorescence if that varies significantly from day to day) rather than the corrected calcium values. However, it is worthwhile attempting to calibrate in order to give an estimate of the absolute changes in $[Ca^{2+}]_i$ involved in the responses and the resting $[Ca^{2+}]_i$ level. If calibrated traces, however, are the preferred

means of expressing results, then it is certainly necessary to measure R_{max}, R_{min}, and autofluorescence every day.

5. Washing equipment thoroughly: Traces of ionomycin and detergents can lead to cell damage and high basal $[Ca^{2+}]_i$ levels. Therefore, it is generally advisable to rinse cuvets, coverslip holders, stirrer bars, and so forth, for at least 20 min under a running tap prior to use.

References

1. Ridgway, E. B. and Ashley, C. C. (1967) Calcium transients in single muscle fibres. *Biochem. Biophys. Res. Commun.* **29,** 229–234.

2. Campbell, A. K., Lea, T. J., and Ashley, C. C. (1979) Coelenterate photoproteins, in *Detection and Measurement of Free Ca²⁺ in Cells* (Ashley, C. C. and Campbell, A. K., eds.), Elsevier, North-Holland, Amsterdam, pp. 13–72.

3. Amman, D., Meier, P. C., and Simon, W. (1979) Ca^{2+} measurements using microelectrodes, in *Detection and Measurement of Free Ca²⁺ in Cells* (Ashley, C. C. and Campbell, A. K., eds.), Elsevier, North-Holland, Amsterdam, pp. 117–129.

4. Murphey, E., Coll, K., Rich, T. L., and Williamson, J. R. (1980) Hormonal effects on calcium homeostasis in isolated hepatocytes. *J. Biol. Chem.* **255,** 6600–6608.

5. Tsien, R. Y. (1981) A non-disruptive technique for loading buffers and indicators into cells. *Nature* **290,** 527,528.

6. Tsien, R. Y., Pozzan, T., and Rink T. J. (1982) Calcium homoeostasis in intact lymphocytes: cytoplasmic free calcium monitored with a new, intracellularly trapped fluorescent indicator. *J. Cell Biol.* **94,** 325–334.

7. Tsien, R. Y. (1980) New calcium indicators and buffers with high selectivity against magnesium and protons: design, synthesis, and properties of prototype structures. *Biochemistry* **19,** 2396–2404.

8. Cobbold, P. H. and Rink, T. J. (1987) Fluorescence and bioluminescence measurement of cytoplasmic free calcium. *Biochem. J.* **248,** 313–328.

9. Tsien, R. Y., Rink, T. J., and Poenie, M. (1985) Measurement of cytosolic free Ca^{2+} in individual cells using fluorescence microscopy with dual excitation wavelengths. *Cell Calcium* **6,** 145–157.

10. Gynkiewicz, G., Poenie, M., and Tsien, R. Y. (1985) A new generation of Ca^{2+} indicators with greatly improved fluorescence properties. *J. Biochem. Sci.* **260,** 3440–3450.

11. Iredale, P. A., Martin, K. F., Hill, S. J., and Kendall, D. A. (1992) Agonist-induced changes in $[Ca^{2+}]_i$ in N1E-115 cells: differential effects of bradykinin and carbachol. *Eur. J. Pharmacol.* **26,** 163–168.

12. Iredale, P. A., Martin, K. F., Alexander, S. P. H., Hill, S. J., and Kendall, D. A. (1992) Inositol 1,4,5 trisphosphate generation and calcium mobilisation via activation of an atypical P_2-receptor in the neuronal cell line N1E-115. *Br. J. Pharmacol.* **107,** 1083–1087.

13. Iredale, P. A., Martin, K. F., Alexander, S. P. H., Hill, S. J., and Kendall, D. A. (1992) Qualitative differences in $[Ca^{2+}]_i$ and $InsP_3$ generation following stimulation of N1E-115 cells with micromolar and millimolar ATP. *Biochem. Pharmacol.* **44,** 1479–1487.

14. Iredale, P. A., Martin, K. F., Hill, S. J., and Kendall, D. A. (1993) The effects of B-phorbol-12,13 dibutyrate on agonist-induced InsP$_3$ generation and $[Ca^{2+}]_i$ increases in N1E-115 cells: differential modulation of responses to angiotensin II and brady-kinin. *Biochem. Pharmacol.* **45,** 611–617.

15. Dickenson, J. M. and Hill, S. J. (1991) Histamine stimulated increases in intracellular calcium in the smooth muscle cell line, DDT$_1$MF-2. *Biochem. Pharmacol.* **42,** 1545–1550.

16. Dickenson, J. M. and Hill, S. J. (1992) Histamine H$_1$-receptor-mediated calcium influx in DDT$_1$MF-2 cells. *Biochem. J.* **284,** 425–431.

17. Dickenson, J. M. and Hill, S. J. (1993) Adenosine A$_1$-receptor stimulated increases in intracellular calcium in the smooth muscle cell line, DDT$_1$MF-2. *Br. J. Pharmacol.* **108,** 85–92.

18. Dickenson, J. M. and Hill, S. J. (1993) Intracellular cross talk between receptors coupled to phospholipase C via pertussis toxin-sensitive and insensitive G proteins in DDT$_1$MF-2 cells. *Br. J. Pharmacol.* **109,** 719–724.

19. Dickenson, J. M., White, T. E., and Hill, S. J. (1993) The effects of elevated cyclic AMP levels on histamine H$_1$-receptor-stimulated inositol phospholipid hydrolysis and calcium mobilization in the smooth muscle cell line, DDT$_1$MF-2. *Biochem. J.* **292,** 409–417.

CHAPTER 18

Measurement of Ca^{2+} Fluxes in Permeabilized Cells Using $^{45}Ca^{2+}$ and Fluo-3

*Robert A. Wilcox, James Strupish,
and Stefan R. Nahorski*

1. Introduction

Many cell-surface receptors via G-proteins activate phosphoinositidase C, which catalyzes the hydrolysis of phosphatidylinositol 4,5-bis-phosphate to produce the second messengers, *myo*-inositol 1,4,5-trisphosphate [Ins(1,4,5)P_3 or IP$_3$] and diacylglycerol *(1)*. Ins(1,4,5)P_3 interacts with a specific receptor populations of ligand-gated channels to mobilize non-mitochondrial intracellular Ca^{2+} stores *(1)*. Because Ins(1,4,5)P_3 is plasma-membrane-impermeant, this phenomenon was first demonstrated in permeabilized pancreatic acinar cells *(2)*, and all subsequent studies in cells have involved introduction of Ins(1,4,5)P_3 by rendering a cell population permeable *(3)*, using microinjection techniques *(4)*, or by the presentation of chemically modified membrane-permeable Ins(1,4,5)P_3 analogs, such as photolabile "caged-IP$_3$" *(5)*. An alternative approach involves disruption of the plasma membrane and preparation of microsomes from the intracellular vesicular Ca^{2+} stores *(6,7)*. However, these preparations exhibit a loss of Ins(1,4,5)P_3-responsiveness compared to cells. Here we describe a $^{45}Ca^{2+}$-release assay and a fluo-3 assay, two methods we use to monitor Ins(1,4,5)P_3-induced Ca^{2+} mobilization from nonmitochondrial intracellular Ca^{2+} stores using cytosol-like buffer (CLB) and permeabilized SH-SY5Y neuroblastoma cell populations.

From: *Methods in Molecular Biology, Vol. 41: Signal Transduction Protocols*
Edited by: D. A. Kendall and S. J. Hill Copyright © 1995 Humana Press Inc., Totowa, NJ

The $^{45}Ca^{2+}$-release assay was similar to those previously described *(3,8)*, and involves preloading a "spike" of $^{45}Ca^{2+}$ into the intracellular Ca^{2+} stores of permeabilized cells and then monitoring the resultant release of $^{45}Ca^{2+}$ induced by concentrations of $Ins(1,4,5)P_3$ and other agonists. We have utilized this assay to assess the intrinsic activity of $Ins(1,4,5)P_3$ a wide range of $Ins(1,4,5)P_3$ analogs, and other Ca^{2+}-mobilizing agents. The assay can be undertaken at a low cell density to minimize cellular metabolism of the inositol polyphosphates, or at high cell density to allow metabolic time-courses to be associated with Ca^{2+} mobilization.

The fluo-3 assay utilizes the highly fluorescent Ca^{2+}-indicator dye fluo-3 to detect the release and uptake of native Ca^{2+} from the intracellular Ca^{2+} stores. Fluo-3 fluorescence at resting cellular $[Ca^{2+}]$ (100–200 n*M*) increases 5- to-10 fold (> 1 µ*M*) following stimulation of Ca^{2+}release from intracellular stores *(9)*. Using $[^{32}P]$-$Ins(1,4,5)P_3$, we have confirmed that a tight correlation exists between $Ins(1,4,5)P_3$ concentrations and the extent and duration of Ca^{2+} mobilization *(10)*. Therefore, the time-course of $Ins(1,4,5)P_3$-induced Ca^{2+} release and metabolism can be assessed as $Ins(1,4,5)P_3$ undergoes active metabolism. Indeed, fluo-3 has recently been used to monitor Ca^{2+} release successfully from permeabilized cells *(11,12)* and cerebellar microsomes *(13)*.

The methods described here for monitoring inositol polyphosphates-induced Ca^{2+} fluxes in permeabilized SH-SY5Y cells can be readily modified for other cell types and for testing a wide range of different Ca^{2+} mobilizing agents.

2. Materials

2.1. Reagents

1. Ultrapure H_2O derived from a specialist unit (Elgastat, High Wycombe, Bucks, UK; or Millipore, Watford, Herts, UK). Prepare all solutions in plasticware, and use ultrapure water unless otherwise stated.
2. Fura-2 free acid and fluo-3 free acid (Calbiochem-Novabiochem, San Diego, CA; Molecular Probes, Eugene, OR): Prepare as 1-m*M* stock solutions.
3. $^{45}CaCl_2$ solution, 74 MBq/mL (CES3, Amersham, Amersham, UK).
4. 200 m*M* $CaCl_2$ and 200 m*M* EGTA in 400 m*M* KOH—the latter may require sonication for solubilization.
5. HEPES-buffered saline (HBS)/EDTA solution: 0.02 % (w/v) Na_2EDTA, 0.9 % (w/v) NaCl, and 20 m*M* HEPES-free acid, corrected to pH 7.2–7.4 using 20% (w/v) NaOH.
6. Adenosine 5'-triphosphate (ATP) disodium salt (Na$_2$ ATP; Grade 1, Sigma A-2383, Sigma, St. Louis, MO).

7. CLB: Prepare as a fivefold stock solution in ultrapure H_2O to yield final concentrations of 120 mM KCl, 2 mM KH_2PO_4, 5 mM $(CH_2COONa)_2$, 2 mM $MgCl_2$, and 20 mM HEPES-free acid. Because the ATP required is relatively unstable in alkaline solutions, the working CLB solution is prepared just prior to use.
8. 10 mM EGTA in 20 mM KOH.
9. Saponin or β-escin (Sigma): 1 mg/mL solution prepared fresh in working stock CLB.
10. Ionomycin from *Streptomyces conglobatus:* calcium salt (Sigma I-0634, Fwt 747.1) or free acid (Calbiochem-Novabiochem) and antibiotic A23187 (Calbiochem-Novabiochem) prepared as 2 mM 10-μL stock aliquots in ethanol or dry dimethyl sulfoxide (DMSO).
11. Silicon oil mixture for separation of the aqueous phase and cell pellet: Dow Corning (BDH, Poole, Dorset, UK) silicone fluids 556 (0.98 g/mL; 60%) and 550 (1.07 g/mL; 40%) shaken thoroughly. The efficiency of the silicon oil mix is batch-dependent and must be empirically tested since other mixing ratios can be more effective. Alternatively, if the cells are pelleted first, then a 1/1 (v/v) mixture will effectively separate the pellet and aqueous phase during the second spin.
12. 0.4% Azur A (Fluka, Gillingham, Dorset, UK) stock in ATP-fee CLB, pH 7.2.
13. Oligomycin (Sigma): 5 mg/mL solution made up in ethanol (EtOH). This acts to exclude Ca^{2+} uptake into the mitochondrial pool. However, the addition of oligomycin is not critical since at $[Ca^{2+}] < 1$ μM, mitochondrial Ca^{2+} uptake ($K_m \approx 10$ μM) is minimal *(14)*.
14. *myo*-Inositol 1,4,5-trisphosphate [Ins(1,4,5)P$_3$] (University of Rhode Island, Chemical Foundation, USA): 20 μL of a 1–2 mM stock solution.
15. Lumasolve (Lumac, Landgraaf, The Netherlands) and Emulsifier-safe scintillation cocktail (Canberra-Packard, Pangbourne, Berks, UK).

2.2. Equipment

1. Fluorescence spectrophotometer (fluorimeter) with variable excitation and emission optics ideally with a chart recorder or computer interface for fluo-3 experiments. We utilize the model LS5β and LS50β fluorimeters (Perkin-Elmer, Beaconsfield, Bucks, UK).
2. Magnetic stirrer and control unit for cuvets (Rank Bros., Bottingham, Cambridgeshire, UK), magnetic stirrer bars for cuvets (Perkin-Elmer), and disposable fluorimetry cuvets (Sarstedt, Leicester, Leicestershire, UK).
3. Bench-top microcentrifuge, with a capacity for 24 standard 1.5-mL microcentrifuge tubes (Heraeus, Sepatech, Osterode, Germany; Desaga, Sarstedt).
4. Electroporator (Gene-Pulser, BioRad, Hamel Hempstead, Herts, UK).

5. IBM 486 PC or compatible with Prism version 1.0 (GraphPad Software, San Diego, CA) for computer-assisted curve-fitting calculations. Many other similar curve-fitting programs are available.

3. Methods
3.1. Preparation of Calcium-Free Stock Solutions

1. The most critical consideration is the reduction of Ca^{2+} contamination during the preparation of all solutions, especially CLB. Reagents used must be the highest purity available, especially divalent metal salts, such as $MgCl_2 \cdot 6H_2O$. Ultrapure water is essential, and all solutions must be prepared and stored in plastic containers, never glass, which leaches significant quantities of metal cations.
2. Stock solutions of stimulating agonists, such as inositol polyphosphates, must be checked for calcium contamination, especially when conducting fluo-3 experiments, where Ca^{2+} contamination can be misinterpreted as Ca^{2+} release. Stock solutions can be readily pretreated to remove Ca^{2+} contamination with solid-phase Ca^{2+} chelating agents, such as chelex-100 (BioRad) or immobilized BAPTA-based "Ca^{2+} sponges" (Molecular Probes). We have treated a wide range of inositol polyphosphates with these chelators and have detected no loss owing to adsorption.

3.2. Preparation of CLB Working Solution and Buffering of Ca^{2+}

1. The fivefold CLB stock (1 vol) is added to a container with preweighed Na_2 ATP (2 mM final), diluted to 5 vol with H_2O, and the pH ($\approx$5.3) corrected to 7.2 using 20% (w/v) KOH and 1 mM HCl.
2. The contaminating free Ca^{2+} ($[Ca^{2+}]_{free}$) is buffered to final concentration of between 6 and 120 nM by addition of an appropriate aliquot of EGTA (10 mM in 20 mM KOH) usually 10–20 µL/100 mL of CLB (1–2 µM EGTA final).
3. The $[Ca^{2+}]_{free}$ of the CLB is confirmed using a tracer concentration of fura-2 (100 nM). Briefly, a sample of EGTA-buffered CLB (2 mL) is placed in a fluorimeter cuvet with 4 µL of 50 µM fura-2. The fluorescent intensity (F_{CLB}) of the buffer is determined at excitation and emission wavelengths of 340 and 510 nm, respectively. Recording the resultant F values, add 5 µL of the 200 mM $CaCl_2$ solution (F_{max}), followed by 30 µL of 200 mM EGTA (F_{min}), and then calculate $[Ca^{2+}]$ from the equation of Grynkiewicz et al. *(14)*, where the K_d for fura-2 is about 139 nM at 25°C and 224 nM at 37°C:

$$[Ca^{2+}]_{free} \ (nM) = [(F_{CLB} - F_{min}) / (F_{max} - F_{CLB})] \times K_d$$

Ideally, the $[Ca^{2+}]_{free}$ should be 60–120 nM, certainly <200 nM, to avoid significant mitochondrial uptake *(15)* and Ca^{2+} inhibition of Ins(1,4,5)P$_3$ receptor responses *(16)*. If it exceeds this, then add more 10 mM EGTA stock to the CLB and repeat the procedure. Once determined, the $[Ca^{2+}]_{free}$ value should remain relatively stable for a given batch of fivefold CLB stock.

4. For fluo-3 experiments, CLB is prepared in an identical fashion, except EGTA (1–2 µM) is replaced by fluo-3 (1 µM) and the solution protected from light. $[Ca^{2+}]_{free}$ of the fluo-3-supplemented CLB is measured using the same approach as for fura-2. Fluo-3 has a K_d of approx 400 nM at 22°C and 864 nM at 37°C *(17)*, and respective excitation and emission wavelengths of 505 and 530 nm. Typically, $[Ca^{2+}]_{free}$ is 150–200 nM in CLB supplemented with 1 µM fluo-3 at 22°C.

3.3. Permeabilization of Cells

1. We routinely utilize electropermeabilization *(18)* or the cholesterol-specific detergents saponin *(3)* and β-escin *(19)* to produce Ins(1,4,5)P$_3$-responsive permeabilized cells. Both of these methods require empirical determination of optimal conditions for each cell type used. The DNA dye Azur A (0.025% [w/v] CLB) can be used quantify permeabilization because it is excluded from intact cells, but stains the nuclei of permeabilized cells blue. We obtain >99% Azur A-stained SH-SY5Y neuroblastoma cells when permeabilized with 25–100 µg/mL saponin or β-escin, or via optimized electropermeabilization (*see* Note 1). Electropermeabilization of SH-SY5Y cells in 0.8-mL vol is accomplished in a Gene Pulser (Bio-Rad) using 3–12 discharges of a 3-µF capacitor with a field strength of 1.5 kV/cm and time constant of 0.1 ms.

2. Electropermeabilization is rapid and efficiently generates plasma membrane pores, which allow free passage of inositol polyphosphates. However, under some conditions, the process can generate significant concentrations of Al^{3+} from the disposable cuvets used, and this may effect cellular metabolism of inositol polyphosphates *(20)*. Consequently, we have favored saponin and β-escin permeabilization, which generate large plasma membrane pores and offer the flexibility of initiating an experiment with the permeabilization event.

3.4. The ⁴⁵Ca²⁺-Release Assay

1. Produce an inositol polyphosphate dilution series in CLB that is twofold the final required concentration. For example, Ins(1,4,5)P$_3$ concentrations (µM) of 30, 10, 3, 1, 0.3, 0.1, 0.03, 0.01, and 0.003 fully cover the Ca^{2+}-release dose–response range in most cell types we have tested. Ionomycin (5 µM final, free acid or calcium salt) and Ins(1,4,5)P$_3$ (20–30 µM)

prepared as twofold stocks in CLB are used as internal controls to define, respectively, the total mobilizable and $Ins(1,4,5)P_3$-sensitive cellular Ca^{2+} pools in permeabilized SH-SY5Y cells. All agents are added as duplicate 50-µL aliquots to decapped 1.5-mL microcentrifuge tubes, and two duplicate blanks (CLB alone) define the baseline calcium release. Thus, a typical experiment consists of 22–24 tubes that are stored on ice, and then allowed to come to room temperature for 15 min prior to addition of the permeabilized cells.

2. A postconfluent SH-SY5Y monolayer (175-cm^2 flask) is washed once and then harvested using the HBS/EDTA solution (*see* Note 2). Cell clumping occurs particularly with fibroblasts and can cause reduced precision. If necessary, titurate the cells two to three times through a 19-gage needle to obtain a single-cell suspension. Alternatively, trypsinized cells can also be used in the assay following centrifugation through fetal calf serum to deactive the trypsin.

3. Resuspend and centrifuge to wash the cell suspension twice in 5 mL of CLB (500g, 1 min).

4. Discard the supernatant and add 1.8 mL of CLB to resuspend the cells. Add 200 µL of saponin stock solution (1 mg/mL, ≈100 µg/mL final) to the cells, mix gently, and leave for 40–60 s, and then centrifuge (500g, 2 min).

5. Resuspend the pellet in about 5–6 mL CLB. For SH-SY5Y cells, this corresponds to a cell density of about 2–4 x 10^6 cells/mL, 0.5–0.8 mg/mL protein *(21)*, or a cellular DNA content of 100–200 µg/mL *(22)*. Then add 1 µL oligomycin and 1 µL $^{45}Ca^{2+}$/2 mL of permeabilized cells. Quickly, gently vortex the cells and $^{45}Ca^{2+}$, and leave for 15–20 min at 18–20°C, with periodic vortexing. In SH-SY5Y cells, $^{45}Ca^{2+}$ loading into intracellular Ca^{2+} stores proceeds rapidly and reaches equilibrium within 10 min (*see* Note 3).

6. Using a repeating pipet (Eppendorf, Hamburg, Germany), titurate the cell suspension, and then add 50 µL of the cell suspension to 50 µL of the agonist in the decapped 1.5-mL microcentrifuge tubes. Immediately add 300–400 µL of silicon oil mixture (*see* Note 4), and after 1–1.5 min, load the tubes into a microcentrifuge and centrifuge the cells through the oil (16,000g, 3 min). Alternatively, the cells can be pelleted after 1–1.5 min (16,000g, 2 min), the silicon oil then added, and a second centrifugation performed (16,000g, 1 min). This latter procedure is not ideal for time-course studies. However, once optimal stimulation times are determined, this procedure yields identical data, and the solid and reproducible pellets produce more consistent data values.

7. Remove the aqueous phase and most of the silicon oil, but avoid disturbing the cell pellet or "smear" on the opposite side of the tube. Invert the tubes on paper underlayered with foil for 30 min to remove the oil.

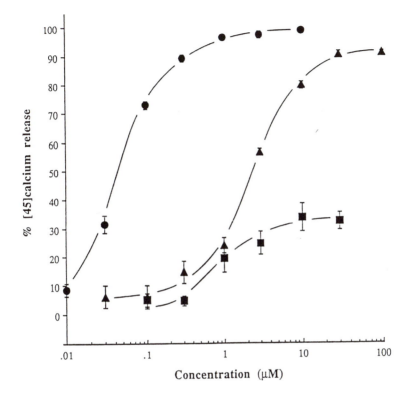

Fig. 1. Ins(1,4,5)P$_3$-, L-*chiro*-Ins(2,3,5)PS$_3$-, and Ins(1,3,4,6)P$_4$-induced ^{45}Ca^{2+} release. The data indicate the percentage ^{45}Ca^{2+} released at 20–22°C from the Ins(1,4,5)P$_3$-sensitive intracellular stores of saponin-permeabilized SH-SY5Y cells ($n \geq 4$) by Ins(1,4,5)P$_3$ (●), and the partial agonists L-*chiro*-Ins(2,3,5)PS$_3$ (■) and Ins(1,3,4,6)P$_4$ (▲). (These compounds were kindly provided by B. V. L. Potter, University of Bath, Bath, UK and R. Gigg, National Institute for Medical Research, Mill Hill, UK, respectively).

8. Add 1 mL of scintillant, and when the pellets are solubilized, vortex the tube and count the retained ^{45}Ca^{2+}. Large cell pellets may require solubilization in 100 μL of lumasolve, prior to the addition of scintillation cocktail.

9. The percentage release of ^{45}Ca^{2+} was calculated using the average value of the four CLB blanks. These define the zero ^{45}Ca^{2+}-release value. In SH-SY5Y cells, maximally effective concentrations of ionomycin (10 μ*M*) and Ins(1,4,5)P$_3$ (10–30 μ*M*) mobilize 90 and 70%, respectively, of the total preloaded ^{45}Ca^{2+}. Figure 1 shows typical ^{45}Ca^{2+}-release dose–response curves in SH-SY5Y cells, for Ins(1,4,5)P$_3$ and the partial agonists, *myo*-

inositol 1,3,4,6-tetrakisphosphate [Ins(1,3,4,6)P$_4$] *(23)* and L-*chiro*-inositol 2,3,5-trisphosphothioate L-*chiro*-[Ins(2,3,5)PS$_3$] *(24)*.

10. The EC$_{50}$ (concentration producing half-maximal stimulation) values and slope factors for ^{45}Ca^{2+} release were calculated using the computer-assisted curve-fitting programs, such as Prism version 1.0. For example, in SH-SY5Y cells, Ins(1,4,5)P$_3$, Ins(1,3,4,6)P$_4$, and L-*chiro*-Ins(2,3,5)PS$_3$ have respective EC$_{50}$ values of 52 ± 2, 1980 ± 137, and 1595 ± 574 nM ($n \geq 4$). The mean slope factors were always >1 (*see* Fig. 1).

3.5. The Use of Fluo-3 to Assess Ca^{2+} Uptake and Ins(1,4,5)P$_3$-Mediated Ca^{2+}-Release

1. Ca^{2+}-mobilizing agents: Prepare as 100-fold stocks in H$_2$O, and precheck for Ca^{2+}-contamination and fluorescence quenching in fluo-3-supplemented CLB (fluo-3/CLB).

2. Magnetic stirrer bars ("fleas") are washed using 7X detergent (Flow Labs), followed by an overnight soak in the HBS/EDTA, and are extensively washed with ultrapure water to remove any contaminating metal salts.

3. Warm the cuvet holder, the 3-mL plastic fluorimetry cuvet, and about 5–10 mL of fluo-3/CLB to 37°C. Add a clean flea to the cuvet, and wash the cuvet out once with 3 mL of warm fluo-3/CLB.

4. Wash a postconfluent SH-SY5Y monolayer (175-cm^2 flask) once, and then harvest using the HBS/EDTA solution. Centrifuge the cells (500g, 1 min), resuspend the pellet, and centrifuge wash twice in 5 mL of CLB (500g, 1 min).

5. Discard the supernatant, and resuspend the cells in 1.5 mL to a density of ≈6–12 x 10^6 cells/mL in fluo-3/CLB at 37°C. For SH-SY5Y cells, this corresponds to 2–2.7 mg/mL protein *(21)* or a cellular DNA content of about 325–750 µg/mL *(22)* (*see* Notes 5 and 8).

6. Then transfer the cells to the prewarmed cuvet containing a clean magnetic flea, and allow 2–3 min, with stirring, to come to equilibrium. Then add 15 µL of 2.5 mg/mL saponin or β-escin solution (25 µg/mL final) to initiate permeabilization. This results in a rapid uptake of Ca^{2+}, which reaches a steady state after about 2–3 min. Then a 1-µM spike of Ca^{2+} can be added to the cell suspension to ensure the intracellular calcium stores are not depleted. When this is taken up and a steady baseline obtained, the experiment is commenced. Ins(1,4,5)P$_3$ doses (100-fold stocks added in 15-µL vol) produce rapid calcium release responses without significant lag periods and decline as the permeabilized cell preparations actively metabolize the Ins(1,4,5)P$_3$ via 5-phosphatase and 3-kinase activities to form Ins(1,4)P$_2$ and Ins(1,3,4,5)P$_4$, respectively *(25,26)*.

7. We have largely utilized this assay to compare Ins(1,4,5)P$_3$ with various synthetic inositol polyphosphate analogs in order to obtain real-time esti-

mates of the kinetics of Ca^{2+} mobilization and the metabolic susceptibility of the analogs. Figure 2 shows a typical trace generated using the fluo-3 method (*see* Notes 6 and 7).

8. The assay is calibrated at the end by adding 20–30 μM of A23187 to define the total mobilizable Ca^{2+}. Then additions of 500 μM Ca^{2+} (F_{max}) followed by 2 mM EGTA (F_{min}) allow actual Ca^{2+} concentrations to be calculated for experimental fluorescence intensity (F) values, using the equation of Grynkiewicz et al. *(14)*, where fluo-3 has K_d values of 400 and 864 nM at 22 and 37°C, respectively *(17)*.

4. Notes

4.1. Optimizing the ⁴⁵Ca²⁺-Release Assay for Different Cell Lines

We have successfully performed $^{45}Ca^{2+}$ assays using many adherent and nonadherent cell lines, such as Rat-1, Swiss 3T3, NIH-3T3, Yac-1, and 1321N1 cells, as well as primary bovine adrenal chromaffin cells. However, the conditions used for each cell type must be optimized empirically. It is essential to:

1. Use Azur A staining to confirm that >95% of cells are permeabilized, and choose the minimal insult to the cells that can produce efficient permeabilization.
2. Select an appropriate cell density, where $^{45}Ca^{2+}$ release is readily detected, but where cellular metabolism of inositol polyphosphate is minimal over 2–3 min. For SH-SY5Y cells, we have selected a cell density where 30–50% of the $^{45}Ca^{2+}$ spike ($\approx$2–3 μM) is incorporated into the intracellular stores. Utilizing fura-2, we have confirmed that the $\approx$2.4 μM of $^{45}Ca^{2+}$ added to CLB supplemented with 2 μM EGTA produces an initial increase in total $[Ca^{2+}]_{free}$ from approx 150–350 nM.
3. Produce a time-course for $^{45}Ca^{2+}$ loading, and identify the plateau where an equilibrium loading state is obtained and maintained. For most cells, this occurs between 10 and 40 min.
4. The density of the silicon fluid mixture varies between batches and according to temperature. Thus, empirically select an appropriate mixture of the silicon fluids that efficiently and reproducibly separates the cells and aqueous phase.

4.2. Optimizing the Fluo-3 Assay

5. Permeabilized cells lose their functional Ca^{2+} responses following prolonged exposure to micromolar levels of Ca^{2+}. Consequently, when using the fluo-3 assay to follow cellular metabolism of inositol-polyphosphates

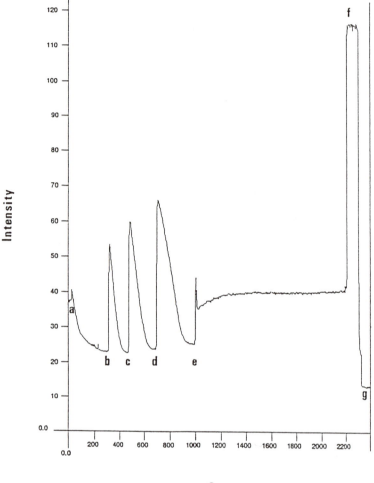

Fig. 2. Profiles of Ins(1,4,5)P$_3$- and L-*chiro*-Ins(2,3,5)PS$_3$-induced Ca^{2+} mobilization using fluo-3 detection. SH-SY5Y cells were permeabilized by addition of 2.5 µg/mL β-escin (a). The cells were then exposed to Ins(1,4,5)P$_3$ at concentrations of 1 µM (b), 3 µM (c), and 10 µM (d), followed by the metabolically resistant partial agonist L-*chiro*-Ins(2,3,5)PS$_3$ (e). F$_{max}$ (f) and F$_{min}$ (g) values are determined as described in Section 3.6., step 8.

(or other Ca^{2+}-mobilizing agents), it is essential to have a sufficiently high cell density to ensure active metabolism, such that concentration of Ca^{2+}-mobilizing agent is rapidly reduced, allowing net reuptake of released Ca^{2+}.

6. The fluo-3 assay is not ideal for experiments involving caffeine, because this Ca^{2+}-mobilizing agent potently quenches fluo-3 fluorescence in a dose-dependent fashion. It is important to test all agents for both Ca^{2+} contamination and fluorescence quenching and enhancement to avoid production of artifactual fluorescence intensity data.

7. Autofluorescence and quenching problems can be overcome by using Ca^{2+}-sensitive minielectrodes, the use and manufacture of which have been recently and comprehensively reviewed *(27,28)*. Only slight modifications to the fluo-3/CLB method are required. The fluo-3 in the CLB is replaced with 1 μM EGTA, and the cells resuspended in a final vol of 0.8–1 mL prior to permeabilization. Because the Ca^{2+}-sensitive minielectrode assays are conducted at 20–22°C, increased cell density produces metabolic time-courses for the inositol polyphosphates that are comparable to the fluo-3 assay.

8. The fluo-3 assay using low cell densities can be used to observe rapid Ca^{2+}-mobilization events in the absence of active inositol polyphosphate catabolism. For an example of this approach, *see* ref. *11*.

References

1. Berridge, M. J. (1993) Inositol trisphosphate and calcium signalling. *Nature* **361,** 315–325.
2. Streb, H., Irvine, R. F., Berridge, M. J., and Schulz, I. (1983) Release of Ca^{2+} from a nonmitochondrial intracellular store in pancreatic acinar cells by inositol-1,4,5-triphosphate. *Nature* **306, 67**–69.
3. Strupish, J., Cooke, A. M., Potter, B. V. L., Gigg, R., and Nahorski, S. R. (1988) Stereospecific mobilization of intracellular Ca^{2+} by inositol 1,4,5-triphosphate. *Biochem. J.* **253,** 901–905.
4. Berridge, M. J. (1988) Inositol triphosphate-induced membrane potential oscillations in Xenopus oocytes. *J. Physiology* **403,** 589–599.
5. Walker, J. W., Somlyo, A. V., Goldman, Y. E., Somlyo, A. V., and Trentham, D. R. (1987) Kinetics of smooth and skeletal muscle activation by laser pulse photolysis of caged inositol 1,4,5-triphosphate. *Nature* **327,** 249–252.
6. Ely, J. A., Hunyady, L., Baukal, A. J., and Catt, K. J. (1990) Inositol 1,3,4,5-tetrakisphosphate stimulates Ca^{2+} release from bovine adrenal microsomes by a mechanism independent of the inositol 1,4,5-trisphosphate receptor. *Biochem. J.* **268,** 333–338.
7. Joseph, S. K., Hansen, C. A., and Williamson, J. R. (1989) Inositol tetrakisphosphate mobilizes calcium from cerebellum microsomes. *Mol. Pharmacol.* **36,** 391–397.
8. Gershengorn, M. C., Geras, E., Purrello, V. S., and Rebeccchi, M. J. (1984) Inositol trisphosphate mediates thyrotrophin-releasing hormone mobilization of nonmitochondrial calcium in rat mammotrophic pituitary cells. *J. Biol. Chem.* **259,** 10,675–10,681.
9. Haugland, R. P. (1992) *Molecular Probes. Handbook of Fluorescent Probes and Research Chemicals,* 5th ed., Molecular Probes, Eugene, OR.

10. Safrany, S. T., Wojcikiewicz, R J. H., Strupish, J., McBain, J., Cooke, A. M., Potter, B. V. L., and Nahorski, S. R. (1991) Synthetic phosphorothioate-containing analogues of inositol 1,4,5-trisphosphate mobilize intracellular Ca^{2+} stores and interact differentially with inositol 1,4,5-trisphosphate 5-phosphatase and 3-kinase. *Mol. Pharmacol.* **39,** 754–761.

11. Meyer, T. and Stryer, L. (1990) Transient calcium release induced by successive increments of inositol 1,4,5-trisphosphate. *Proc. Natl. Acad. Sci. USA* **87,** 3841–3845.

12. Missiaen, L., Taylor, C. W., and Berridge, M. J. (1991) Spontaneous calcium release from inositol trisphosphate-sensitive calcium stores. *Nature* **352,** 241–244.

13. Michelangeli, F. (1991) Measuring calcium uptake and inositol 1,4,5-trisphosphate-induced calcium release in cerebellum microsomes using fluo-3. *J. Fluorescence* **1,** 203–206.

14. Grynkiewicz, G., Poenie, M., and Tsien, R. Y. (1985) A new generation of Ca^{2+} indicators with greatly improved fluorescent properties. *J. Biol. Chem.* **260,** 3440–3449.

15. Carafoli, E. (1987) Intracellular calcium homeostasis. *Annu. Rev. Biochem.* **56,** 395–433.

16. Danoff, S. K., Suppattapone, S., and Snyder, S. H. (1988) Characterization of a membrane protein from brain mediating the inhibition of inositol 1,4,5-trisphosphate receptor binding by calcium. *Biochem. J.* **254,** 701–705.

17. Merrit, J. E., McCarthy, S. A., Davies, M. P. A., and Moores, K. E. (1990) Use of fluo-3 to measure cytosolic Ca^{2+} in platelets and neutrophils. *Biochem. J.* **269,** 513–519.

18. Knight, D. E. and Scrutton, M. C. (1986) Gaining access to the cytosol: the technique and some applications of electropermeabilization. *Biochem. J.* **234,** 497–506.

19. Kobayashi, S., Kitazawa, T., Somlyo, A. V., and Somlyo, A. P. (1989) Cytosolic heparin inhibits musarinic and α-adrenergic Ca^{2+} release in smooth muscle. *J. Biol. Chem.* **264,** 17,997–18,004.

20. Loomis-Husselbee, J. W., Cullen, P. J., Irvine, R. F., and Dawson, A. P. (1991) Electroporation can cause artefacts due to solubilization of cations from electrode plates. Aluminum ions enhance conversion of inositol 1,3,4,5-tetrakisphosphate into inositol 1,4,5-trisphosphate in electroporated L1210 cells. *Biochem. J.* **277,** 883–885.

21. Bradford, M. M. (1976) A rapid and sensitive method for the quantitation of microgram quantities of protein utilising the principle of protein-dye binding. *Anal. Biochem.* **72,** 248–254.

22. Labarca, C. and Paigen, K. (1980) A simple, rapid and sensitive DNA assay procedure. *Anal. Biochem.* **102,** 344–352.

23. Gawler, D. J., Potter, B. V. L., Gigg, R., and Nahorski, S. R. (1991) Interactions between inositol tris- and tetrakis-phosphates. Effects on intracellular Ca^{2+} mobilization in SH-SY5Y cells. *Biochem. J.* **276,** 163–167.

24. Safrany, S. T., Wilcox, R. A., Liu, C., Dubreuil, D., Potter, B. V. L., and Nahorski, S. R. (1993) Identification of partial agonists with low intrinsic activity at the inositol 1,4,5-trisphosphate receptor. *Mol. Pharmacol.* **43,** 499–503.

25. Downes, C. P. and MacPhee, C. H. (1990) *myo*-Inositol metabolites as cellular signals. *Eur. J. Biochem.* **193**, 1–18.
26. Shears, S. B. (1991) Regulation of the metabolism of 1,2-diacylglycerols and inositol phosphates that respond to receptor activation. *Pharmacol. Ther.* **49**, 79–104.
27. Fry, C. H. (1990) Ion-selective electrodes. Manufacture and use in biological systems, in *Methods in Experimental Physiology and Pharmacology: Biological Measuring Techniques,* vol. 7, Biomesstechnik-Verlag, Berlin, Germany.
28. Orchard, C. H., Boyett, M. R., Fry, C. H., and Hunter, M. (1991) The use of electrodes to study cellular Ca²⁺ metabolism, in *Cellular Calcium: A Practical Approach* (McCormack, J. G. and Cobbold, P. H., eds.), IRL, Oxford, UK, pp. 83–114.

CHAPTER 19

Single-Cell Calcium Imaging

Katrina A. Marsh

1. Introduction

It is widely accepted that calcium ions are the primary regulators of smooth muscle contraction and relaxation; however, the mechanisms by which changes in intracellular calcium ion concentration ($[Ca^{2+}]_i$) are obtained within the cell are incompletely understood. Increases in $[Ca^{2+}]_i$ can be mediated either via a release of calcium from intracellular stores *(1–3)* or via an influx of calcium ions from the extracellular fluid *(4,5)*.

Many studies that investigate agonist-induced changes in $[Ca^{2+}]_i$, for example, in smooth muscle tissue, have concentrated on isolated tissue *(6)* or populations of cultured cells *(5,7)*. With the advent of single-cell imaging techniques, it has been possible to study changes in $[Ca^{2+}]_i$ of individual cultured cells as a function of time *(8,9)*. Investigators are therefore now in a position to study the temporal and spatial aspects of the response of single cultured cells to various agonists and to study the effect of stimulating other second-messenger systems on the calcium response.

Single-cell calcium imaging has recently revealed an inherent heterogeneity within a population of smooth muscle cells taken from the bovine trachea. In this case, the number of individual cells responding to the nonapeptide, bradykinin, was found to increase in a concentration-dependent manner *(10)*. This discovery has only been made possible by studying changes within individual cells, rather than a masked overview of the response of the population of cells as a whole. The involvement of calcium influx vs intracellular calcium release can also be investigated in this system by utilizing calcium-rich and calcium-free extracellular media.

From: *Methods in Molecular Biology, Vol. 41: Signal Transduction Protocols*
Edited by: D. A. Kendall and S. J. Hill Copyright © 1995 Humana Press Inc., Totowa, NJ

This chapter aims to give the reader some basic background knowledge behind the theory of single-cell calcium imaging, and also to provide instructions on how to put this into practice and to analyze the results obtained. The methodology concentrates on the methods used in my laboratory using the "MagiCal" hardware system and "TARDIS" software (supplied by Applied Imaging International Ltd., Hylton Park, Sunderland, Tyne & Wear, UK) to image fura-2-loaded bovine tracheal smooth muscle cells. Several basic steps are involved in single-cell calcium imaging regardless of the manufacturer of the system, and these are as follows:

1. Cells are loaded with calcium-sensitive fluorochrome and mounted on a fluorescence microscope receiving light of the appropriate wavelength(s).
2. Light emitted from the cells is passed through an image-intensifying camera to produce an electrical signal corresponding to the image that has been scanned in a raster format.
3. The electrical signal undergoes analog-to-digital conversion, so that the image is converted to a series of pixels each of which represents the light intensity at that particular part of the image.
4. Arithmetic processing of these images is performed to reveal a series of images that have been corrected for background fluorescence and represent calcium ion concentrations rather than light intensity.
5. Processing of the resulting data is performed to extract the relevant information from the experiment.

1.1. The Calcium Indicator, Fura-2

In order to visualize intracellular calcium ions, the ratiometric fluorescent dye fura-2 is used in my laboratory. Fura-2 binds calcium in a 1:1 ratio, and the excitation wavelength of the dye changes when it binds to free calcium ions within the cell. In its free form, fura-2 has a high excitation efficiency at 380 nm and a low excitation efficiency at 340 nm. When fura-2 binds to calcium ions, the excitation spectrum of the dye shifts to shorter wavelengths. Therefore, the peak excitation wavelength of fura-2 in its free form is 380 nm, and the peak excitation wavelength of fura-2 when it is bound to calcium is 380 nm. With increasing concentrations of calcium, the excitation efficiency at 380 nm is decreased, whereas, simultaneously, that of 340 nm is increased. By measuring the fluorescence emitted (at 500 nm) by the fura-2 when separately illuminated by both 340- and 380-nm wavelengths, the amount of fura-2 that is bound to calcium can be gaged and from this the unknown calcium ion concentration.

It is the ratio of fluorescence obtained at 340-nm excitation to that at 380-nm excitation that can be used to measure the free calcium ion concentration in a cell according to the relationship described by Grynkiewicz et al. in 1985 *(11)*, whereby:

$$[Ca^{2+}] = [K_d \beta (R - R_{min}) / (R_{max} - R)]$$

In this instance, K_d = the dissociation constant for fura-2 and calcium under the conditions used, R = the measured ratio for which the $[Ca^{2+}]_i$ is to be determined, R_{max} = the maximum fluorescence ratio obtained in the presence of an excess of calcium ions, R_{min} = the minimum fluorescence ratio obtained in the absence of calcium ions, and β = the fluorescence ratio at 380 nm of the minimum fluorescence to maximum fluorescence.

In practice, a particular cell type and analysis system require calibration with respect to calcium ion concentration. This is done using the calcium ionophore ionomycin in the presence of excess calcium which allows the flooding of the cell with calcium ions, thus yielding the maximum fluorescence ratio. The minimum fluorescence ratio is calculated after the chelation of free calcium ions with EGTA. Once obtained, these parameters are entered into the equation along with K_d, β, and the measured ratio, R, to reveal the unknown calcium ion concentration. The use of this equation in calculating $[Ca^{2+}]_i$ depends on several assumptions:

1. The intracellular dye is fully de-esterified;
2. The only source of fluorescence within the cell comes from the fura-2; and
3. The spectral properties of the dye are the same inside and outside the cell.

Fura-2 as a free acid is unable to cross the cell membrane. However, it is commercially available as a cell-permeant acetoxymethylester (Fura-2 AM). Once the compound has entered the cell, it is then hydrolized by intracellular esterases effectively trapping the free acid within the confines of the cell membrane. However, for all its attributes, fura-2 is not infallible in its assistance in calculating $[Ca^{2+}]_i$. Several artifacts can arise in calcium imaging, which are discussed in detail in other publications, for example, excessive calcium buffering *(12)*, the effect of path length on the determination of fluorescence *(9)* and the incomplete hydrolysis of ester bonds *(13)*. However, despite these potential problems, a reliable and reproducible method for calculating $[Ca^{2+}]_i$ in single cultured cells has been developed.

2. Materials

1. Physiological saline solution (PSS): 145 mM NaCl, 5 mM KCl, 1 mM MgSO$_4$, 10 mM N-2-hydroxyethylpiperazine-N'-2-ethanesulfonic acid (HEPES), 10 mM glucose, and either 2 mM CaCl$_2$ (PSSA), 10 mM CaCl$_2$ (PSSB), or 0.1 mM EGTA (PSSC).
2. Fetal calf serum (FCS).
3. Fura-2 acetoxymethylester: 5 μM in PSSA containing 10% FCS.
4. Ionomycin stock solution: 800 μM in PSSB.
5. EGTA stock solution: 250 mM in PSSC.
6. Cell-culture materials.
7. 22-mm Round glass coverslips, washed, dried, and sterilized.
8. High-vacuum silicone grease.
9. Forceps.
10. Lens tissue.
11. Stainless-steel coverslip holder consisting of two interlocking cylindrical pieces of metal between which the coverslip is placed and held.
12. Image analysis hardware and software.

2.1. Hardware Components (Fig. 1)

1. Light source: Excitation wavelengths of 340 and 380 nm are required for the ratiometric measurements of calcium using fura-2. The equipment utilizes a 75-W xenon arc lamp that emits an even spectrum of illumination between 300 and 800 nm.
2. Filter wheel: A rotating filter wheel is positioned between the UV light source and the fluorescence microscope, which holds four separate filters that rotate on an axis. The filter wheel controller (FWC) unit controls the movement of the wheel and dictates which filter (either 340 or 380 nm) is positioned in the light path at any one time. For single-cell imaging of smooth muscle cells, the FWC is programmed such that incident light falls on the specimen in alternate wavelengths of 340 and 380 nm every 1.5 s.
3. Heated chamber: In order to maintain the cells under investigation at 37°C throughout the duration of the experiment, a temperature-controlled chamber is available. This piece of equipment is constructed from an aluminum alloy, in a doughnut-style shape so that the coverslip holder and contents sit in the center. This configuration allows the cells to be viewed through the objective lens located directly below the glass coverslip.
4. Microscope: The Nikon Diaphot inverted microscope (Nikon UK Ltd., Telford, Shropshire, UK) with epifluorescence attachment is used for the visualization of fura-2-loaded cells using quartz objective lenses. The longer wavelength of emitted fluorescence is separated from the excitation

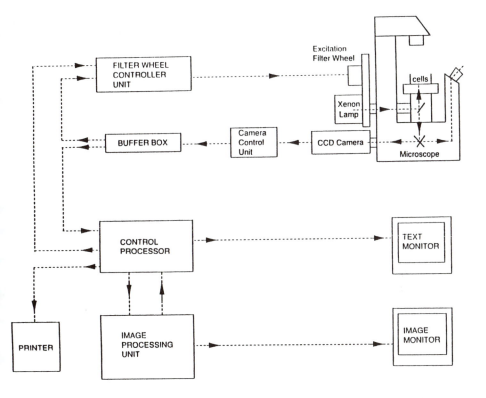

Fig. 1. Schematic diagram demonstrating the various interlinking hardware components of the single-cell calcium imaging system.

wavelengths by passing through a dichroic mirror and then a 510-nm barrier filter. Emitted light then passes into the eyepiece or the chosen camera equipment.

5. Camera: The system I use houses a charge-coupled device (CCD) image-intensifying camera (Photonic Science, Mountfield, East Sussex, UK). Light emitted from the specimen through the microscope is focused on a high-gain hybrid image intensifier. The output from this is then passed via a fiber-optic taper to a CCD image sensor and then into the video buffer box. The camera control unit basically houses the camera power supply and intensifier, and video gain controls.

6. Video buffer box: This component distributes the incoming video signal to other parts within the system, such as the filter wheel controller unit, image processing unit, and monitor.

7. Control processor: The control processor is the controlling computer within the MagiCal system handling the general control of the system. The processor communicates with the filter wheel controller, video buffer box, printers, monitor, keyboards, and image-processing unit (IPU).

8. IPU: This unit performs all the manipulations on captured images that it has stored in memory. In the system described, there is an 8-Mbyte image memory that will store up to 120 images of 256×256 pixel size, although a 32-Mbyte image memory is also available. The IPU also contains a 2-D look-up table for the conversion of ratioed values into calibrated calcium ion concentrations.

9. Monitors: Two-color monitors are often used in an image analysis system, one for text where commands are given and one displaying the images in memory. In addition, I have a monochrome monitor that displays the image obtained directly from the image-intensifying camera.

10. Printers: Once images have been processed and analyzed, they can then be converted to hard copy prints. The MagiCal system can (and in my case does) output to a monochrome printer, a color printer, and a Polaroid freeze-frame video recorder.

2.2. Image Digitization

The first stage of image processing is usually to average each frame as it is collected by the camera using analog hardware averaging. This is performed in order to reduce camera noise before continuing to process further. The signal that emerges from the image-intensifying CCD camera is an analog signal that is a voltage varying continuously with time. This voltage represents the light intensity of each frame that has been scanned in a raster pattern. However, digital computers work with numbers rather than analog quantities, and therefore, the signal must undergo an analog-to-digital conversion. This is done by sampling the analog voltage at regular time intervals along each scan line, which in effect divides each line into small picture elements (pixels). The voltage at each pixel is then read as an 8-bit number that is assigned a gray level, correlating to the light intensity of that particular pixel, in the range of 0 to 255. The final image is then subject to pseudocoloring of each pixel, where each gray level is assigned a different color making differences in intensity more apparent to the human eye.

3. Methods

3.1. Cell Culture

Standard cell-culture methods are used *(14)* to provide viable cells for image analysis, which will depend on the cell type used. For the purpose of calcium imaging, cells are sparsely seeded onto clean (*see* Note 1), sterile glass coverslips 2–3 d prior to analysis. It may be necessary to treat coverslips with factors that increase the adhesion of the cells to glass.

3.2. Cell Loading

1. Coverslips with attached cells are gently washed three times in PSSA (*see* Note 2).
2. Cells are then loaded by incubating at 37°C for 30 min in a 5 μM solution of fura-2 AM made up in a 10% solution of FCS in PSSA.
3. Excess dye is then removed by washing with PSSA three more times.
4. Cells remain in PSSA for 15 min to allow for complete de-esterification of the fura-2 AM molecule.

3.3. Image Capture

1. Stainless-steel coverslip holders are greased with high-vacuum silicone grease (*see* Note 3) to stop leakage from the resulting chamber.
2. A single coverslip is placed (*see* Note 4) in the metal holder, which is then mounted in the temperature-controlled chamber on the microscope. Excess buffer present on the underside of the coverslip is wiped off carefully using lens tissue.
3. The PSS solution (900 µL) required for the experiment is placed in the coverslip chamber.
4. By running through the software program, cells are then illuminated in order to visualize and select an appropriate field of view of healthy cells.
5. The intensifier gain of the CCD camera is then set to a level, such that the image in view displays the widest possible range of the 256 gray levels available. In practice, the pixels representing the highest light intensity at 380 nm should be set at a gray level of 256.
6. The field of view is then moved to an area that is devoid of any cells for the purpose of measuring background fluorescence.
7. An averaged image of each of the 340- and 380-nm image types is captured and stored.
8. The field of view is moved back to the area of interest.
9. A sequence of up to 118 averaged images (*see* Note 5) of alternating 340-(A) and 380-nm (B) types is captured and stored.

10. Agents to be added are placed directly onto the cells in a volume of 100 μL (*see* Note 6).

11. The image sequence is then subjected to background subtraction, which is performed by the image analysis software on command. The 340-nm background is subtracted from each of the 340-nm images on a pixel-by-pixel basis, and similar processing is performed with the 380-nm background and images.

3.4. Calibration

1. Coverslips are mounted as in Section 3.3., step 2, and 950 μL of PSSB are placed in the coverslip chamber.
2. Several (approx 20) basal images are captured.
3. Ionomycin (25 μL of stock solution; final concentration of 20 μ*M*) is placed onto the coverslip, and a further 40 images are collected.
4. EGTA (25 μL of stock solution; final concentration of 6 m*M*) is added to the bathing media, and the remainder of the sequenced is captured.
5. Graphical analysis (*see* Section 3.6.) is performed to obtain the maximum and minimum fluorescence at 340 (A) and 380 nm (B).
6. From these raw values, the R_{min} (A/B in the presence of EGTA), R_{max} (A/B in the presence of ionomycin), and β values can be obtained.
7. The calculated calibration parameters are then entered into the 2-D lookup table contained within the environment menu for use in the conversion of ratioed values to calcium ion concentrations.

3.5. Image Ratioing

1. The 2-D look-up table corresponding to the cell type under investigation is loaded into the calcium-imaging program.
2. On command, the image analysis software then ratioes background-corrected image pairs (A/B) on a pixel-by-pixel basis that are automatically converted into calcium ion concentrations.
3. There now remains a sequence of 59 ratioed images that correspond to the changes in intracellular calcium ion concentration occurring as a function of time under the chosen experimental conditions.

3.6. Graphical Analysis of Single Cells

1. The cell of interest is outlined using a light pen on the image monitor screen.
2. The mean intracellular calcium ion concentration of the cell outlined is then calculated on command, and a graphical representation of the changes in intracellular calcium ion concentration is constructed.
3. This process is repeated for each cell in the field of view.
4. Particular areas of interest within a cell can also be analyzed in the same way.

5. Several other methods of presenting data and analyzing results are available depending on the system used and the software available. These include the display of a sequence of images in a montage arrangement, and 3-D histogram analysis of the calcium content of a particular cell revealing spatial characteristics of the calcium response; interrogation of the pixel profile of a line drawn through the cell.

3.7. Storage of Images

Several methods for long-term storage of data are available. A sequence of images can be archived and stored on floppy disk. However, this is rather time-consuming and also requires a large number of disks. Tape streamer cartridges have a much larger capacity, and copying sequences are in general a faster procedure than the previous method. Optical disks are also available far data storage, and optical disk storage is often a much more convenient method for the user. In addition to a large capacity on each disk, this method also demonstrates the fastest time for writing sequences of images to disk that can be read many times.

4. Notes

1. Since the cells for image analysis are sparsely seeded, it is essential that the coverslips are cleaned thoroughly to avoid fluorescent particles present on the glass interfering with the images. Overnight soaking in a weak solution of nonfluorescent detergent followed by rinsing in running tap water and then distilled water will suffice.
2. It is beneficial to prepare calcium-containing buffers on the day of use to avoid precipitation of salts or growth of microorganisms, which again will interfere with the fluorescent images.
3. The use of excessive amounts of grease will not only reduce the area of the coverslip available for viewing, but can also smear across the remainder of the coverslip when wiping the underside of the coverslip with lens tissue, thus blurring the image.
4. Watchmaker forceps, either curved or straight, are a good investment for the handling of coverslips with the minimum of accidental damage to the coverslip.
5. The maximum number of images that can be captured in one sequence depends on the size of the image being captured and also the size of the image memory available.
6. Care must be taken when adding agents directly to the cell chamber, since too much force when pipeting can result in the cells under observation being expelled from the field of view.

References

1. Somlyo, A. P., Walker, J. W., Goldman, Y. E., Trentham, D. R., Kobayashi, S., Kitazawa, T., and Somlyo, A. V. (1988) Inositol triphosphate, calcium and muscle contraction. *Phil. Trans. R. Soc. Lond. B.* **320,** 399–404.

2. Berridge, M. J. and Irvine, R. F. (1989) Inositol phosphates and cell signalling. *Nature* **341,** 197–205.

3. Irvine, R. F. (1990) "Quantal" Ca^{2+} release and the control of Ca^{2+} entry by inositol phosphates—a possible mechanism. *FEBS Lett.* **263,** 5–9.

4. Benham, C. D. and Tsien, R. W. (1987) A novel receptor-operated Ca^{2+}-permeable channel activated by ATP in smooth muscle. *Nature* **328,** 275–278.

5. Murray, R. K. and Kotlikoff, M. I. (1991) Receptor-activated calcium influx in human airway smooth muscle cells. *J. Physiol.* **435,** 123–144.

6. Takuwa, Y., Takuwa, N., and Rasmussen, H. (1987) Measurements of cytoplasmic-free Ca^{2+} concentration in bovine tracheal smooth muscle using aequorin. *Am. J. Physiol.* **253,** C817–C827.

7. Felbel, J., Trockur, B., Ecker, T., Landgraf, W., and Hofmann, F. (1988) Regulation of cytosolic calcium by cAMP and cGMP in freshly isolated smooth muscle from bovine trachea. *J. Biol. Chem.* **263,** 16,764–16,771.

8. Cheek, T. R., Jackson, T. R., O'Sullivan, A. J., Moreton, R. B., Berridge, M. J., and Burgoyne, R. D. (1989) Simultaneous measurement of cytosolic calcium and secretion in single bovine adrenal chromaffin cells by fluorescent imaging of Fura-2 in cocultured cells. *J. Cell Biol.* **109,** 1219–1227.

9. Neylon, C. B., Hoyland, J., Mason, W., and Irvine, R. F. (1990) Spatial dynamics of intracellular calcium in agonist-stimulated vascular smooth muscle cells. *Am. J. Physiol.* **259,** C657–C686.

10. Marsh, K. A. and Hill, S. J. (1993) Characteristics of the bradykinin-induced changes in intracellular calcium ion concentration of single bovine tracheal smooth muscle cells. *Br. J. Pharmacol.* **110,** 29–35.

11. Grynkicwicz, G., Poenie, M., and Tsien, R. Y. (1985) A new generation of Ca^{2+} indicators with greatly improved fluorescence indicators. *J. Biol. Chem.* **260,** 3440–3450.

12. Moore E. D. W., Becker, P. L., Fogarty, K. E., Williams, D. A., and Fay, F. S. (1990) Calcium imaging in single living cells: theoretical and practical issues. *Cell Calcium* **11,** 157–179.

13. Roe, M. W., Lemasters, J. J., and Herman, B. (1990) Assessment of Fura-2 for measurements of cytosolic free calcium. *Cell Calcium* **11,** 63–73.

14. Marsh, K. A. and Hill, S. J. (1992) Bradykinin B_2 receptor-mediated phosphoinositide hydrolysis in bovine cultured tracheal smooth muscle cells. *Br. J. Pharmacol.* **107,** 443–447.

CHAPTER 20

Characterization of Calcium/Calmodulin-Stimulated Protein Kinase II

Paula E. Jarvie and Peter R. Dunkley

1. Introduction

The enzyme Ca^{2+}/calmodulin-stimulated protein kinase II (CaM-PK II) belongs to a family of calmodulin-stimulated serine/threonine protein kinases that are widely distributed in nature, but are especially enriched in the brain *(1–3)*. CaM-PK II is also found in a number of subcellular fractions, where it may be associated with specific binding proteins *(3)*. CaM-PK II has a number of unique characteristics that allow it to be identified *(see* Section 1.1.), and have its activity *(see* Section 1.2.) and level determined *(see* Section 1.3.).

1.1. General Methods Used in Characterization of CaM-PK II

CaM-PK II always has one or more subunits with molecular masses in the range of 45–65 kDa. The α subunit has an apparent molecular weight on sodium dodecyl sulfate-polyacrylamide gel electrophoresis (SDS-PAGE) of 50 kDa, the β subunit 60 kDa, the β' subunit 58 kDa, the δ subunit 60 kDa, and the γ subunit 59 kDa *(1–3)*. The proportion of each of these subunits varies considerably depending on the tissue, but the α and β subunits predominate in brain *(3)*. The form of CaM-PK II in vivo is not known, but the isolated enzyme is generally a multimeric complex of 500–700 kDa, which has an α-to-β ratio of 3:1 *(1–3)* in forebrain.

From: *Methods in Molecular Biology, Vol. 41: Signal Transduction Protocols*
Edited by: D. A. Kendall and S. J. Hill Copyright © 1995 Humana Press Inc., Totowa, NJ

An SDS-PAGE *(4)* technique used to fractionate the subunits of CaM-PK II according to their apparent molecular weights is described in Section 3.1. and is the basis of a number of other procedures described in this chapter. The subunits of CaM-PK II have isoelectric points of approx 6.8 and can be fractionated on two-dimensional PAGE, but their relative insolubility often causes them to streak on isoelectric focusing *(5,6)*. For more conclusive identification of the enzyme CaM-PK II, subunits may be cut from a polyacrylamide gel and partially digested with *Staphylococcus aureus* strain V8 protease into peptides, which can be fractionated on a second gel *(5,7,8)*. This technique is described in Section 3.2.

CaM-PK II subunits have characteristic mobilities on SDS-PAGE and patterns of peptides after protease digestion, which can be detected with a protein stain when purified enzyme is used *(9)*. However, since CaM-PK II is able to autophosphorylate itself *(see* Section 1.2.), the preferred method for detecting and quantitating CaM-PK II, especially with subcellular fractions and impure enzyme, is by autoradiography of radiolabeled protein. Autoradiography is described in Section 3.3. and autophosphorylation in Sections 3.4. and 3.5.

1.2. Activity of CaM-PK II

The protein kinase activity of CaM-PK II can be studied in a number of ways *(1–3,10–12)*. The ability to autophosphorylate itself is most often used, and this method is described in Sections 3.4. and 3.5. Its ability to phosphorylate either exogenous substrates, including peptides, or endogenous substrates present in the tissue of interest is described in Sections 3.6. and 3.7. Autophosphorylation and endogenous substrate phosphorylation can be used with both intact and lysed tissue, whereas exogenous substrate assays require lysed tissue. Certain precautions need to be observed during the preparation of tissue or during the purification of the enzyme and during incubations to assess CaM-PK II activity. This is because the enzyme is particularly sensitive to high temperatures *(11,13)*.

1.2.1. Autophosphorylation

Each subunit of CaM-PK II is capable of undergoing autophosphorylation in the presence of calcium and calmodulin *(1–3,10–12)*. Three modes of autophosphorylation exist. Primary autophosphorylation leads to an autonomous form of the enzyme that retains activity in the absence of calcium and calmodulin, and this unique feature of the kinase is fundamental to its characterization *(11)*. Secondary autophosphorylation irreversibly decreases the activity of the enzyme, whereas tertiary autophosphorylation inhibits the activity of the enzyme and blocks the binding of calmodulin *(11)*.

The ability of CaM-PK II to autophosphorylate itself can be measured in two ways. In lysed tissue, $[\gamma\text{-}^{32}P]$ATP is used with a short incubation to label the enzyme, whereas $^{32}P_i$ is employed in a long prelabeling period with intact tissue. Labeled samples are then run on SDS-PAGE, before or after immunoprecipitation, and the activity can be quantitated by densitometry of the resulting autoradiograph.

1.2.1.1. $[\gamma\text{-}^{32}P]$ATP Labeling of Lysed Tissue

CaM-PK II activity can be assessed by autophosphorylation in homogenates and subcellular fractions *(1–3,10–12,14,15)*, as well as in purified soluble fractions *(1–3,9,10)*. In subcellular fractions, there are many protein kinases and phosphatases, and it is sometimes difficult to be sure of which phosphoprotein bands on a gel are CaM-PK II. Weinberger and Rostas *(16)* developed a method using zinc that allows autophosphorylation reactions to proceed in the virtual absence of endogenous substrate phosphorylation, or the activity of other protein kinases or phosphatases. This procedure allows a significant enrichment of CaM-PK II phosphorylation relative to the phosphorylation of other proteins *(8)* and is described in Section 3.4.

1.2.1.2. $^{32}P_i$ Labeling of Intact Tissue

CaM-PK II activity is difficult to measure directly by autophosphorylation in intact tissue, because only a very small proportion of the enzyme becomes labeled *(11,13)*. However, if antibodies are available, immunoprecipitation can be used to enrich CaM-PK II and autophosphorylation can then be assessed *(12,17)*.

1.2.2. Enzyme Assays

The substrate specificity of CaM-PK II is very broad, but the enzyme has particularly high activity against synapsin I and microtubule-associated protein-2 *(1,2)*. *In situ*, the major substrates of CaM-PK II that have been identified to date, in addition to the aforementioned, are tyrosine and phenylalanine hydroxylases, pyruvate kinase, glycogen synthase, myosin light chain, and ion channels *(1–3,10)*.

CaM-PK II readily phosphorylates a number of these exogenous substrates in both lysed tissue and as the purified enzyme *(1,2)*. Specificity has been improved by selecting a peptide substrate that is phosphorylated only by CaM-PK II *(18)*. CaM-PK II is also able to phosphorylate endogenous substrates. In lysed brain tissue, CaM-PK II has synapsin I as its major substrate *(5)*. This protein has two sites phosphorylated by CaM-PK II at one end of the protein, and these can be separated from a peptide phosphorylated by protein kinase A and CaM-PK I at the other end of the protein *(5)*. Since proteins and peptides do not penetrate cell membranes, activity is best measured in intact tissue by analyzing the phosphorylation of endogenous substrates, such as synapsin I in synaptosomes *(1,2,5,10,19,20)*, or tyrosine hydroxylase in PC12 cells and bovine adrenal chromaffin cells *(21–23)*. Tyrosine hydroxylase (TOH) is phosphorylated on four sites *(23)*, one of which, serine 19, is labeled by CaM-PK II. Trypsin digestion of the TOH bands and fractionation of the resulting peptides by high-performance liquid chromatography (HPLC) separates out two phosphopeptides containing only serine 19, and the level of phosphate in these peaks can be quantitated *(22–24)*.

1.2.2.1. CALCIUM-INDEPENDENT ACTIVITY

When CaM-PK II is preincubated with Ca^{2+}, calmodulin, and ATP, autophosphorylation occurs. If the enzyme is now incubated with a peptide substrate and radiolabeled ATP, calcium-independent CaM-PK II activity can be determined *(1,2,10,11,20)*. When intact tissues are stimulated, the level of calcium-independent CaM-PK II activity increases, and this increase is used to provide an indication of the extent of autophosphorylation of CaM-PK II in intact tissues *(13)*.

1.3. Levels of CaM-PK II

CaM-PK II is not a very antigenic protein, and it has proven difficult to produce polyclonal antibodies, but a number of laboratories now have monoclonal antibodies to particular subunits *(1,2,10,12)*. CaM-PK II binds calmodulin to each of its subunits, but the enzyme requires high concentrations of calmodulin relative to other calmodulin-activated kinases *(1,2)*.

The level of CaM-PK II protein can be measured using antibody binding, as described in Section 3.9., or by calmodulin binding, as described in Section 3.10. Both these procedures require initial transfer of the protein to nitrocellulose membrane *(25,26)*, as described in Section 3.8.

2. Materials

All solutions referred to in Section 3. are listed in alphabetical order in Section 2.1. Equipment is similarly listed in Section 2.2.

2.1. Solutions (see Note 1)

1. Absolute ethanol.
2. Acetic acid: 7% in H_2O.
3. Acrylamide stock: 30% acrylamide, 0.8% bis-acrylamide, filtered under vacuum using a prefilter grade filter.
4. Adenosine-5'-triphosphate (ATP) disodium salt: 1 mM in 30 mM Tris, pH 7.4, at 20°C.
5. Ammonium persulfate: 10% in H_2O.
6. Antibody buffer: 1% gelatine in Tris-buffered saline tween (TBST).
7. Assay buffer A: 30 mM Tris, 10 mM $MgSO_4$, and 10 mM EGTA, pH 7.4.
8. Assay buffer B: 30 mM Tris, 4 mM $MgSO_4$, and 4 mM EGTA, pH 7.4.
9. Assay buffer C: 30 mM Tris, 20 mM $MgSO_4$, and 2 mM EGTA, pH 7.4.
10. Assay buffer D: 30 mM Tris and 18 mM $MgSO_4$, pH 7.4.
11. [γ-^{32}P]ATP radioisotope, 10 mCi/mL, ≥3000 Ci/mmol.
12. Biocam solution:
    a. Stock biotinylated calmodulin: 2 mg/mL in H_2O (*see* Note 2).
    b. To use: Dilute stock biotinylated calmodulin 1/400 with blocking solution 3. Allow 4–8 mL for one 15 × 11 cm membrane.
13. Blocking solution 1: 5% skim milk powder in Tris-buffered saline (TBS).
14. Blocking solution 2: 2% gelatine in TBS. Prepare fresh and cool.
15. Blocking solution 3: 5% skim milk powder and 1 mM $CaCl_2$ in TBS.
16. $CaCl_2$ (calcium chloride dihydrate): 12 mM in 30 mM Tris-HCl, pH 7.4.
17. $CaCl_2$: 7.5 mM in 30 mM Tris-HCl, pH 7.4.
18. $CaCl_2$: 2 mM in 30 mM Tris-HCl, pH 7.4.
19. Calmodulin (bovine brain): 1 mg/mL in 30 mM Tris-HCl, pH 7.4.
20. Color development solution: 5-bromo-4-chloro-3-indolyl phosphate/nitro blue tetrazolium (BCIP/NBT) alkaline phosphatase:
    a. Stock NBT solution: 50 mg/mL NBT in 70% dimethyl formamide.
    b. Stock BCIP solution: 25 mg/mL BCIP in 100% dimethyl formamide.
    c. Alkaline phosphatase buffer: 100 mM Tris-HCl, 100 mM NaCl, and 5 mM $MgCl_2$, pH to 9.5 with NaOH.
    Just prior to color development, mix NBT and BCIP stock solutions with alkaline phosphatase buffer, so that concentrations of NBT and BCIP are 300 and 150 μg/mL, respectively (*see* Note 3).
21. Coomassie blue stain: 25% propan-2-ol, 10% acetic acid, and 0.4% coomassie blue.

22. Depolarizing buffer: 46 mM NaCl, 77 mM KCl, 25 mM NaHCO$_3$, 1.2 mM MgCl$_2$, 100 µM CaCl$_2$, and 10 mM glucose. Bubble with 5% CO$_2$ in oxygen for 30–60 min. Make fresh.
23. Destain solution: 25% propan-2-ol and 10% acetic acid.
24. Diluted resolving buffer: 25% resolving buffer and 0.1% SDS.
25. Diluted stacking buffer: 25% stacking buffer and 0.1% SDS.
26. Drying solution: 50% methanol and 1% glycerol.
27. EDTA (disodium salt): 25 mM in 30 mM Tris-HCl, pH 7.4.
28. Fixing solution: 53% methanol and 7% acetic acid.
29. Krebs-like buffer: 118 mM NaCl, 5 mM KCl, 25 mM NaHCO$_3$, 1.2 mM MgCl$_2$, 100 µM CaCl$_2$, and 10 mM glucose. Bubble with 5% CO$_2$ in oxygen for 30–60 min. Make fresh.
30. Krebs-like buffer: 10X concentrated.
31. Molecular-weight protein standards: Dissolve in diluted solubilizing buffer (one third concentration), so that the concentration of combined protein is approx 1 µg protein/3 µL.
32. Orthophosphoric acid: 75 mM in H$_2$O.
33. ^{32}P$_i$ radioisotope: 10 mCi/mL, ≥8500 Ci/mmol.
34. Peptide substrate: 1 mM in 30 mM Tris-HCl, pH 7.4 (*see* Note 4).
35. Primary antibody solution: Species-specific primary antibody is diluted to appropriate dilution with antibody buffer. Allow 4–8 mL for one 15 × 11 cm membrane.
36. Protease solution (*Staphylococcus aureus* strain V8):
    a. Enzyme diluent: 125 mM Tris-HCl, 0.1% SDS, 2% β-mercaptoethanol, and 30% glycerol, pH 6.8. Add minimal bromophenol blue.
    b. Stock protease solution: Dissolve V8 protease in enzyme diluent to a concentration of 500–1000 U/mL. Store at –20°C.
    c. To use: Dilute sufficient stock protease solution 1 in 100 with enzyme diluent. Allow 50 µL/band (finally 0.25–0.5 U/band).
37. Rehydrating buffer: 125 mM Tris-HCl, 0.1% SDS, and 2% β-mercaptoethanol, pH 6.8.
38. Resolving buffer: 1.5M Tris-HCl, pH 8.8.
39. Running buffer: 25 mM Tris-HCl, 1.44% glycine, and 1% SDS. The pH of this solution should be 8.6. No adjustment should be necessary.
40. SDS: 10% in H$_2$O.
41. Secondary antibody solution A: Goat antimouse IgG or goat antirabbit IgG alkaline phosphatase conjugate. The conjugates are stored at 4°C and diluted 1/3000 in antibody buffer for use. Allow 25 mL for one 15 × 11 cm membrane.
42. Secondary antibody solution B: Stock Strepavidin alkaline phosphatase conjugate: 1 mg/mL. Dilute stock 1 in 1000 with wash solution. Allow 25 mL for one 15 × 11 cm membrane.

43. Solubilizing buffer:
    a. Stock solution: Make the following for 100 mL, but only dilute to 88 mL: 200 mM Tris-HCl, 28% glycerol, 8% SDS, 6 mM EGTA, and minimal bromophenol blue, pH 6.8.
    b. Mix 4.4 mL above stock solution and 0.6 mL β-mercaptoethanol.
44. Stacking buffer: 500 mM Tris-HCl, pH 6.8.
45. TBS: 10 mM Tris-HCl and 150 mM NaCl, pH to 8.0.
46. TBST: 0.05% Tween in TBS.
47. Tetraethylmethylethylenediamine (TEMED).
48. Transfer buffer: 50 mM boric acid, 2 mM EDTA, and 4 mM β-mercaptoethanol, pH to 8.9 with NaOH.
49. Trifluoperazine: 500 μM in 30 mM Tris-HCl, pH 7.4.
50. Tris buffer: 30 mM in H$_2$O, pH 7.4, with HCl at 20°C.
51. Wash solution: 0.05% Tween and 1 mM CaCl$_2$ in TBS.
52. X-Ray developer and fixer appropriate for the film used.
53. Zinc sulfate (ZnSO$_4$): 50 mM in 30 mM Tris-HCl, pH 7.4.

2.2. Equipment

This list of equipment only includes major or specialized items. All items of equipment generally found in a well-equipped laboratory are not included.

1. Autoradiography film.
2. Densitometer.
3. Gel running apparatus.
4. Glass gel plates, spacers, and combs.
5. Intensifying screens suitable for enhancing β emissions. Preferably these are fixed inside a light-proof cassette.
6. Nitrocellulose membrane.
7. P81 phosphocellulose paper.
8. Perspex sheets (approx 6 mm thick) cut to the same size as the autoradiography film.
9. Power supply: Capable of 3000 V and having the ability to vary both current and voltage.
10. Refrigerated water cooling system.
11. Western blot transfer apparatus.

3. Methods
3.1. SDS-PAGE

1. Assemble gel plate mold (commonly 1-mm spacers between the plates).
2. Resolving gel (10% acrylamide): A constant percentage resolving gel (commonly 10% acrylamide) is preferred for the study of CaM-PK II. Mix the reagents together in the following ratio to begin polymerization. Make

enough to fill 4/5 of the gel mold. For 30 mL: Mix 10 mL acrylamide stock, 12.2 mL H$_2$O, 7.5 mL resolving buffer, 150 µL ammonium persulfate (10%), 300 µL SDS (10%), and 20 µL TEMED. Pour immediately into assembled gel plate mold. Gently overlay the resolving gel with diluted resolving buffer and leave to set undisturbed for at least 30 min.

3. Stacking gel: Pour off the overlay solution. Mix enough stacking gel to more than fill the gel plate mold. For 10 mL: Mix 1.5 mL acrylamide stock, 5.9 mL H$_2$O, 2.5 mL stacking buffer, 100 µL ammonium persulfate (10%), 100 µL SDS (10%), and 10 µL TEMED. Use a little of this stacking gel to wash the top surface of the set resolving gel, and then pour the remaining solution onto the top of the resolving gel to fill the mold completely. Remove any air bubbles, insert a comb with the required number of sample wells, and allow to set for at least 20 min. When set, remove the comb, wash off any unset stacking solution with diluted stacking buffer, straighten any bent wells with a long syringe needle, and overlay with diluted stacking buffer. Gels may be used immediately, stored at room temperature overnight, or stored in the cold for up to a week.

4. Running a gel:
   a. Gels that have been stored in the cold should be equilibrated at room temperature for approx 1 h before use.
   b. Discard the overlay solution from the stacking gel by pouring off as much as possible and using a long syringe to remove the remainder.
   c. Carefully remove the bottom spacer (if any), and wipe out any vaseline or similar product used to seal the gel mold.
   d. Fill the lower chamber of the electrophoresis bath with running buffer to the required level, and place the gel in the chamber. Dislodge any air bubbles from underneath the gel by squeezing running buffer with a Pasteur pipet, and then clamp in position.
   e. Protein samples containing autophosphorylated CaM-PK II (*see* Section 3.4.) should be made up in solubilizing buffer in a ratio of 2 parts protein sample to 1 part solubilizing buffer and defrosted 2 min at 70°C if frozen. Ideally, load 40–100 µg protein/well in a volume of 50–100 µL.
   f. Fill the upper chamber with running buffer, and underlay the protein samples into the wells using a 100-µL glass syringe.
   g. Reserve the two end wells for molecular weight protein standards made up similarly in solubilizing buffer. Load the same volume of standard per well as for the protein samples.
   h. Connect the power supply leads to the electrodes with the positive lead at the bottom. Set the variable voltage and power controls to maximum, turn on the power, and set the current as required (*see* Note 5). Some form of cooling mechanism (e.g., a fan or cooling coil) should be used if the current is >5 mA.

   i. As the run progresses, a blue dye front will be clearly visible. Allow this front to just run off the bottom of the gel.

   j. When the run is completed, dismantle the apparatus and the gel mold, and place the gel in a container.

5. Staining a gel:
   a. Cover the gel with coomassie blue stain, leave on a gently shaking water bath for 30 min, and then remove the stain.
   b. Rinse gently and cover with destain solution. Leave shaking for approx 60 min. Repeat once or twice using 7% acetic acid until the gel is sufficiently destained and the protein bands are clearly visible.

6. Drying a gel: Soak the gel in fixing solution for 30 min and then drying solution for 30 min. Place the gel between thoroughly wet sheets of cellophane, and tape securely to a glass plate. Leave to dry for approx 2 h with an infrared lamp warming from behind or overnight without a lamp. An autoradiograph (*see* Section 3.3.) may be obtained from the dried gel.

3.2. Protease Digestion

1. First gel: Run labeled samples on a gel (*see* Section 3.1.) and obtain an intensified autoradiograph overnight (*see* Section 3.3.). The bands of interest can be cut from the gel using a scalpel or razor blade. These pieces should be marked in some manner (e.g., snipping of the bottom left-hand corner) to enable correct orientation of the bands on the second gel.

2. Second gel: Prepare a 15% acrylamide gel, 1.5 mm thick, with the top 30% being stacking gel. Use a 1.5-mm thick comb having a normal design for the wells when pouring the stacking gel.

3. Place the gel pieces in individual vials or tubes with enough rehydrating buffer to cover them for approx 10 min. During this time, the cellophane will loosen and can be removed from the gel band.

4. Using tweezers, a fine spatula, or other suitable implement, place the gel pieces vertically in the wells of the second gel in such a manner that they are all aligned the same way.

5. Add 50 µL diluted protease solution to each well around the gel piece.

6. These gels are run overnight at a constant voltage. First run for 60–90 min at 100 V (until dye front has run well into the stacking gel). Switch off the power for 30–60 min to allow digestion of the proteins by the enzyme, then switch the power back on, and run overnight at a suitable voltage (*see* Note 6) until the dye front reaches the bottom of the gel.

7. Stain the gel with coomassie blue stain if desired (*see* Section 3.1.) or simply soak in 7% acetic acid for 30 min.

8. Dry the gel ready for autoradiography (*see* Section 3.3.).

3.3. Autoradiography and Densitometry

1. Run labeled samples on a gel, and dry the gel (*see* Sections 3.1. and 3.2.).
2. Either of the following methods may be used to obtain an autoradiograph:
   a. Intensified autoradiograph: Place a sheet of autoradiography film next to an intensifying screen, and then place the dried gel face down on top of the film. Close the cassette, and keep at –80°C for the required time (overnight is often suitable).
   b. Normal exposure: Make a "sandwich" in the following sequence: perspex sheet, autoradiography film, dried gel (face down), perspex sheet. Place this sandwich inside a light-proof bag and clamp firmly. Keep in the dark for a few days as required.
3. Remove the film from the cassette or package, and process according to directions for the X-ray developer and fixer used. Wash well and allow to dry.
4. If quantitation is desired, densitometry may be performed only on autoradiographs produced by normal exposure, since the response obtained using intensified screens is generally not linear at the high or low ends of the grey scale.
5. Following instructions for the densitometer, scan the tracks or sections of tracks on the gel. The relative peak heights or areas under the peak of the CaM-PK II subunits or phosphopeptides can be compared from track to track. If these values are first normalized with respect to a standard control track on the same gel, comparisons can be made between samples run on other gels and between experiments.

3.4. Autophosphorylation in Lysed Tissue

Final conditions required in assay (*see* Notes 7, 8, 9, and 10):

Standard method	Zinc method
30 mM Tris-HCl	30 mM Tris-HCl
1 mM Mg^{2+}	1 mM Mg^{2+}
1 mM EGTA	1 mM EGTA
±1.2 mM Ca^{2+}	±5 mM Zn^{2+}
±50 µg/mL calmodulin	100 Ci/mL, 40 µM ATP
±50 µM trifluoperazine	1 mg/mL protein
100 µCi/mL, 40 µM ATP	Total vol 100 µL, pH 7.4
1 mg/mL protein	Incubation temperature: 37°C
Total vol 100 µL, pH 7.4	Incubation time: 5 min
Incubation temperature: 37°C	
Incubation time: 15 s	

1. Prepare sample containing CaM-PK II (*see* Notes 11 and 12), diluted to 2.5 mg/mL in 30 mM Tris pH 7.4, for subcellular fractions or a suitable dilution for purified enzyme.

2. Prepare assay tubes as follows (use disposable plastic and keep on ice): 10 μL assay buffer A, ± 10 μL CaCl$_2$ (12 mM), ± 5 μL calmodulin (1 mg/mL) ± 10 μL trifluoperazine (500 μM) and enough 30 mM Tris to bring the volume in each tube to 40 μL. Appropriate control tubes should be prepared.

3. Calculate required amounts of [γ-^{32}P]ATP radioisotope and "cold" ATP (*see* Note 13a). Make "cold" ATP solution in Tris, and put into a separate tube. Add the calculated amount of [γ-^{32}P]ATP radioisotope, and mix carefully. Add 20 μL of this ATP mixture to each assay tube (*see* Note 14).

4. Prewarm assay tubes for 1 min at 37°C in a water bath, then initiate reaction by adding 40 μL sample (2.5 mg protein/mL) to the appropriate assay tube in the water bath, and mix. Stop the reaction after 15 s by the addition of 70 μL solubilizing buffer. Keep on ice until all tubes have been completed.

5. Heat all tubes at 100°C for 2 min. Samples are now ready to be loaded onto a gel or frozen until required. The autophosphorylation of CaM-PK II can be quantitated by densitometry of the resulting autoradiograph.

6. The zinc procedure is essentially the same as the standard method, except that the incubations are extended to 5 min, and 10 μL ZnSO$_4$ (50 mM) is used instead of Ca^{2+} and calmodulin.

3.5. Autophosphorylation in Intact Tissue

1. Prepare synaptosomes *(27)*, wash, and resuspend in Krebs-like buffer so that the overall protein concentration in the next step will be 5 mg/mL (*see* Appendix).

2. ^{32}P$_i$ preincubation: final conditions required: 750 μCi/mL ^{32}P$_i$, 25 mM NaHCO$_3$, 5 mg/mL protein, 1.2 mM MgCl$_2$, 118 mM NaCl, 100 μM CaCl$_2$, 5 mM KCl, and 10 mM glucose. Incubate 150 μL synaptosome sample with enough ^{32}P$_i$ radioisotope to allow a concentration of 750 μCi/mL (*see* Appendix), for 45 min at 37°C in a disposable plastic tube, shaking frequently to keep the sample suspended.

3. Depolarization: Incubate 70-μL aliquots from the preincubation with an equal volume of prewarmed Krebs-like buffer or an equal volume of prewarmed depolarizing buffer for 5 s at 37°C.

4. Stop the incubations by lysis in detergent and subsequent immunoprecipitation of CaM-PK II *(12,17)* followed by gel electrophoresis.

3.6. Standard Peptide Assay

Final conditions required in assay (*see* Notes 9 and 15): 30 mM Tris-HCl, 1 mM Mg^{2+}, 1 mM EGTA, ±1.2 mM Ca^{2+}, ±50 μg/mL calmodulin, ±100 μM peptide substrate, 50 μCi/mL, 1 mM ATP, 0.25 mg/mL protein, total vol: 100 μL, pH 7.4, incubation temperature: 30°C, incubation time: 45 s.

1. Prepare sample containing CaM-PK II (*see* Notes 11 and 12), dilute to 1 mg/mL in 30 m*M* Tris-HCl, pH 7.4, for subcellular fractions or a suitable dilution for purified enzyme.

2. Using a ruler and pencil, divide 13.5 × 9.0 cm sheets of P81 paper into sufficient 1.5-cm squares to allow triplicate samples of all sample tubes and five squares for blanks. Prepare a separate sheet for total radioactivity counts, one square for each tube. Each square should be appropriately numbered in pencil in the corner, and the sheets arranged conveniently on a polystyrene board with a pin through each corner to keep paper raised above the board.

3. Prepare assay tubes as follows (use disposable plastic and keep on ice): 25 μL assay buffer B, ±10 μL CaCl₂ (12 m*M*), ±5 μL calmodulin (1 mg/mL), ±10 μL peptide substrate (1 m*M*) and enough 30 m*M* Tris to bring the volume in each tube to 55 μL. Appropriate control tubes should be prepared.

4. Calculate required amounts of [γ-^{32}P]ATP radioisotope and "cold" ATP (*see* Note 13b). Make "cold" ATP solution in Tris and put into a separate tube. Add the calculated amount of [γ-^{32}P]ATP radioisotope, and mix carefully. Add 20 μL of this ATP mixture to each assay tube (*see* Note 14).

5. Prewarm assay tubes for 1 min at 30°C in a water bath, then initiate reaction by adding 25 μL protein sample (1.0 mg/mL) to the appropriate assay tube in the water bath, and mix. Stop the reaction after 45 s by spotting triplicate 20-μL aliquots onto the numbered P81 paper squares. Place the sheets of P81 paper in a container with approx 1 L of 75 m*M* orthophosphoric acid.

6. Also place the P81 paper squares that were set aside for blanks in the container of orthophosphoric acid.

7. Keep the rest of the sample in the assay tube until the end of the assay, and then spot a 20 μL sample from each tube onto the P81 papers for a total count of radioactivity in each tube. Leave to dry in air. Do not put these into the container of orthophosphoric acid.

8. When all the assay tubes have been processed and all the paper squares are in the container with orthophosphoric acid, wash the papers three times for 10 min in fresh changes of orthophosphoric acid (75 m*M*) on an orbital shaker (use approx 1 L each wash for 100 P81 squares).

9. Rinse the papers twice in absolute ethanol, and then leave under a flow of compressed air until dry.

10. Holding the dried P81 sheets with tweezers, cut out the numbered squares containing the blank and assay samples, place in glass scintillation vials, and count in a β scintillation spectrometer using the Cherenkov method *(24)* for 2–10 min each.

11. Count the papers used for the total radioactivity in a similar manner. Calculate CaM-PK II activity by comparing the counts obtained for samples containing Ca^{2+} and calmodulin with those obtained for basal levels.

3.7. Peptide Assay for Calcium-Independent Activity

1. Follow Section 3.6., steps 1 and 2 except that protein concentration should be 2.5 mg/mL for subcellular fractions.
2. Preincubation to prepare a primary *(11)* autophosphorylated CaM-PK II sample: final conditions required (*see* Notes 16, 17, and 18): 30 m*M* Tris-HCl, 5 m*M* Mg^{2+}, 0.5 m*M* EGTA, ±0.75 m*M* Ca^{2+}, ±50 μg/mL calmodulin, 200 μ*M* ATP, 1 mg/mL protein, total vol: 100 μL, pH 7.4, incubation temperature: 30°C, incubation time: 15 s.
   a. Prepare assay tubes as follows (use disposable plastic and keep on ice): 25 μL assay buffer C, ±10 μL $CaCl_2$ (7.5 m*M*), ±5 μL calmodulin (1 mg/mL), 20 μL ATP (1 m*M*), and enough 30 m*M* Tris to bring the volume in each tube to 60 μL. Appropriate control tubes should be prepared.
   b. Prewarm assay tubes for 1 min at 30°C in a water bath, then initiate reaction by adding 40 μL protein sample (2.5 mg/mL) to the appropriate assay tube in the water bath, mix, and incubate for 15 s at 30°C (*see* Note 14).
   c. Stop preincubation by adding 150 μL diluted EDTA (25 μL EDTA [25 m*M*] mixed with 125 μL Tris [30 m*M*]) (*see* Note 16).
3. Peptide incubation: final conditions required: 30 m*M* Tris-HCl, 5 m*M* Mg^{2+}, 0.05 m*M* EGTA, ±0.275 m*M* Ca^{2+}, ±50 μg/mL calmodulin, 50 μCi/mL, 100 μ*M* ATP, 0.1 mg/mL protein, total vol: 100 μL, pH 7.4, incubation temperature: 30°C, incubation time: 45 s.
   a. Prepare assay tubes as follows (use disposable plastic and keep on ice): 25 μL assay buffer D, ±10 μL $CaCl_2$ (2 m*M*), ±4.5 μL calmodulin (1 mg/mL), ±10 μL peptide substrate (1 m*M*), and enough 30 m*M* Tris to bring the volume in each tube to 55 μL. Appropriate control tubes should be prepared.
   b. Calculate required amounts of [γ-^{32}P]ATP radioisotope and "cold" ATP (*see* Note 13c). Make "cold" ATP solution in Tris and put into a separate tube. Add the calculated amount of [γ-^{32}P]ATP radioisotope, and mix carefully. Add 20 μL of this ATP mixture to each assay tube.
   c. Prewarm assay tubes for 1 min at 30°C in a water bath, then initiate reaction by adding 25 μL diluted autophosphorylated sample from step 2c to the appropriate assay tube in the water bath, and mix. Stop the reaction after 45 s by spotting triplicate 20-μL aliquots onto the numbered P81 paper squares.
4. Complete as for standard assay (*see* Section 3.6., step 5).

3.8. Transfer to Nitrocellulose

Samples are first run on a polyacrylamide gel (*see* Section 3.1.) and then transferred to nitrocellulose membrane *(26,27)*. CaM-PK II can be readily transferred *(8)*.

1. After Section 3.1., step 4, equilibrate the gel in transfer buffer for approx 30 min.
2. CaM-PK II is electrophoretically transferred from the polyacrylamide gel to a nitrocellulose membrane. Prior to transfer, cut a piece of nitrocellulose membrane and several pieces of filter paper to the same size as the gel to be transferred. Equilibrate these, together with the fiber pads from the transfer apparatus in transfer buffer for approx 30 min.
3. Prepare a "sandwich" in the gel holder as follows: fiber pad, filter paper, gel, nitrocellulose membrane, filter paper, and fiber pad, beginning on the negative side of the gel holder. Fill the transfer cell with precooled transfer buffer (4°C), and position the negative side of the gel holder on the negative side of the apparatus. Carry out the transfer with the current set between 0.6 and 1.0 A for a minimum of 6 h (usually overnight) using a cooling coil connected to a water bath at 4°C. Place the transfer cell on a magnetic stirrer, so that the buffer can be stirred continually during the transfer.
4. When transfer is completed, dismantle the apparatus and the "sandwich." Part of the gel can be stained and dried (*see* Section 3.1.) and then autoradiography (*see* Section 3.3.) performed if desired, in order to check that the transfer has been successful (*see* Fig. 6 in ref. *8*).

3.9. Antibody Binding

This method of characterization relies on the availability of a specific antibody to CaM-PK II *(1,2,10,12)*.

1. Protein samples containing CaM-PK II are first run on a gel and transferred to nitrocellulose membrane (*see* Section 3.8.).
2. Mark the nitrocellulose membrane, cut into sections if planning to use more than one primary antibody, and soak in 100 mL blocking solution 1 with gentle shaking for 45–60 min. Wash the membrane three times in TBST over a period of 15 min with gentle shaking. Use 100 mL each time.
3. Soak the nitrocellulose membrane in 100 mL blocking solution 2, and agitate gently for 45–60 min. Again, wash the membrane three times in TBST over a period of 15 min with gentle shaking.

4. Incubation with primary antibody: Incubate the washed membrane with primary antibody solution within a sealed plastic bag (*see* Note 19). Rock the membrane gently for at least 1 h (3 h or more may be required for some antibodies). Repeat the washing procedure with TBST.

5. Secondary antibody incubation: Incubate the membrane for 1 h with secondary antibody solution A. As with the previous step, seal the membrane in a plastic bag, and place on a rocking platform. Final washes: wash the membrane twice with TBST and then once with TBS alone to remove Tween from the membrane.

6. Color development: Pour color development solution over the washed membrane, and gently agitate while color is appearing. Stop the color development by rinsing with distilled water for 10 min. The time for color development varies with concentration of proteins from immediate to 4 h. Average time used is 15–30 min (*see* Note 20). Dry the membranes between filter paper, and store in sealed plastic bags. Photography is recommended in case colors fade.

3.10. Calmodulin Binding Assay

The amount of CaM-PK II present in lysed tissue can also be measured using biotinylated calmodulin.

1. Lysed protein samples are first run on a gel and transferred to nitrocellulose membrane (*see* Section 3.8.).

2. Mark the nitrocellulose membrane, immerse in 100 mL blocking solution 3, and agitate gently for 45–60 min. Wash the membrane three times in wash solution over a period of 30 min with gentle shakng. Use 100 mL for each wash.

3. Incubation with Biocam: Incubate the washed membrane with Biocam solution within a sealed plastic bag, and rock gently for 30 min. Repeat the washing procedure as previously.

4. Secondary antibody incubation: Incubate the membrane for 2 h with secondary antibody solution B. As with the previous step, seal the membrane in a plastic bag, and place on a rocking platform. Final washes: wash the membrane twice with wash solution and then once with TBS alone to remove Tween from the membrane.

5. Color development: Pour color development solution over the washed membrane, and gently agitate while color is appearing. Stop the color development by rinsing with distilled water for 10 min. Biocam color development is almost immediate. Dry the membrane between filter paper, and store in sealed plastic bags. Photography is recommended in case colors fade.

4. Notes

1. All H_2O used should be doubled-deionized or distilled.
2. Biotinylated calmodulin is available from Calbiochem-Novabiochem (San Diego, CA).
3. NBT and BCIP powders used in the color development solution should be stored at –80°C, desiccated, and away from light. Stock solutions should be stored at 4°C in amber vials and may be used for up to 1 mo. After mixing with alkaline phosphatase buffer, protect from light and use as quickly as possible. Use 20 mL of color development solution for one 15 × 11 cm membrane.
4. A suitable peptide substrate is available from Auspep (Melbourne, Australia): Cat. 2025, Calmodulin Dependent Protein Kinase Substrate. Alternative substrates are available from other sources.
5. Time for running gels varies with the size of the gel and the applied current. Examples are as follows:

For 18 × 20 cm gels:	For 16 × 16 cm gels:
7–8 mA approx 16 h	4 mA approx 16 h
20 mA approx 6 h	20 mA approx 4 h

6. A gel measuring 18 × 20 cm wlll take approx 16 h to run at 70 V.
7. CaM-PK II can exist in more than one autophosphorylated form depending on the exact incubation conditions used *(11)*. If low ATP conditions (<100 μM) and long times (>30 s) are used, then the mobility of CaM-PK II on SDS-PAGE is decreased *(27)*. The conditions chosen here do lead to a proportion of CaM-PK II decreasing mobility, but are a compromise to minimize the amount of radiolabeled ATP required. The mobility shift is, however, a useful diagnostic tool for characterizing CaM-PK II.
8. CaM-PK II is not autophosphorylated in the absence of Ca^{2+}. In subcellular fractions, it may be autophosphorylated in the presence of Ca^{2+} alone because of endogenous calmodulin, but if purified enzyme is used, calmodulin alone will not activate it. Trifluoperazine blocks the formation of a calcium–calmodulin complex, inhibits autophosphorylation even in the presence of Ca^{2+} and calmodulin, and is therefore useful for inhibiting CaM-PK II autophosphorylation, while allowing phosphorylation due to other kinases to proceed.
9. Up to 10 mM Mg^{2+} can be used *(28)*, but this is not essential.
10. In the presence of high concentrations of zinc, most protein kinase and protein phosphatase activity is inhibited, but CaM-PK II autophosphorylation is preserved. Essentially, the same results are found if labeling is undertaken with zinc in the presence or absence of Ca^{2+} and/or calmodulin *(8)*. It must be noted that after autophosphorylation in the presence of zinc, the mobility of CaM-PK II is slightly increased on SDS-PAGE *(16)*.

11. Degradation of CaM-PK II occurs during isolation, and can be minimized by the addition of protease inhibitors and especially by removal of Ca^{2+}, which alters CaM-PK II conformation and makes it more susceptible to thermal inactivation *(11)*. Initial steps in the isolation of tissue, such as homogenization and centrifugation, should therefore be undertaken in a buffer containing 1 mM EDTA.

12. If possible, tissue should be freshly prepared at 4°C and used immediately, since freezing (either –20 or –80°C) or long-term storage lead to a significant loss of CaM-PK II activity *(11)*.

13. [γ-^{32}P]ATP/ATP mixture in 30 mM Tris-HCl, pH 7.4: Prepare enough for the number of tubes to be labeled plus two more. Final concentration in the assay tubes should be:
    a. 50 μCi/mL and 40 μM ATP (i.e., 5 μCi/tube and 4 nmol/tube in a volume of 20 μL/tube for a total incubation volume of 100 μL).
    b. 50 μCi/mL and 1 mM ATP (i.e., 5 μCi/tube and 100 nmol/tube in a volume of 20 μL/tube for a total incubation volume of 100 μL).
    c. 50 μCi/mL and 100 μM ATP (i.e., 5 μCi/tube and 10 nmol/tube in a volume of 20 μL/tube for a total incubation volume of 100 μL).

14. Preincubation of the enzyme with Ca^{2+} and calmodulin leads to thermal inactivation, and enzyme reactions should, in general, be initiated by addition of tissue sample (or purified enzyme) rather than ATP.

15. Lower ATP levels can be used to increase the specific activity of the ATP, but when using subcellular fractions, ATPase activity can decrease ATP levels significantly over a 45-s period.

16. Increasing the concentration of Mg^{2+} favors primary autophosphorylation *(11,28)* and, hence, the formation of calcium-independent CaM-PK II. However, when the EDTA is added at the end of the incubation, the solution becomes very acidic owing to the release of H^+ ions on binding of Mg^{2+} to the EDTA. To minimize this, the Mg^{2+} concentration should not be more than 5 mM.

17. Higher concentrations of ATP also favor calcium-independent activty *(11)*, but decrease the specific activity of ATP in the subsequent peptide incubation owing to necessary carryover of ATP.

18. The levels of Ca^{2+} and EGTA are reduced to minimize carryover into the peptide assay and to avoid further acid generation by Ca^{2+} binding to the EDTA.

19. The type of plastic used for sealing into plastic bags for primary antibody incubation is important. Thin bags stick to the membranes, producing poor results, so a reasonably heavy plastic bag is required.

20. Background purple coloration has been found to depend on the dilution of the primary antibody and the length of the incubation. Both of these factors need to be determined for individual antibodies to arrive at the desired sensitivity, yet minimize background.

References

1. Schulman, H. (1988) The multifunctional Ca^{2+}/calmodulin-dependent protein kinase, in *Advances in Second Messenger and Phosphoprotein Research*, vol. 22 (Greengard, P. and Robinson, G., eds.), Raven, New York, pp. 39–112.

2. Sodeling, T. R., Fukunaga, K., Rich, D. P., Fong, Y. L., Smith, K., and Colbran, R. J. (1990) Regulation of brain Ca^{2+}/calmodulin-dependent protein kinase II, in *The Biology and Medicine of Signal Transduction* (Nishizuka, Y., ed.), Raven, New York, pp. 206–211.

3. Rostas, J. A. P. and Dunkley, P. R. (1992) Multiple forms and distribution of calcium/calmodulin-stimulated protein kinase II in brain. *J. Neurochem.* **59,** 1191–1202.

4. Laemmli, U. K. (1970) Cleavage of structural proteins during the assembly of bacteriophage T4. *Nature* **227,** 680–685.

5. Dunkley, P. R., Baker, C. M., and Robinson, P. J. (1986) Depolarization dependent protein phosphorylation in rat cortical synaptosomes. Characterization of active protein kinases by phosphopeptide analysis of substrates. *J. Neurochem.* **46,** 1692–1703.

6. Rodnight, R., Zamoni, R., and Tweedle, A. (1988) An investigation of experimental conditions for studying protein phosphorylation in microslices of rat brain by two dimensional electrophoresis. *J. Neurosci. Methods* **24,** 27–38.

7. Cleveland, D. W., Fischer, S. G., Kirschner, M. W., and Laemmli, U. K. (1977) Peptide mapping by limited proteolysis in sodium dodecyl sulphate and analysis by gel electrophoresis. *J. Biol. Chem.* **252,** 1102–1106.

8. Jeitner, T., Jarvie, P., Costa, M., Rostas, J. A. P., and Dunkley, P. R. (1991) Calmodulin kinase II, C kinase and cyclic AMP kinase in guinea-pig myenteric ganglia: presence of the enzymes and characterisation of their substrates. *Neuroscience* **40,** 555–569.

9. Rostas, J. A. P., Brent, V., Seccombe, M., Weinberger, R. P., and Dunkley, P. R. (1989) Purification and characterisation of calmodulin stimulated protein kinase II from two day and adult chicken forebrain. *J. Mol. Neurosci.* **1,** 93–104.

10. Kelly P. T. (1991) Calmodulin-dependent protein kinase II: multifunctional roles in neuronal differentiation and synaptic plasticity. *Mol. Neurobiol.* **5,** 153–177.

11. Dunkley, P. R. (1991) Autophosphorylation of neuronal calcium/calmodulin-stimulated protein kinase II. *Mol. Neurobiol.* **5,** 179–202.

12. Molloy, S. S. and Kennedy, M. B. (1991) Autophosphorylation of type II Ca^{2+}/calmodulin-dependent protein kinase in cultures of postnatal rat hippocampal slices. *Proc. Natl. Acad. Sci. USA* **88,** 4756–4760.

13. Dunkley, P. R., Côté, A., and Harrison, S. (1991) Autophosphorylation of calmodulin-stimulated protein kinase II in intact synaptosomes. *J. Mol. Neurosci.* **2,** 193–201.

14. Rostas, J. A. P., Weinberger, R. P., and Dunkley, P. R. (1986) Multiple pools and multiple forms of calmodulin stimulated protein kinase during development. Relationship to post-synaptic densities. *Prog. Brain Res.* **69,** 355–371.

15. Dunkley, P. R., Jarvie, P. E., and Rostas J. A. P. (1988) Distribution of calmodulin—and cyclic—AMP stimulated protein kinases in synaptosomes. *J. Neurochem.* **51,** 57–68.

16. Weinberger, R. P. and Rostas, J. A. P. (1991) The effect of zinc on protein phosphorylation in rat cerebral cortex. *J. Neurochem.* **57,** 605–614.

17. Gorelick, F. S., Wang, J. K. T., Lai, Y., Nairn, A. C., and Greengard, P. (1988) Autophosphorylation and activation of Ca-calmodulin dependent protein kinase II in intact nerve terminals. *J. Biol. Chem.* **263,** 17,209–17,212.

18. Pearson, R. B., Woodgett, J. R., Cohen, P., and Kemp, B. E. (1985) Substrate specificity of a multifunctional calmodulin-dependent protein kinase. *J. Biol. Chem.* **260,** 14,471–14,476.

19. Presek, P., Jessen, S., Dreyer, J., Jarvie, P., Findik, D., and Dunkley, P. R. (1992) Tetanus Toxin inhibits depolarization stimulated protein phosphorylation in rat cortical synaptosomes: effect on synapsin I phosphorylation and translocation. *J. Neurochem.* **59(4),** 1336–1343.

20. Rostas, J. A. P., Brent, V., and Dunkley, P. R. (1987) The effect of calmodulin on the activity of calmodulin-stimulated protein kinase II. *Neurosci. Res. Commun.* **1,** 3–8.

21. Bunn, S. J., Harrison, S. M., and Dunkley, P. R. (1992) Protein phoshorylation in bovine adrenal medullary chromaffin cells: histamine stimulated phosphorylation of tyrosine hydroxylase. *J. Neurochem.* **59,** 164–174.

22. Haycock, J. W. (1990) Phosphorylation of tyrosine hydroxylase in situ at serine 8, 19, 31 and 40. *J. Biol. Chem.* **265,** 11,682–11,691.

23. Haycock, J. W. (1992) Multiple signalling pathways in bovine adrenal chromaffin cells regulate tyrosine hydroxylase phosphorylation at ser[19], ser[31] and ser[40]. *Neurochem. Res.* **18,** 15–26.

24. Dunkley, P. R., Jarvie, P. E., and Sim, A. T. R. (1995) Second messenger stimulated protein phosphorylation in the nervous system, in *Neurochemistry: A Practical Approach*, 2nd ed. (Turner, A. J. and Batchelard, H. S., eds.), IRL, Oxford, in press.

25. Towbin, H., Staehelin, J., and Gordon, J. (1979) Electrophoretic transfer of proteins from polyacrylamide gels to nitrocellulose sheets: procedures and some applications. *Proc. Natl. Acad. Sci. USA* **76,** 4350–4354.

26. Biorad Transblot Cell Operating Instructions. Biorad Laboratories, Hercules, CA.

27. Dunkley, P. R., Heath, J. W., Harrison, S. M., Jarvie, P. E., Glenfield, P. J., and Rostas, J. A. P. (1988) A rapid percoll gradient procedure for isolation of synaptosomes directly from an S1 fraction: homogeneity and morphology of subcellular fractions. *Brain Res.* **441,** 59–71.

28. Lou, L. L., Lloyd, S. J., and Schulman, H. (1986) Activation of the multifunctional Ca^{2+}/calmodulin-dependent protein kinase by autophosphorylation: ATP modulates production of an autonomous enzyme. *Proc. Natl. Acad. Sci. USA* **83,** 9497–9501.

Appendix
Calculations for Autophosphorylation in Intact Tissue

1. Because $^{32}P_i$ has a very short half-life, the decay factor is substantially different each day. This requires that different volumes of $^{32}P_i$ be used in the preincubation from day to day. To enable consistency between experiments, a calculation is given (step 3) for the volumes of $^{32}P_i$ and 10X concentrated Krebs buffer, and for the protein concentration of the tissue sample to use in the preincubation. A little 10X concentrated Krebs buffer is added to the preincubation to compensate for the lack of buffer salts (and thus decreased ionic strength) in the $^{32}P_i$ addition.

2. Assumptions:
   a. Stock $^{32}P_i$ is 10 mCi/mL.
   b. Required $^{32}P_i$ concentration in incubation is 0.75 µCi/mL.

3. Derivation:

   (10 mCi/mL × decay factor × volume $^{32}P_i$) =
   $\qquad\qquad\qquad\qquad$ (0.75 mCi/mL × total vol preincubation)

   Total vol preincubation = $V + P + K$
   where V = vol sample (µL), P = vol $^{32}P_i$ (µL), K = vol 10X Krebs (µL), and D = decay factor.

   Therefore:

   $$10 \times D \times P \;=\; 0.75\,(V + P + K)$$
   $$=\; 0.75\,V + 0.75P + 0.75K \qquad (1)$$

   and

   $$K = 1/10\,(P + K)$$
   $$10K = P + K$$
   $$9K = P \qquad (2)$$

   Subtitute for P in Eq. (1):

   $$10 \times D \times 9K \;=\; [0.75V + (0.75 \times 9K) + 0.75K]$$

   Rearranging:

   $$[(10 \times D \times 9K) - (0.75 \times 9K) - 0.75K] = 0.75V$$
   $$K[(90D - 0.75) \times (9 + 1)] = 0.75V$$
   $$K(90D - 7.5) = 0.75V$$
   $$K = \frac{0.75V}{(90D - 7.5)}$$

   and

   $$P = 9K \text{ (from Eq. [2])} \qquad (3)$$

4. In general:

$$\text{Vol (10X) Krebs} = [(\text{conc.} \,^{32}\text{P}_i \text{ required} \times \text{vol sample})/(90 \times \text{decay factor})] - (10\text{X conc.} \,^{32}\text{P}_i \text{ required})$$

$$\text{Vol} \,^{32}\text{P}_i \text{ required} = 9 \times \text{vol (10X) Krebs} \qquad (4)$$

5. Concentration of protein required in sample: To obtain a final protein concentration of 5 mg/mL in the preincubation:

$$(\text{Vol sample} \times \text{required protein concentration}) = (\text{total vol} \times 5 \text{ mg/mL})$$

$$\text{Required protein conc.} = [(\text{total vol} \times 5 \text{ mg/mL})/\text{vol sample}] = [5 \,(\text{vol sample} + \text{vol} \,^{32}\text{P}_i + \text{vol (10X) Krebs})]/\text{vol sample} \qquad (5)$$

CHAPTER 21

Protein Kinase C

*Measurement of Translocation, Activation,
and Role in Cellular Responses*

Sandra E. Wilkinson and Trevor J. Hallam

1. Introduction

The serine/threonine kinase protein kinase C (PKC) is regarded as playing a key role in stimulation of cellular responses in many different cells and tissues. There are now known to be at least 10 different isoenzymes of PKC (α, β_I, β_{II}, γ, δ, ϵ, ζ, η, θ, and λ) which vary in their cofactor requirement for activation and also in their substrate specificity. The conventional PKCs (cPKC) (α, β_I, β_{II}, and γ) require Ca^{2+}, phospholipid, and diacylglycerol (DAG) for activation. The novel PKC (nPKC) enzymes (δ, ϵ, η, and θ) are Ca^{2+}-independent, whereas members of the atypical PKCs (aPKC), ζ and λ, lack the binding site for DAG. Phorbol esters, such as 12-*O*-tetradecanoylphorbol-13-acetate (TPA), also bind to and activate those enzyme isotypes that are sensitive to DAG and are used as exogenously applied activators of isolated enzymes or of cells to mimic the actions of DAG *(1)*.

The classical route leading to the activation of PKC results from stimulation of cell-surface receptors by binding of a variety of hormones, growth factors, and neurotransmitters, which leads to the activation of one of several forms of phospholipase C via G-proteins or on

From: *Methods in Molecular Biology, Vol. 41: Signal Transduction Protocols*
Edited by: D. A. Kendall and S. J. Hill Copyright © 1995 Humana Press Inc., Totowa, NJ

phosphorylation by a tyrosine kinase. This leads to the generation of inositol 1,4,5-trisphosphate, which stimulates the release of Ca^{2+} from intracellular stores, and DAG. Ca^{2+} is believed to bind to PKC and cause it to translocate to the cell membrane, where it is fully activated by the binding of phosphatidylserine and DAG *(2,3)*. Members of the nPKC and aPKC families must, however, recognize different signals, because these isoforms all lack the Ca^{2+} binding domain and the aPKC isoenzymes also lack the DAG binding site.

The amount of PKC that translocates to the cell membrane is regarded as a measure of enzyme activation. However, it is probable that translocation and activation of enzyme activity, although they do occur simultaneously with some isotypes, are separate events. We describe a number of techniques by which this can be evaluated. Before embarking on any of this work, however, there are a number of caveats of which the researcher should be aware (for full review, *see* ref. *4*). PKC is now not the only known receptor for phorbol esters *(5)*, and there may be other, as yet undiscovered, receptors for this family of nonphysiological activators. Members of the aPKC family lack the common DAG/phorbol ester binding site *(6,7)* and presumably should not be activated by these compounds, although reports in the literature are less than clear on this point *(7,8)*. It is also becoming apparent that some PKC isoenzymes are activated by products of lipid metabolism other than DAG, which may mean that the pattern of enzyme activation induced by phorbol esters bears little resemblance to that stimulated by a more physiological stimulus *(9,10)*. Finally, it must be remembered that the phorbol esters are metabolically stable and can be found in the cell several hours after initial treatment, unlike DAG, which is rapidly degraded by the enzymes DAG lipase and DAG kinase. Again, this may result in a false picture emerging of the range of isoenzymes that may translocate to the plasma membrane on stimulation of the cell by physiological activators, and the length of time that these isotypes would normally remain in the cell membrane. Having said this, new and more selective agents are beginning to appear, for example, Sapintoxin A, which activates PKC-α, β_I, γ, and ϵ, but not PKC-δ *(11)*, and many of the techniques described in this chapter can be used both with such compounds as these and with more physiological stimuli.

2. Materials

2.1. Phorbol Ester-Induced Translocation of Human Neutrophil PKC and Measurement of Enzyme Activity in Both Particulate and Cytosolic Fractions

2.1.1. Isolation of Human Neutrophils

1. Heparin (50 µL/50 mL blood).
2. Lymphocyte separating media (Ficoll-Hypaque, Pharmacia LKB, Uppsala, Sweden).
3. 1.8% Dextran in phosphate-buffered saline (PBS).
4. 0.21% NaCl.
5. $1M$ KCl.
6. PBS: 10 mM Na$_2$HPO$_4$, and 0.15M NaCl, pH 7.2.
7. Buffer A: 0.25M sucrose, 10 mM HEPES, 10 mM β-mercaptoethanol, 1 mM phenylmethylsulfonylfluoride (PMSF), and 1 mM sodium ortho-vanadate, pH 7.4.

2.1.2. Phorbol Ester Stimulation and Cell Fractionation

1. Buffer A: 0.25M sucrose, 10 mM HEPES, 10 mM β-mercaptoethanol, 1 mM PMSF, and 1 mM sodium orthovanadate, pH 7.4.
2. Buffer B: Buffer A + 1% Triton X-100.
3. TPA (or phorbol ester of choice).
4. Dimethyl sulfoxide (DMSO).

2.2. In Vitro Assay of PKC Activity

1. Assay buffer: 25 mM Tris-HCl, 5 mM MgNO$_3$, pH 7.5.
2. Histone IIIs (Sigma nomenclature, Sigma, St. Louis, MO): 5 mg/mL in assay buffer.
3. CaCl$_2$: 5 mM in assay buffer.
4. Phosphatidylserine: 200 µg/mL in assay buffer, sonicated for 2×20 s to induce vesicle formation.
5. [γ-^{32}P] Adenosine-5'-triphosphate (ATP) (Amersham, Amersham, UK) in 100 µM cold ATP.
6. Trichloroacetic acid (TCA): 10% (v/v) (BDH, Poole, Dorset, UK).
7. 20 mM Tetra-sodium pyrophosphate in 5% TCA (v/v).
8. Enzyme.

2.3. The Direct Imaging
of PKC Using a Fluorescent Probe

1. Cells.
2. Culture medium (as appropriate to cell type).
3. 12-(1,3,5,7-tetramethylBODIPY-2-propionyl)phorbol-13-acetate (Molecular Probes, Eugene, OR).
4. DMSO.
5. Microscope with x40 objective or x100 oil-immersion objective lens.
6. Silicon-intensified video camera.

2.4. The Assessment
of Phorbol Ester-Induced p47 Phosphorylation
in Human Platelets

1. 100 mL Blood from healthy human volunteers.
2. 77 mM EDTA, pH 7.4.
3. Tris-buffered saline (TBS): 15 mM Tris-HCl, 140 mM NaCl, pH 7.4 at 4°C.
4. Carrier-free [^{32}P]orthophosphate (Amersham) in dilute HCl.
5. Phorbol ester.
6. DMSO.
7. Sodium dodecyl sulfate polyacrylamide gel electrophoresis (SDS-PAGE) equipment.
8. Phosphoimager or autoradiographic equipment.

3. Methods
3.1. Phorbol Ester-Induced Translocation
of Human Neutrophil PKC, Measurement
of Activity in Both Particulate
and Cytosolic Fractions

3.1.1. Isolation of Human Neutrophils

1. Collect blood from source, e.g., healthy human volunteers and layer onto lymphocyte separating media (1:1 ratio).
2. Spin samples at 1000g for 30 min at room temperature.
3. Discard plasma layer, and mix pelleted granulocytes and red blood cells with an equal volume of 1.8% dextran. Mix gently, and stand for 40 min.
4. Remove upper layer, and spin at 1500g for 10 min at room temperature.
5. Discard upper layer, and lyse red blood cells by osmotic shock. Mix cell pellet with 5.8 mL of 0.21% NaCl for 2 min, then add 0.5 mL 1M KCl, and stand for 5 min. Add PBS to a volume of 20 mL, and pellet cells by centrifugation (1500g for 10 min).
6. Repeat step 5 if red cell lysis is not complete.
7. Resuspend in 10 mL buffer A.

3.1.2. Phorbol Ester Stimulation and Cell Fractionation

1. Divide neutrophils into appropriate number of aliquots for experiment.
2. Add $10^{-8}M$ TPA (or agonist of choice) (*see* Note 1) in DMSO, such that final DMSO concentration is not >0.1%. Appropriate solvent-only controls should be included in every experiment. Incubate cells at 37°C for 0–20 min.
3. Sonicate, on ice, for 6 × 10 s bursts at high power. Centrifuge lysate at 100,000g for 1 h.
4. Remove supernatant (cytosol), and store at 4°C (*see* Note 2). Resuspend pellet (membrane fraction) in buffer B, and repeat step 3. Remove detergent-solubilized extract. The cytosolic and detergent-solubilized fractions may then be assayed for enzyme activity as detailed in Section 3.2 (*see* Note 3).

3.2. In Vitro Assay of PKC Activity

1. Mix 40 μL assay buffer, 10 μL histone (*see* Note 4) (final concentration, 0.2 mg/mL), 10 μL CaCl$_2$ (final concentration, 0.5 mg/mL), 20 μL PS (final concentration, 40 μg/mL), and 10 μL [γ-^{32}P] ATP (*see* Note 7) (Amersham; final concentration, 10 μM)/assay tube (*see* Note 5).
2. Preincubate tubes for 5 min at 30°C. Start reaction by addition of enzyme (appropriately diluted), and allow to run for 10 min.
3. Stop reaction by addition of 1 mL ice-cold 10% TCA. Allow protein to precipitate for at least 30 min at 4°C, and then separate acid-precipitable material from unreacted [γ-^{32}P] ATP by filtration through glass-fiber disks (Whatman, GF/B, Maidstone, Kent, UK) (*see* Note 6).
4. Wash precipitated protein with 4 × 20 mM tetra-sodium pyrophosphate in 5% TCA and then 1X ethanol. Dry disks, and determine incorporated radioactivity by liquid scintillation spectrometry.

3.3. Assessment of the Translocation of PKC Isoenzymes by Western Blot

Although evaluation of PKC activity will give a measure of the proportion of total enzyme that is induced to translocate by treatment with various agonists, no information is derived as to the relative amounts of the different isoenzymes that are involved. Use of the pseudosubstrate site-derived peptides, as substrates in the assay, will do little to clarify the situation, because all these compounds exhibit a certain degree of crossreactivity for the other isoforms. The intracellular distribution of the isotypes, and any redistribution subsequent to treatment with an agonist, can be assessed by Western blot analysis of the proteins found in the subcellular fractions before and after treatment. Antibodies have been raised against peptide sequences unique to the various isoenzymes, and many of these preparations are now commercially available (*see* Note 8). The dilution of primary antibody most suitable for use in a Western blot should be provided by the manufacturers. Any standard procedure for Western blot development and analysis can then be followed.

3.4. The Use of a Fluorescent Probe for PKC to Visualize Enzyme Localization and Translocation

The direct imaging of the subcellular distribution of PKC in a living cell was described by Khalil and Morgan in 1991 *(12)* who used a fluorescent probe for the enzyme to demonstrate that contraction of vascular smooth muscle cells was concurrent with Ca^{2+}-dependent translocation of PKC. This methodology allowed a more accurate determination of the intracellular location of the enzyme before activation, but gave no information as to the degree of involvement of the different isoenzymes. As the range and selectivity of probes increases, however, use of the following techniques could yield valuable information.

3.5. The Direct Imaging of PKC Using a Fluorescent Probe

1. Wash cells and resuspend in appropriate medium.
2. Incubate in the presence of 100 nM PKC probe, 12-(1,3,5,7-tetramethyl-BODIPY-2-propionyl)phorbol-13-acetate, for 10–60 s. The probe should be diluted from a stock solution dissolved in DMSO to a final solvent concentration of 0.1%.
3. Agonist-induced alterations in cellular morphology can be viewed microscopically. Fluorescent images can be recorded with a silicon-enhanced video camera—to improve signal-to-noise ratio, at least 20 consecutive frames should be analyzed and averaged. In the work described by Khalil and Morgan, an excitation filter at 485 ± 5.2 nM, a dichromic filter at 500 nM, and a long-pass filter at 530 nM proved most suitable (*see* Note 9).

3.6. Substrate Phosphorylation

A useful technique for assessing the activity of PKC in intact cells is the measurement of substrate phosphorylation. Although the treatment of [^{32}P] orthophosphate-labeled cells with phorbol esters, or other stimuli of signal transduction pathways, can lead to a large number of proteins being identified as either direct or indirect substrates of PKC, there are several prominent substrates that are useful markers of activation of the enzyme. Following the time-course of phosphorylation of these proteins can give useful information as to the effectiveness of both agonists and inhibitors of PKC in intact cells. Examples include the myristolated, alanine-rich C kinase substrate (MARKS), which is phosphorylated by PKC during macrophage activation and growth factor-dependent mitogenesis *(13)*, an 80-kDa protein present in human fibroblasts *(14)*, and the 47-kDa protein pleckstrin in human platelets *(15)*.

3.7. The Assessment of Phorbol Ester-Induced p47 Phosphorylation in Human Platelets

1. Collect blood using 77 mM EDTA as anticoagulant (3.75 mL/50 mL blood).
2. Spin at room temperature for 20 min at 140g to obtain platelet-rich plasma.
3. Spin plasma at room temperature for 10 min at 600g to deposit platelets as a soft pellet (*see* Note 11).
4. Resuspend platelets in 2.5 mL TBS, add 300 µCi [^{32}P]orthophosphate (Amersham), then incubate at 37°C for 90 min to allow enzyme activities to recover from centrifugation, and return to basal levels.
5. Stimulate cells with $10^{-8}M$ TPA (or appropriate concentrations of other phorbol esters) for desired length of time (*see* Note 12), and then stop reaction by the addition of protein disruption buffer followed by boiling for 2 min.
6. Separate proteins by SDS-PAGE, and visualize labeled proteins either by use of a phosphoimager or by exposure to X-ray film. In the latter case, quantitation of ^{32}P incorporation can be performed by densitometry of the autoradiograph produced.

3.8. Use of Inhibitors to Define a Role for PKC in Eliciting a Cellular Response

Much of the evidence for the physiological role of PKC relies on the ability of phorbol esters to mimic the response to a physiological ligand and on evidence derived from the use of nonselective protein kinase inhibitors, such as H7 or staurosporine. These data have to be viewed with caution *(16,17)*. Some relatively selective inhibitors of PKC based on indolocarbazole or bis-indolylmaleimide structures derived from staurosporine and K252a as starting structures are now being reported, and their careful use will undoubtedly lead to greater clarification of the cellular role of PKC *(18–20)*.

4. Notes

4.1. Phorbol Ester-Induced Translocation of Human Neutrophil PKC and Measurement of Activity Both in Particulate and Cytosolic Fractions

1. The methodology is suitable for a wide range of cell types and agonists. The period over which cells are challenged by agonist will have to be re-established following any change in the protocol.
2. If subcellular fractions are to be stored for more than 48 h, glycerol should be added to a final concentration of at least 10% to retain enzyme activity. Freezing of samples should be avoided, because this is also associated with a loss in enzyme activity.
3. If required, PKC may be partially purified by ion-exchange chromatography, as described in ref. *20*, which will increase the length of time over which the enzyme remains viable. Enzyme in the particulate fraction should be solubilized in 0.5% Triton X-100 for 60 min, followed by centrifugation at 100,000*g* prior to application onto the column.

4.2. In Vitro Assay of PKC Activity

4. This method, using histone as substrate, is only suitable for assessing the activity of members of the cPKC family. Histone is a very inefficient substrate for nPKCs and aPKCs; for these isoforms, an appropriate peptide based on the pseudosubstrate site should be used (*see* ref. *21* and references therein), and the assay methodology described in ref. *22* adopted.
5. The assay, as described, contains saturating levels of Ca^{2+} and phospholipid, ensuring that enzyme activity is maximal. In order to demonstrate that the enzyme is activated by a particular agonist, Ca^{2+} should not be present in the assay, and the concentration of phospholipid dropped to 5 µg/mL.
6. This method is suitable for use in a high-throughput screen for the assessment of the effects of a range of agonists and/or inhibitors on enzyme activity. Acid-precipitated protein may be collected and washed using a cell harvester. However, the machine may have to be adapted if TCA is to be used in it for any period of time.
7. In the interest of safety, [γ-^{33}P] ATP can be substituted for [γ-^{32}P] ATP.

4.3. Assessment of the Translocation of PKC Isoenzymes by Western Blot

8. The crossreactivity of commercial antibodies should always be checked, if possible, because these do not always live up to manufacturers' claims.

4.4. The Direct Imaging of PKC Using a Fluorescent Probe

9. According to Khalil and Morgan *(12)*, pretreatment of the cells with the kinase inhibitor H7 delayed the onset, magnitude, and time of the shortening of the fluorescent signal caused by changes in cell geometry in contracting cells. The inhibitor did not effect translocation or distribution of the enzyme.

10. The specificity of the probe can be confirmed by the demonstration that it is displaceable by the nonfluorescent phorbol ester 12-deoxyphorbol 13-isobutyrate 20-acetate.

4.5. The Assessment of Phorbol Ester-Induced p47 Phosphorylation in Human Platelets

11. Cells should be handled as gently as possible during the entire procedure to minimize both nonspecific enzyme activation and cell clumping.

12. The cells will tolerate concentrations of DMSO up to 1% over a 2–10 min time period. If experiments run over a longer time-course, the concentration of DMSO should be dropped to 0.1%.

References

1. Gschwendt, M., Kittstein, W., and Marks, F. (1991) Protein kinase C activation by phorbol esters: do cysteine-rich regions and pseudosubstrate motifs play a role? *Trends Biochem. Sci.* **16,** 167–169.

2. Nishizuka, Y. (1988) The molecular heterogeneity of protein kinase C and its implications for cellular regulation. *Nature* **334,** 661–665.

3. Berridge, M. J. and Irvine, R. F. (1984) Inositol trisphosphate, novel second messenger in cellular signal transduction. *Nature* **312,** 315–321.

4. Wilkinson, S. E. and Hallam, T. J. (1993) Protein kinase C: is its pivotal role in cellular activation over-stated? *Trends Pharmacol. Sci.* **15(2),** 53–57.

5. Ahmed, S., Kozma, R., Monfries, C., Hall, C., Lim, H. H., Smith, P., and Lim, L. (1990) Human brain n-chimaerin cDNA encodes a novel phorbol ester receptor. *Biochem. J.* **272,** 767–773.

6. Ono, Y., Fujii, T., Ogita, K., Kikkawa, U., Igarashi, K., and Nishizuka, Y. (1988) The structure, expression and properties of additional members of the protein kinase C family. *J. Biol. Chem.* **263,** 6927–6932.

7. Liyanage, M., Frith, D., Livneh, E., and Stabel, S. (1992) Protein kinase C group B members PKC-δ, -ε, -ζ, -η. Comparison of properties of recombinant proteins *in vitro* and *in vivo*. *Biochem. J.* **283,** 781–787.

8. Borner, C., Nichols-Guadagno, S., Fabbro, D., and Weinstein I. B. (1992) Expression of four protein kinase C isoforms in rat fibroblasts. Differential alterations in *ras-*, *src-* and *fos-*transformed cells. *J. Biol. Chem.* **267,** 12,892–12,899.

9. Nakanishi, H. and Exton, J. H. (1992) Purification and characterization of the zeta-isoform of protein kinase C from bovine kidney. *J. Biol. Chem.* **267**, 16,347–16,354.

10. Lee, M.-H. and Bell, R. M. (1991) Mechanism of protein kinase C activation by phosphatidylinositol 4,5-bisphosphate. *Biochem.* **30**, 1041–1049.

11. Ryves, W. J., Evans, A. T., Olivier, A. R., Parker, P. J., and Evans, F. J. (1991) Activation of the PKC-isotypes -α, -β$_I$, -γ, -δ, -ε by phorbol esters of different biological activities. *FEBS Lett.* **288**, 5–9.

12. Khalil, R. A. and Morgan, K. G. (1991) Imaging of protein kinase C distribution and translocation in living vascular smooth muscle cells. *Circ. Res.* **69**, 1626–1631.

13. Hartwig, J. H., Thelen, M., Rosen, A., Janmey, P. A., Nairn, A. C., and Aderem, A. (1992) MARKS is an actin filament cross-linking protein regulated by protein kinase C and calcium-calmodulin. *Nature* **356**, 618–622.

14. Isacke, C. M., Meisenhelder, J., Brown, K. D., Gould, K. L., Gould, S. J., and Hunter, T. (1986) Early phosphorylation events following the treatment of Swiss 3T3 cells with bombesin and the mammalian bombesin-related peptide, gastrin-releasing peptide. *EMBO J.* **5**, 2889–2898.

15. Tyers, M., Rachubinski, R. A., Sterwart, M. I., Varrichio, A. M., Shorr, R. G., Haslam, R. J., and Harley, C. B. (1988) Molecular cloning and expression of the major protein kinase C substrate of platelets. *Nature* **333**, 470–473.

16. Bradshaw, D., Hill, C. H., Nixon, J. S., and Wilkinson, S. E. (1993) Therapeutic potential of protein kinase C inhibitors. *Agents and Actions*, **38**, 135–147.

17. Nixon, J. S., Wilkinson, S. E., Davis, P. D., Sedgwick, A. D., Wadsworth, J., and Westmacott, D. (1991) Modulation of cellular processes by H7, a non-selective inhibitor of protein kinases. *Agents and Actions* **32**, 188–192.

18. Birchall, A. M., Bishop, J., Bradshaw, D., Cline A., Coffey, J., Elliott, L. H., et al. (1994) Ro 32-0432, a selective and orally active inhibitor of protein kinase C prevents T-cell activation. *J. Pharm. Exp. Ther.* **268(2)**, 922–929.

19. Toullec, D., Pianetti, P., Coste, H., Bellevergue, P., Grand, T., Ajakane, M., et al. (1991) The bisindolylmaleimide GF 109203X is a potent and selective inhibitor of protein kinase C. *J. Biol. Chem.* **266**, 15,771–15,781.

20. Davis, P. D., Hill, C. H., Keech, E., Lawton, G., Nixon, J. S., Sedgwick, A. D., et al. (1989) Potent selective inhibitors of protein kinase C. *FEBS Lett.* **259**, 61–63.

21. Hug, H. and Sarre, T. F. (1993) Protein kinase C isoenzymes: divergence in signal transduction? *Biochem. J.* **291**, 329–343.

22. Wilkinson, S. E., Parker, P. J., and Nixon, J. S. (1993) Isoenzyme specificity of bisindolylmaleimide, selective inhibitors of protein kinase C. *Biochem. J.* **294**, 335–337.

CHAPTER 22

The Measurement
of Phospholipase D-Linked Signaling
in Cells

Michael J. O. Wakelam, Matthew Hodgkin,
and Ashley Martin

1. Introduction

There is now increasing evidence that phospholipase D (PLD) activity can be stimulated by a range of hormones, growth factors, and neurotransmitters in a range of cell types *(1)*. The enzyme generally catalyzes the hydrolysis of phosphatidylcholine (PtdCho) to produce phosphatidate (PtdOH) and choline (Cho), although the hydrolysis of phosphatidylethanolamine has been reported. PtdOH can be converted to diacylglycerol (DAG) by the action of phosphatidate phosphohydrolase, and the PLD pathway has thus been suggested to provide a source of DAG in cells. This reaction clearly takes place, however, whether it provides a DAG, which has a signaling function, remains unclear. This chapter describes methods to determine PLD activity and also to examine the levels of both PtdOH and DAG.

There are two major methods for determining PLD activity. First, one can determine the generation of the cellular products (i.e., Cho and PtdOH or use can be made of the "transphosphatidylation" assay. The second assay is based on the mechanism of the reaction catalyzed by PLD; this involves the formation of a putative enzyme-phosphatidyl intermediate that, under normal conditions, utilizes water as a nucleophilic acceptor to generate PtdOH. In the presence of a short-chain primary aliphatic alcohol, which

From: *Methods in Molecular Biology, Vol. 41: Signal Transduction Protocols*
Edited by: D. A. Kendall and S. J. Hill Copyright © 1995 Humana Press Inc., Totowa, NJ

acts as a stronger acceptor, a phosphatidylalcohol is generated that, because it is a poor substrate for phosphatidate phosphohydrolase, accumulates within the cells.

2. Materials

1. [^{3}H]palmitate.
2. HHBG: 1.26 mM calcium chloride, 0.5 mM magnesium chloride, 0.4 mM magnesium sulphate, 5.37 mM potassium chloride, 137 mM sodium chloride, 4.2 mM sodium hydrogen carbonate, 0.35 mM sodium dihydrogen phosphate, 10 mM HEPES, pH 7.4, 1% (w/v) bovine serum albumin (BSA), 10 mM glucose.
3. Butan-1-ol.
4. Methanol.
5. Chloroform.
6. Phosphatidylbutanol (Avanti Polar Lipids, Alabaster, AL).
7. Whatman LK5DF plates (Whatman, Maidstone, UK).
8. 2,2,4 Trimethylpentane (isooctane).
9. Ethylacetate.
10. Acetic acid.
11. [^{3}H]Choline chloride.
12. DMBGH: DMEM containing 10 mM glucose, 20 mM HEPES, pH 7.4, and 1% (w/v) BSA.
13. Dowex-50-H$^+$.
14. [^{3}H]Choline, [^{14}C]choline phosphate, or [^{3}H]glycerophosphocholine, KCl.
15. Triton X-100, phosphatidylserine (Lipid Products, Nutley, UK), imidazole, DTT, *E. coli* Diacylglycerol kinase (Calbiochem, Nottingham, UK), [^{32}P]-γ-ATP, silica thin-layer chromatography (TLC) plates (Merck, 5714, 5 x 20 cm 60F$_{254}$), 1-stearoyl, 2-arachidonyl glycerol (Sigma or Avanti).
16. Plastic backed silica gel 60 TLC plates (Whatman), phosphatidic acid, hexane, diethyl ether, ammonium hydroxide, Coomassie blue R250.

3. Methods

3.1. Assay of PLD by the Transphosphatidylation Assay

The methodology described here has been developed for fibroblast cells grown in monolayer culture *(2)*, it may be necessary to adapt this for other cell types.

1. Cells are grown in 24-well plates until about 70% confluent, the medium is then changed to one containing 4 μCi/mL [^{3}H]palmitic acid, and the cells are then cultured for another 24–36 h until confluent. This is also the required time for radioactive equilibrium labeling to be achieved of the phospholipids, in particular PtdCho.

2. The medium is removed and the cells are washed in 0.5 mL HHBG for 20 min.

3. The cells are then incubated in HHBG containing 0.3% (v/v) butanol (3μL/mL = 30 m*M*). This is then aspirated and the stimulation is initiated by the addition of 0.2 mL HHBG containing 0.3% butan-1-ol (*see* Note 1) and the stimulant (*see* Note 2). After the required time, the incubation is terminated by aspirating the medium and adding 0.5 mL ice-cold methanol (AR grade). The plates are then stood on ice until scraped (*see* Note 1).

4. The wells containing the cell debris in 0.5 mL CH_3OH are scraped into a glass vial (screw top with good seal, e.g., Trident vials). Each well is washed with an additional 0.2 mL CH_3OH, which is then added to the appropriate vial.

5. Add 0.7 mL $CHCl_3$ to each vial, which is then capped prior to vortex mixing. The vials are then left to stand at room temperature for about 15 min.

6. Add 585 μL of distilled water to give a final $CHCl_3$:CH_3OH:H_2O ratio of 1:1:0.8. After vortex mixing, the vials are centrifuged at 1200g for 5 min in order to promote phase separation.

7. As much of the upper phase as possible is discarded without removing any of the lower organic phase which contains the labeled phospholipids, including [^{3}H]phosphatidylbutanol (PtdBut).

8. The lower phase in the glass vial is dried by vacuum centrifugation. This is then resuspended in 2×25 μL $CHCl_3$:CH_3OH (19:1 [v/v]) together with 10 μg unlabeled PtdBut (Avanti Polar Lipids) for spotting onto Whatman LK5DF TLC plates using a positive displacement pipet. The sample is applied to the adsorbent strip at the bottom of each lane no lower than 1 cm from the bottom or no higher than 0.5 cm from the top of the strip.

9. Develop the plates to 1–2 cm from the top (for about 1.5 h) in the organic (upper) phase of 2,2,4 trimethylpentane (isooctane):ethylacetate:acetic acid:water (50:110:20:100). The solvent is prepared in a separating funnel ensuring good mixing; all traces of the aqueous lower phase must be removed. The solvent should be made and used on the day of the experiment. The TLC tank is used unlined, thus the plate is developed under nonequilibrium conditions. The effect of this is that the PtdBut runs further up the plate, thus giving better resolution.

10. PtdBut has an R_f of 0.36–0.4 that can vary between each batch of developing solvent. The major phospholipids remain at or near the origin. The exact position of PtdBut is determined by iodine staining, the positions are marked with a soft pencil, the silica spots are scraped from the plate into a scintillation vial containing 0.5 mL H_2O, scintillant is added, and the radioactivity is determined by scintillation counting after leaving overnight to extract (*see* Note 2).

3.2. Determination of Choline and Choline Phosphate Generation

Methods exist in the literature for separating choline metabolites by TLC, however ion-exchange separation *(3)* is quick, simple, and reproducible.

1. Cells, cultured in 24-well plates, are labeled for 36–48 h with 2 µCi/mL [³H]choline chloride in DMEM containing 2% serum. The radiolabel concentration and the length of labeling may require variation depending on cell type.
2. The radiolabeled medium is replaced with 0.5 mL of fresh unlabeled serum-free DMEM and cells returned to the incubator 2 h prior to the experiment.
3. After the 2 h preincubation, the cells are washed for 5, 10, and 30 min in DMEM containing 10 mM glucose, 20 mM HEPES, pH 7.4, and 1% (w/v) BSA (DMBGH). This can be done in HHBG.
4. The cells are then stimulated by aspirating the wash buffer and adding 150 µL agonist in DMBGH.
5. The incubations are terminated by the direct addition of 0.5 mL ice-cold methanol to the well. This permits the determination both of those choline metabolites associated with the cell, and those released into the medium. If only cell-associated metabolites are to be analyzed, the medium is aspirated prior to methanol addition and fresh buffer added to maintain the methanol:water ratio.
6. The contents of each well is scraped into an insert vial and the well washed with another 0.2 mL of methanol.
7. Add 310 µL CHCl₃ to each tube. The tubes are vortex mixed and stood for 20 min at room temperature or overnight at 4°C.
8. 390 µL CHCl₃ and 480 µL H₂O are added to the tubes, which are mixed and then centrifuged at 1200g for 5 min to split the phases.
9. The upper aqueous phase is taken (0.8 mL) and made to 5 mL with H₂O. This is then loaded onto a 1 mL Dowex-50 H⁺ column (*see* Note 3). The run through is collected together with a water wash of 5–8 mL (volume determined to elute a glycerophosphocholine standard) as the glycerophosphocholine fraction. A 2-mL aliquot of this fraction is transferred to a 20-mL scintillation vial, scintillant is added, and the radioactivity determined.
10. Another volume of water (usually 10–15 mL for a 1-mL column, again as determined from the characterization profile) is added to the column and collected as the choline phosphate fraction. A 2-mL aliquot of this fraction is transferred into a 20-mL scintillation vial and the scintillant added and counted.
11. The choline fraction is then eluted with 5–7 mL of 1M KCl (depending on column characteristics). Scintillant is added to the whole sample and this is counted.

3.3. Characterization
of Dowex 50 H⁺ Ion-Exchange Resin

1. Extracts are prepared as in Section 3.2. from cells which are nonradio-labeled, then [³H]choline, [¹⁴C]choline phosphate, or [³H]glycerophos-phocholine is added.
2. The column is eluted with 25 × 1 mL H₂O additions, each of which is collected and the associated radioactivity determined by scintillation count-ing. The column is then eluted with 15 x 1 mL additions of 1*M* KCl (again each fraction is collected and the radioactivity determined). Each sample is counted using a dual label counting program (³H/¹⁴C) and because the choline is only eluted following the addition of KCl, it is possible to char-acterize the elution profiles of each metabolite on the one column.

3.4. Measurement of DAG Mass
in Cell and Tissue Extracts

This method relies on the conversion of unlabeled DAG to ³²P-labeled PtdOH, which can then be quantified by scintillation counting or phos-phorimaging (*see* ref. *4* for a discussion of the limitations of the assay).

1. Chloroform extracts of the samples are prepared. This is routinely obtained as the lower phase of a standard Bligh and Dyer extraction. For cells in culture, the washing and stimulation are performed as described in Section 3.1. for the transphosphatidylation assay, except radiolabeling is not required and no butanol is added. Ice-cold methanol (400 μL) is used to stop the reaction, and the cells are scraped from the dish into a glass tube. The well is then washed with 200 μL of ice-cold methanol, which is also added to the tube. Chloroform (700 μL) is then added to the tube, which is left to stand on ice for 10 min. After 600 μL of 2*M* sodium chloride is added, the tubes are then mixed well and centrifuged at 1200*g* for 5 min. This procedure promotes phase splitting; the lower (chloroform) phase contains lipids and the upper (aqueous) phase contains soluble metabolites (such as inositol trisphosphate, which can also be mass assayed).

    For tissue samples, 100 mg of finely powdered frozen tissue is extracted as just described. The lower phase can be stored at –70°C when covered with aqueous solvent. In our laboratory, 50% of the chloroform extract from a 3.5-cm dish of quiescent Rat-1 fibroblasts is sufficient for DAG mass determination. For tissue extracts we use three dilutions of the extract.
2. The chloroform extracts are dried *in vacuo* in the original or new glass tubes (the extracts cannot be stored once dry). In our experience the inter-facial material does not affect the assay, so this can be dried. Lipids are

solubilized in a Triton X-100/Phosphatidylserine mixture. Phosphatidyl-serine (30 µL; supplied as 25 mM stock from Lipid Products) is dried into a glass tube under nitrogen and then sonicated in 2.5 mL of 10 mM Imadazole, pH 6.6, 0.6% (w/v) Triton X-100 until the solution is optically clear. Add 50 µL of this to the tubes which are then sonicated in a bath for 30 min. A small amount of ice is added to the sonicating bath to keep the temperature low.

3. Reactions: Once sonicated, 20 µL of 250 mM Imadazole, pH 6.6, 250 mM NaCl, 62.5 mM MgCl$_2$, 5 mM EGTA is added to the solubilized lipid. Add 10 µL of freshly prepared 100 mM DTT. *E. coli* DAG kinase (Calbiochem) is added to a final concentration of 5 mU/tube. This is usually equivalent to 0.5 µL/tube, which we add in 10 µL of 10 mM Imadazole, pH 6.6. The reaction is started by the addition of 10 µL of 5 mM ATP containing 1 µCi of [^{32}P]-γ-ATP made up in 100 mM Imadazole, pH 6.6; this results in a final ATP concentration of 0.5 mM in a final reaction vol of 100 µL. The tubes are incubated at 30°C for 30 min.

4. Extraction of [^{32}P]-labeled lipids: The reaction is stopped by addition of 1 mL of chloroform:methanol:conc. HCl (150:300:2). After 10 min, 300 µL of chloroform and 400 µL of H$_2$O are added. The tubes are vortexed and centrifuged at 1200g for 5 min to promote phase splitting. The upper (aqueous) phase containing unreacted [^{32}P]ATP is removed and replaced with 1 mL of a synthetic upper phase. The tubes are mixed and centrifuged as before. Once this wash has been removed the tubes may be dried *in vacuo*.

5. Quantitation of phosphorylated lipids: Dried phosphorylated extracts are solubilized in 40 µL of chloroform:methanol (19:1) and 20 µL is spotted onto a silica TLC plate (Merck, 5714, 5 x 20 cm 60F$_{254}$). The plates are developed in chloroform:methanol:acetic acid (38:9:4.5) to within 5 mm of the top of the plate. Radiolabeled bands are located by autoradiography and the PtdOH band (at the top of the plate) is scraped into scintillation vials, scintillant is added, and the associated radioactivity is determined by liquid scintillation counting. Phosphorimaging can also be used to locate and quantify the bands.

The mass of PtdOH is related to the mass of DAG by means of a standard curve. 1-Stearoyl,2-arachidonyl glycerol (SAG) (available from Sigma or Avanti) is used in the concentration range 0–2000 pmol/tube. The SAG standard is dried *in vacuo* and solubilized in parallel to test samples. To ensure the reaction proceeds to completion and that the micellar structure is not affected by the amount of lipid in the samples, defined quantities of cellular lipid extract should be added to the DAG standards and conversion percentage determined *(4)* (*see* Note 4).

3.5. Assay of PtdOH by Staining with Coomassie Blue

This method involves the chromatographic separation of phosphatidate from other lipids and the staining of the lipid by Coomassie blue which can then be quantitated by densitometry.

1. Cells are cultured in 10-cm diameter plates in order to provide sufficient lipid for reliable detection. If the cells are not confluent it may be necessary to pool two or more dishes (*see* Note 5).

2. Following washing and stimulation, as outlined in Section 3.1., the medium is rapidly aspirated and the cells washed twice with ice-cold phosphate-buffered saline. The cells are then scraped in 0.5 mL methanol and the plate washed with 0.5 mL methanol. The two methanol samples are transferred to a glass tube to which is added 0.5 mL chloroform and 0.4 mL water. The phases are split by the addition of 0.5 mL chloroform and 0.5 mL 1*M* NaCl.

3. After a brief centrifugation (1220*g*), 0.8 mL of the lower chloroform phase is transferred to a fresh glass tube and dried under vacuum.

4. A Merck 20-cm silica gel 60 plastic-backed plate with fluorescent indicator is pre-run in hexane/diethyl ether (1:1 [v/v]). All the chromatographic procedures used here employ paper-lined tanks that are equilibrated for at least 1 h.

5. The lipid samples are taken up in chloroform and applied carefully in 0.5-cm strips 10 cm from the bottom of the pre-run plate. Standards of phosphatidate (0.5, 1.0, 2.5, and 10 μg) are applied to each plate.

6. The plates are developed to 95% of their lengths using chloroform/methanol/ammonium hydroxide (65:35:7.5, by vol). The neutral and zwitterionic lipids migrate in this solvent whereas the acidic lipids remain at the origin.

7. After drying for about 15 min, the plate is cut 1.5 cm above the origin to remove those lipids that migrated in the solvent given in step 6.

8. The plate is then developed in the reverse direction for 90–95% of the remaining length using chloroform/methanol/acetic acid (8:2:1, by vol).

9. The dried plates are stained for 1 h using 0.03% Coomassie blue R250 in 20% methanol (v/v) containing 100 m*M* NaCl.

10. After destaining for 10–15 min in 20% (v/v) methanol, the plates are air-dried overnight.

11. The absorbencies of the PtdOH bands are then determined using a densitometer in the reflectance mode at 580 nm. The areas of the peaks produced by the PtdOH standards are plotted to generate a standard curve for each plate and the mass of phosphatidate in the incubations calculated. Duplicate samples should be run on separate plates.

12. If required the data can be normalized by measuring the phospholipid phosphorus mass in the chloroform extract by the method of Bartlett *(5)*.

4. Notes

1. A number of short-chain primary aliphatic alcohols can be utilized in the transphosphatidylation assay (i.e., methanol, ethanol, propanol, and butanol. We use butanol because it can be used at a lower concentration and is less toxic.

2. It is advisable to include an internal positive control in each experiment (e.g., TPA in Swiss 3T3 cells induces a large response). It is preferable to perform the entire assay in one day. However, if it is impossible to run the TLC the same day, then the lipids must be stored dry in sealed tubes purged with N_2 at $-20°C$ overnight. PtdBut in cellular lipid samples is unstable in $CHCl_3$.

3. Although the level of glycerophosphocholine rarely changes in response to acute stimulation, there is often a high level of radioactivity associated with this fraction in some cell types, therefore its separation from the choline phosphate fraction is advisable.

4. Chloroform extracts from cells and tissues contain other lipids that can be phosphorylated by the *E.coli* DAG kinase. This will result in the appearance of other less polar phospholipids on the TLC plate. One of these is probably ceramide phosphate, and this assay can therefore be used to identify and quantify ceramide (with appropriate standards).

5. As with the DAG assay, the PtdOH assay can utilize lipid samples prepared from tissue samples as described in Section 3.4.

Acknowledgments

Work from this laboratory is supported by the Wellcome Trust and the Medical Research Council.

References

1. Cook, S. J. and Wakelam, M. J. O. (1992) Phospholipases C and D in mitogenic signal transduction. *Rev. Physiol. Biochem. Pharmacol.* **119,** 14–45.

2. Cook, S. J., Briscoe, C. P., and Wakelam, M. J. O. (1991) The regulation of phospholipase D activity and its role in sn 1,2-diradylglycerol formation in bombesin- and phorbol 12-myristate 13-acetate-stimulated Swiss 3T3 cells. *Biochem. J.* **280,** 431–438.

3. Cook, S. J. and Wakelam, M. J. O. (1989) Analysis of the water soluble products of phosphatidylcholine breakdown by ion exchange chromatography: bombesin and c-kinase stimulated choline generation in Swiss 3T3 cells. *Biochem. J.* **263,** 581–587.

4. Paterson, A., Plevin, R., and Wakelam, M. J. O. (1991) Accurate measurement of sn-1,2-diradylglycerol mass in cell lipid extracts. *Biochem. J.* **280,** 829–836.

5. Bartlett, G. R. (1959) Phosphorous assay in column chromatography. *J. Biol. Chem.* **234,** 466–468.

CHAPTER 23

Phospholipase A$_2$ Activity

Ronald M. Burch

1. Introduction

Phospholipases A$_2$ (PLA$_2$s) are important enzymes in signal transduction, being responsible for release of arachidonate from membrane phospholipids for the production of prostaglandins, leukotrienes, platelet-activating factors, and other bioactive lipids *(1)*. Phospholipids are composed of a glycerophosphate in which the two hydroxyls are esterified with long-chain fatty acids and the phosphoryl moiety forms a phosphodiester bond with a polar "head group," commonly choline, ethanolamine, serine, and inositol. Phospholipid structures are designated by a *s*tereospecific *n*omenclature (*sn*) based on L-glycerol-3-phosphate. Thus, the fatty acid esterified to carbon-1 of the glycerol backbone is termed the "*sn*-1" fatty acid, and is commonly saturated, e.g., palmitate. The fatty acid esterified to carbon-2 of the glycerol backbone is the *sn*-2 fatty acid, and is often polyunsaturated, e.g., arachidonate.

PLA$_1$s specifically hydrolyze the *sn*-1 fatty acid and yield a lysophospholipid, whereas PLA$_2$s hydrolyze the *sn*-2 fatty acid and yield another lysophospholipid. PLCs yield head group phosphates and diacylglycerols, and PLDs yield the head group and phosphatidic acid. PLA$_2$ enzymes belong to several families; one of these is a group of closely related proteins of about 14,000-Dalton molecular mass and composed of a variety of secreted enzymes similar to snake venoms *(2)*. More distantly related are the secreted pancreatic enzymes and honey bee venom PLA$_2$. Most recently described is a quite genetically distinct enzyme of 100,000-Dalton molecular mass, which is cytosolic and which

From: *Methods in Molecular Biology, Vol. 41: Signal Transduction Protocols*
Edited by: D. A. Kendall and S. J. Hill Copyright © 1995 Humana Press Inc., Totowa, NJ

becomes associated with membranes at free Ca^{2+} concentrations of 300 nM or more *(3)*. This enzyme has caused considerable excitement as a candidate for the "hormonally sensitive" PLA_2. However, many stimuli that increase prostaglandin synthesis also increase transcription and translation of the low-molecular-weight venom-related PLA_2s.

Numerous methods have been described for measuring PLA_2 activity. Many are gathered together in other sources *(2,4)*. Described here are two assays that have proven quite useful in assaying PLA_2 activity in cell extracts. The labeled fatty acid release assay is amenable to simultaneous multiple-tube assays. This assay uses as substrate a phospholipid that contains a radiolabeled fatty acid, usually [^{3}H]arachidonate in signal-transduction studies, which is released by PLA_2 and detected after separation from the substrate. However, the detection of released fatty acid may yield a false conclusion that the enzyme responsible is PLA_2. For example, PLA_1-mediated degradation to yield a lysophospholipid followed by lysophospholipase degradation also yields the free fatty acid. Similarly, degradation of a phospholipid by a PLC followed by diacylglycerol lipase can also yield the free fatty acid. Thus, in order to use the labeled fatty acid release assay, one must be certain that the conditions of the assay do in fact reflect PLA activity.

To assure oneself that PLA_2 activity is being assessed, another assay, which uses a phospholipid radiolabeled in the head group, is used. The product detected in this assay is the radiolabeled lysophospholipid. This assay is much more laborious, requiring extraction of the reaction products, followed by thin-layer chromatography separation of products with comparison of R_f to authentic standards. Often this assay is performed on a few samples when new experimental conditions are used to assure that the easier labeled fatty acid assay is valid.

2. Materials

1. Substrate: A convenient substrate for fatty acid release assays is 1-stearoyl-2[^{3}H]-arachidonyl-phosphatidylcholine, available commercially. This is diluted to an appropriate activity with the same phospholipid, not radioactively labeled, also available commercially. Other phospholipids may be evaluated as substrates for particular PLA_2s, with phosphatidylethanolamines and phosphatidylinositols being available commercially. For lysophospholipid assays, 1-palmitoyl-2-arachidonoyl-phosphatidyl[^{3}H]-choline is commercially available *(see Note 1)*.

2. Reaction buffer: 50 mM Tris-HCl, pH 7.4 or 8.5 (*see* Note 3 for pH), and 5 mM Ca^{2+} (*see* Note 2).

3. Dole's extraction medium *(5)*: Isopropanol:*n*-heptane:1N H$_2$SO$_4$, 40:10:1 (v:v:v).

4. Silica gel, 200 mesh.

5. Formic acid, 16N, diluted to 1N with distilled water.

6. 1-Butanol.

7. Speed-Vac vacuum centrifuge.

8. Standard mix of lipids: 1-Palmitoyl-2-arachidonoyl-phosphatidylcholine, 100 µg, 1-palmitoyl-lysophosphatidylcholine, 200 µg, and arachidonic acid, 100 µg, are placed into a 12 × 75-mm glass tube. Solvents are removed under a stream of nitrogen (Note 1), and the mixture is redissolved in 1 mL chloroform.

9. Lipid solvent: Chloroform:methanol, 2:1.

10. Thin-layer chromatography plates, silica gel G, with Celite preadsorbant zone: If multiple samples are to be run, then 20 × 20-cm plates scored into 19 zones are most convenient. Use appropriate size chromatography development tank.

11. Thin-layer chromatography solvent *(6)*: Chloroform:ethanol:distilled water:triethylamine, 30:34:8:35.

12. Iodine chamber: Chromatography development tank containing several iodine crystals (kept in a fume hood), or a Pasteur pipet containing a small wad of glass wool over which are layered several crystals of iodine, and overlayered with another wad of glass wool. A pipet bulb should be used to wash iodine vapor over plate inside a fume hood.

3. Methods

3.1. PLA₂ Activity
in Cell Membranes or Supernatant Measured
as Fatty Acid Release (See Notes 2 and 3)

1. Substrate is prepared by adding 1-stearoyl-2-[^{3}H]arachidonyl-phosphatidylcholine, 100,000/1 mL of 0.1M Tris-HCl (*see* Note 3) followed by 0.08 mg/mL nonradioactive phospholipid (this yields 10 µM final substrate concentration after dilution in the reaction buffer).

2. Sonicate the phospholipid mixture using a Kontes microtip probe.

3. Distribute the substrate solution to polypropylene tubes, 20 µL.

4. Membrane preparation or cell supernatant, 80 µL, in reaction buffer, is added to each tube (*see* Note 2).

5. Tubes are incubated at 37°C for 1–30 min (include zero time tubes to determine amount of nonesterified fatty acid in the substrate mixture).

6. Reactions are stopped by the addition of 0.5 mL Dole's extraction medium to each tube. Extraction is completed by the addition of 0.3 mL *n*-heptane, and then 0.3 mL distilled water, with vortexing for 15 s after the addition of each reagent. The tubes are finally centrifuged at 500*g* for 5 min to separate the organic (upper) and aqueous phases.

7. From the upper phase (≈0.4 mL), an aliquot of 0.3 mL is taken and placed into a glass tube.

8. Approximately 50 mg of silica gel is added to each tube. The tubes are vortexed to adsorb traces of substrate. From each tube, an aliquot of 200 µL is taken and added to a scintillation vial.

9. Radioactivity is determined by liquid scintillation counting after the addition of nonaqueous counting cocktail.

3.2. PLA$_2$ Activity
in Cell Membranes or Supernatant Measured
as Lysophospholipid Production

1. Prepare substrate as in Section 3.1., step 1, except that labeled phospholipid is 1-palmitoyl-2-arachidonyl-phosphatidyl[^{3}H]choline and unlabeled phospholipid is 1-palmitoyl-2-arachidonyl-phosphatidylcholine.

2. Assay is performed as in Section 3.1., steps 2–5.

3. Reaction is stopped by the addition of 10 µL 1*N* formic acid.

4. 1-Butanol (0.2 mL) is added *(7)* and the mixture is allowed to incubate for 30 min, with vortexing for 15 s every 5 min.

5. The organic layer is collected with a Pasteur pipet and placed into a 12 × 75-mm glass tube.

6. A standard mix solution of phosphatidylcholine, lysophosphatidylcholine, and arachidonic acid (100 µL) is added to the tube, and then the contents of the tube are evaporated to dryness under vacuum using a Speed-Vac.

7. The residue in the tube is redissolved in lipid solvent (50 µL) and vortexed for 15 s.

8. A thin-layer plate is marked with a pencil, 15 cm above the beginning of the silica gel phase. The redissolved lipids are applied in small aliquots about 0.5 cm below the top of the preadsorbant layer of the plate, allowing evaporation to occur prior to adding additional aliquots. Care is taken not to allow the lipid mixture to spread into the silica gel phase. When all phospholipid has been applied, the glass tube is rinsed with 10 µL additional lipid solvent, and this is also applied to the plate. Evaporation of solvent is important and may be hastened using a stream of nitrogen.

9. Thin-layer chromatography solvent is placed into the developing tank (situated in a fume hood) to a depth of 3–4 mm. The thin-layer plate is lowered into the tank, the tank cover replaced, and a piece of polyethylene film

placed over the outside to seal against air leak caused by the fume hood. The plate is allowed to develop until the solvent has reached the pencil line (perhaps 2 h).

10. The plate is removed from the development tank, and solvent is allowed to evaporate in fume hood (approx 15 min). Plate is either placed in iodine tank, or iodine vapor is blown over lanes. Lipids will become brown. A pencil is used to mark upper and lower boundaries of lipids. Lysophosphatidylcholine will have migrated about 2 cm from the preadsorbant/adsorbant interface (R_f = 2/15 = 0.13). Phosphatidylcholine will have migrated about 3 cm, R_f = 0.2.

11. The zones corresponding to the lipids are scraped from the plate with a razor blade into scintillation counting vials, and radioactivity is determined using liquid scintillation counting.

4. Notes

1. Lipids containing unsaturated fatty acids, such as arachidonic acid, are highly susceptible to oxidation. When preparing lipid substrates and standard solutions, it is important to keep exposure to oxygen to a minimum. Lipid stock solutions should always be placed back into –20°C storage conditions as soon as possible.

2. Control of Ca^{2+} concentration is important in PLA$_2$ assays. For secreted PLA$_2$s, optimum activity is usually observed at Ca^{2+} concentrations of 5–10 mM. For some cytosolic PLA$_2$s, similar Ca^{2+} concentrations are optimum, whereas for others, free Ca^{2+} <100 nM is sufficient for activity. Certain PLA$_2$s will associate with membranes at free Ca^{2+} >100 nM. Thus, preparation of membrane pellets should be investigated at very low free Ca^{2+}, by adding 0.5 mM EDTA, or at higher free Ca^{2+} to determine whether PLA$_2$ becomes associated. Similarly, to maximize PLA$_2$ found in supernatants of cell extracts, EDTA should be included in the lysis buffer. It is important to calculate free Ca^{2+} accurately in incubation buffers using appropriate equations (e.g., EQCAL, Biosoft, Cambridge, UK) or direct measurement using a Ca^{2+} electrode.

3. The pH optimum should be determined for the specific PLA$_2$ being investigated. Most secreted PLA$_2$s have optima of 8–8.5, whereas several cytosolic enzymes have optima near 7.

References

1. Burch, R. M., Kyle, D. R., and Stormann, T. (1993) *Molecular Biology and Pharmacology of Bradykinin Receptors.* Landes, Austin, TX.
2. Dennis, E. A. (1983) Phospholipases, in *The Enzymes,* vol. XVI (Boyer, P. D., ed.), Academic, New York, pp. 308–353.

3. Clark, J. D., Lin, L.-L., Kriz, R. W., Ramesha, C. S., Sultzman, L. A., Lin, A. Y., Milona, N., and Knopf, J. L. (1991) A novel arachidonic acid-selective cytosolic PLA$_2$ contains a Ca^{2+}-dependent translocation domain with homology to PKC and GAP. *Cell* **65,** 1043–1051.

4. Reynolds, L. J., Washburn, W. N., Deems, R. A., and Dennis, E. A. (1991) Assay strategies and methods for phospholipases. *Methods Enzymol.* **197,** 3–23.

5. Dole, V. P. (1956) A relation between non-esterified fatty acids in plasma and the metabolism of glucose. *J. Clin. Invest.* **35,** 150–154.

6. Korte, K. and Casey, L. (1982) Phospholipid and neutral lipid separation by one-dimensional thin-layer chromatography. *J. Chromat.* **232,** 47–53.

7. Bjerve, K. S., Daae, L. N. W., and Bremer, J. (1974) The selective loss of lyso-phospholipids in some commonly used lipid extraction procedures. *Anal. Biochem.* **58,** 238–245.

CHAPTER 24

Techniques
for the Measurement
of Nitric Oxide

Anna M. Leone, Peter Rhodes,
Vanessa Furst, and Salvador Moncada

1. Introduction

Nitric oxide (NO) is an important biological mediator involved in control of blood vessel tone, neurotransmission, and immune function (*1–3*). NO has physicochemical properties that make it ideal to function as an intercell communicator, and these include an ability to travel rapidly between cells (*4*), a short half-life, and a rapid oxidation to inactive anions (*2*). Unfortunately, some of these same characteristics make quantification of NO difficult. Most studies both in vivo and in vitro have used indirect indices of NO production, such as breakdown products of NO metabolism, second messengers, or biological events in effector systems. However, recent developments in probe technology have allowed the direct detection of NO in certain defined situations.

The complex fate of NO in many systems helps to confound analysis. NO is metabolized to a range of breakdown products in vivo, only some of which have been rigorously characterized. However, nitrite, nitrate, nitrosyl hemoglobin, nitroso serum albumin, and low-mol-wt nitrosothiols have all been used as indices of NO production. This chapter explores some of the established techniques for using such indices and gives details of some specific applications.

From: *Methods in Molecular Biology, Vol. 41: Signal Transduction Protocols*
Edited by: D. A. Kendall and S. J. Hill Copyright © 1995 Humana Press Inc., Totowa, NJ

2. Direct Measurements

Direct measurements of NO are attractive because they offer the highest specificity. The two most direct approaches are the detection of electric current produced when NO is oxidized and the detection of light produced when NO reacts with ozone.

2.1. Probes

The two main probe systems rely on the electrochemical oxidation of NO to generate electric current. They can be accurately placed in in vitro systems to determine the exact locality at which NO is being produced and to follow in real time the kinetics of NO. Judicial positioning of more than one probe in a system can be used to assess the direction and speed of NO travel *(4)*. The specificity of NO probes depends on their ability to exclude other molecules that might give rise to a signal. Shibuki *(5)* developed an electrochemical sensor based on a modified oxygen electrode (Clark electrode), consisting of platinum wire as the working electrode (anode) and silver wire as the counterelectrode (cathode). The electrodes are mounted in a capillary tube filled with sodium chloride/hydrochloric acid solution and separated from the analytical solution by a gas-permeable membrane of chloroprene rubber. A constant potential is applied and a direct current measured secondary to oxidation of NO on the platinum anode. The response time of the sensor is 3–6 s, the detection limit 5×10^{-7} mol/L, and the range of linearity is 1×10^{-6} to 3×10^{-4} mol/L. The detection limits of some commercial forms of this probe may be no lower than 5×10^{-6} mol/L, and these would be unsuitable for work in many systems, particularly where constitutive NO generation is under investigation. Shibuki applied his electrode to measurement of NO release in the central nervous system *(5)*. Others have used commercially available versions of this probe to detect NO production in various cell-culture preparations *(6)*. Most versions of this probe are several millimeters in diameter and do not allow particularly accurate positioning in a cellular system, but they are quite robust and various models are commercially available.

Malinski *(4)* developed a detection system that also relies on the electrochemical oxidation of NO. The reaction is catalyzed on polymeric metalloporphyrin. The porphyrinic semiconductor is covered with a thin layer of the cation-exchanger Nafion, which eliminates anionic interference from nitrite. The high sensitivity, small diameter (0.2–1.0 µm), and fast response time (10 ms) are useful features for detection of NO in microsystems, such as single cells. The detection limit may be as low as 1×10^{-8} mol/L, and the current–concentration relationship is linear between 1×10^{-8} and 3×10^{-3} mol/L, a wider range than for the Clark electrode. A particular advantage of the Malinski probe is that it can detect NO at the surface of the cell membrane where the concentration is much higher than in the surrounding fluid *(7)*. Malinski used the microsensor to demonstrate travel of NO from endothelium to smooth muscle *(4)* and the release of NO by platelets. There have also been successful in vivo applications of the Malinski probe, including the measurement of NO in rat brain during ischemia *(8)*. Potentially, the probe could be mounted on a 22-gage needle or an intravenous or Swan-Ganz catheter to allow monitoring in tissues and circulating blood *(7)*. No commercial form of the Malinski probe is yet available; it is a delicate system that requires considerable technical skill to build and operate.

2.2. Chemiluminescence

2.2.1. Direct Measurement of NO in Breath

Reaction between NO and ozone leads to the generation of light.

$$NO + O_3 \rightarrow NO_2{}^* + O_2$$

$$NO_2{}^* \rightarrow NO_2 + light$$

Detection of light produced in this way was first applied to measurement of NO as an atmospheric pollutant *(9)*. A variety of chemiluminescence analyzers is now available for quantification of NO, and such equipment can be adapted readily for routine measurement of the endogenous NO that is excreted in expired air *(10,11)*. The technique is highly sensitive (detection limit 1 ppb) and linear over a wide range; 1–1000 ppb can easily be covered without recalibration of equipment. Analyses are both rapid and highly reproducible. Further work is required to establish the exact origins of NO contributing to the various NO profiles that can be recorded with this technology. Both the lower and the upper respiratory tracts release NO into the expirate *(11,12)*. Whether or not useful representation of systemic NO production could be obtained from breath analyses is still to be determined.

The response time of chemiluminescence equipment for the analysis of NO in breath is now rapid (<0.5 s) so that exhaled NO levels can be determined in a single breath *(11)*. The method has been used to detect changes in NO production that seem to occur during exercise and with asthma *(11,13)*. However, the technique is unsuitable for assessment of patients who do not have good control over their respiratory cycle. This is because a slow (4 rpm), regular breathing pattern with prolonged exhalation time (>5 s) is necessary to create the NO plateaux required for standardized measurements. Breathing with normal or low tidal volumes (<11) and normal or high frequency (>8 rpm) does not give rise to a stable end expiratory concentration of NO. Therefore, assessment of NO in the expirate of subjects with chronic dyspnea or neuromuscular disorders would be difficult by this means.

2.2.2. Mass Spectrometry

The presence of NO in breath can be demonstrated using the nitrosation of thioproline and analysis of the nitrosothioproline derivative by gas chromatography-mass spectrometry. This has been applied to validate the use of chemiluminescence for analysis of NO in breath (Fig. 1) *(10)*. Such technology is a powerful analytical tool, particularly when combined with stable isotopes that can be used as tracers in biological systems. However, it is also complex and is not practical for routine use.

2.2.3. Measurement of NO in Liquid Samples

Chemiluminescence can be applied to measurement of NO in the head space above tissue cultures, cell extracts, or other liquid samples *(14)*. This methodology takes advantage of the low solubility of NO in aqueous solutions; NO has a partition coefficient of approx 20. NO dissolved in the liquid phase can be displaced into the head space above a sample by bubbling an inert gas through the specimen. Chemiluminescence can be produced by reaction of a variety of molecules with ozone; a requirement is an unsaturated or strongly polar group, as is found in sulfides, amines, and the solvent dimethyl sulfoxide (DMSO) *(9,14)*. However, the specificity of chemiluminescence analyzers for NO is high in most systems, because the majority of other molecules potentially able to give chemiluminescence with ozone are nonvolatile or do not occur in biological systems.

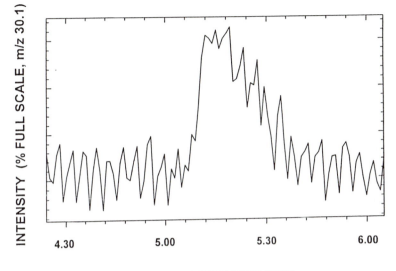

Fig. 1. A mass chromatogram of a direct breath injection monitored at m/z 30.1. A clear peak at the known retention time of NO can be seen at about 5.2 min.

The specificity of a chemiluminescence signal for NO can be demonstrated by passing the sample through a solution of Fe^{2+} before it reaches the reaction chamber, removing the NO contribution to the signal, or by using low-temperature traps to remove contaminants from the sample *(10)*.

3. Indirect Measurements

There is a wide range of indirect indices for NO production. It includes NO metabolites, the second messenger cyclic guanosine monophosphate (cGMP), and effector organ responses, such as vessel dilation and platelet function.

3.1. Bioassay

Some of the earliest measurements of NO production have involved bioassays able to provide units of response in particular biological systems; such bioassay is primarily qualitative *(1,3)*. Many of the biological parameters regulated by NO, such as vessel tone, are also controlled by a variety of other molecules; specificity is therefore a potential problem with bioassay systems, and steps need to be taken to assess the specificity of each system, including inhibition of NO production as well as inhibition of other factors with similar biological actions.

A number of L-arginine analogs are competitive inhibitors for NO synthesis and can be used to assess the influence of NO on biological parameters. They can be used both in vitro and in vivo *(15)*.

3.2. cGMP

NO stimulates the enzyme-soluble guanylate cyclase to produce cGMP, an intracellular messenger that interacts with a range of receptor proteins to produce its effects. Concentrations of cGMP have therefore been used as an index of NO production *(2)*. However, many molecules activate guanylate cyclase, causing increases in cGMP production *(16)*. These include all natriuretic peptides so far identified (atrial, B-type, C-type), carbon monoxide, nitrosothiols, melatonin, and hydroxyl radicals. cGMP concentrations can also be influenced indirectly by substances that can act as phosphodiesterase inhibitors. The specificity of changes in cGMP as an index of NO production may therefore be poor. In contrast with bioassay systems, the low specificity of cGMP is difficult to improve, because it is not possible to inhibit selectively most of the molecules that may activate guanylate cyclase.

3.3. Nitrite and Nitrate

NO generated in most systems is short-lived, being converted rapidly to a range of breakdown products, which includes nitrite and nitrate. The relative stability of these anions and the wide range of analytical techniques available for their measurement have encouraged their use as indices of NO production. This works well in aqueous solution in the presence of oxygen, where the conversion of NO to nitrite is quantitative *(17)*, and other sources of nitrite apart from NO can quite easily be controlled. The use of both nitrite and nitrate as indices for in vivo NO production is not as straightforward because the ratio of nitrite and nitrate may vary considerably; many factors, including hemoglobin, superoxide anions, and hydroxyl radicals, may be involved in NO degradation. In addition, sources of nitrite and nitrate other than NO are difficult to control. Therefore, it is important to measure both anions in most circumstances, and ideally they should be measured simultaneously.

There are specific problems with using nitrite and nitrate concentrations in whole blood to assess NO formation in vivo. Nitrite is not stable in whole blood, being rapidly oxidized to nitrate *(18)*. The time delay before centrifugation of a whole-blood sample influences the plasma nitrite levels subsequently measured. Also, nitrite can probably be formed in plasma after sampling by the breakdown of less-stable NO metabolites, such as *S*-nitrosothiols *(19)*. Nitrate is stable in whole blood and plasma, and because it is present in much higher concentrations than nitrite, small contributions from the breakdown of less-stable NO metabolites are probably not as important as for nitrite. However, the half-life of nitrate in vivo is believed to be several hours *(20)*, and it may therefore be chronologically insensitive to small acute fluctuations in NO production. The ubiquitous nature of these anions calls for specific dietary preparation of subjects before study, and defined methods of venesection, storage, and sample handling.

The most sensitive and specific methods available for the determination of nitrite and nitrate in plasma are gas chromatography-mass spectrometry (GC-MS) *(21)* and gas chromatography-combustion isotope ratio mass spectrometry (GCCIRMS). Such technology has the great strength of providing isotopic analyses, so that studies can be performed with stable isotopes. These are powerful analytical tools that can be used to address specific questions. However, mass spectrometry is not suitable for routine analyses. It is also disadvantageous because nitrite and nitrate must be measured separately. Nitrite must first be converted to nitrate, and both anions must be converted to a nitroaromatic. These processes may introduce sample contamination from reagents and laboratory ware.

Other methods are available that are less sensitive than mass spectrometry. These include diazotization (the Griess reaction—a relatively simple colorimetric assay *[22]*); chemiluminescence, which requires conversion of nitrate into nitrite and conversion of nitrite into NO *(14)*; and high-performance liquid chromatography (HPLC) *(23)*. All of these methods have been usefully applied in NO research. However, they all require extensive pretreatment of samples prior to analysis, which inevitably risks contamination with nitrite and nitrate.

3.3.1. High-Performance Capillary Electrophoresis

Recently, we developed a method using high-performance capillary electrophoresis (HPCE) for the direct and simultaneous analysis of nitrite and nitrate in plasma *(24)*. HPCE is performed by application of high voltages (10–30 kV) across narrow bore fused silica capillaries. It is an analytical separation technique characterized by high resolving power, minimal sample preparation requirements, the ability to operate in aqueous media, rapid analyses, and low sample consumption (1–50 nL injected). Sensitivity has not been a particular strength, and recently there have been some important advances toward improving this aspect of the technology. These include high-sensitivity Z-cell capillaries, dedicated diode-array detectors, and extended light-path capillaries. We have measured basal plasma nitrite and nitrate, as well as pathophysiological changes in the concentrations of these anions using both a Waters Quanta 4000 capillary electrophoresis system fitted with a high-sensitivity Z-cell capillary (Waters Chromatography Division, Millipore Corporation, Milford, MA) and an HP 3D capillary electrophoresis system with diode-array detector fitted with an extended light-path capillary (Hewlett-Packard Ltd., Stockport, Cheshire, UK). The Z-cell capillary has an extended optical path length of 3 mm, a 40-fold increase compared with a standard 75-μm internal diameter capillary; the sensitivity increase for basal plasma nitrite is about 10-fold—not as great as the increase in path length because light is lost through the capillary wall. The extended light path or "bubble" capillary used in the HP 3D system contains a three-times expanded internal diameter section (225 μm) at the point of detection, giving a three-times extended optical path length and a similar 10-fold increase in sensitivity for basal plasma nitrite.

1. Care to avoid sample contamination is important at every stage of the assay, and all sampling syringes and laboratory ware used are thoroughly rinsed with Milli-Q+ water to remove surface nitrite and nitrate (Millipore, Bedford, MA).
2. Samples for analysis are diluted 1:10 with MQ+ water in the insert of Ultrafree MC filters (Millipore) that have a nominal mol wt cutoff of 5 kDa; they are then ultrafiltered at 5000*g*.
3. For analyses on the HP 3D capillary electrophoresis system, we use 72-cm fused silica capillaries of 75-μm internal diameter (extended light path, 225 μm).

4. The electrolyte consists of 25 mM sodium sulfate containing 5% NICE-Pak OFM Anion-BT (Waters proprietary osmotic flow modifier) in Milli-Q+ water.
5. Samples are injected by electromigration for 20 s at –6 kV and analyzed at an applied negative potential of 30 kV.
6. The capillary is purged with electrolyte for 1 min between runs.
7. Data are acquired at a response time of 0.1 s, with 0.01 peak width selection, at 214 nm (band width 2 nm) onto an HP 3D CE Chem Station data system.

Standard curves prepared for nitrite and nitrate added to plasma in the 0–50 and 0–400 μM range, respectively, give regression coefficients of 0.98 for nitrite and 0.99 for nitrate. The intra-assay coefficient of variance (CV) is 10% for basal nitrite, 4.6% for 50 μM nitrite, 6.4% for basal nitrate, and 1.2% for 50 μM nitrate ($n = 10$). Inter-assay CVs for basal and spiked nitrite and nitrate are 9–11% ($n = 14$). The presence of anions in plasma at much higher concentrations than nitrite and nitrate, such as chloride and sulfate, is a cause of interference in many assays. However, the use of direct detection at 214 nm in this assay is advantageous, because neither chloride nor sulfate is observed at this wavelength. The wavelengths at which maximum UV absorbance occurs for nitrite and nitrate are lower, but noise is significantly increased below 210 nm, and therefore a small band width value for the diode-array detector was chosen. The selective nature of the electromigration step, coupled with the high resolving power of the technique, make it possible to keep sample handling to a minimum. In addition, the significant mass-to-charge difference between nitrite and nitrate resulted in good separation between these two anions (Fig. 2).

Analysis of nitrite and nitrate in biological samples has been performed for many reasons (25). Much of the early interest in these anions was related to nitrosamine formation and the potential role of nitrosamines in the etiology of cancer. In recent years, the discovery of NO as an important mediator of cell function has greatly increased interest in the measurement of these anions. In vivo increases in nitrite and nitrate have been found in inflammatory conditions (26–28) and during interleukin-2 therapy (29). These changes are widely believed to originate from increases in NO production. In vivo decreases in nitrite and nitrate concentration as an indication of reduced NO production have not been reported. This is presumably because of the intrinsic problem of measuring a reduction in low-level constitutive NO release above a relatively high background/dietary contribution.

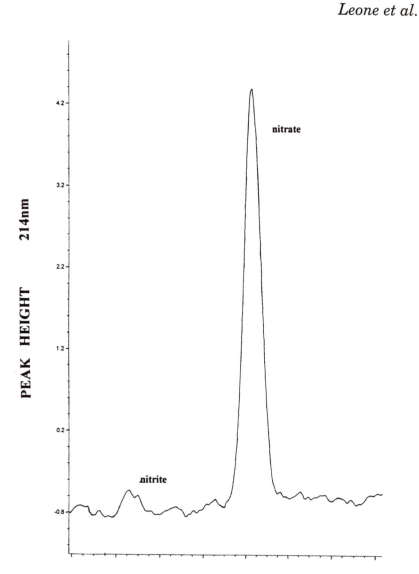

PEAK HEIGHT   214nm

TIME (mins)

Fig. 2. A typical capillary ion analysis for basal plasma nitrite and nitrate.

3.4. Measurement of S-Nitrosothiols

A variety of *S*-nitrosothiols is believed to be present in human plasma. The mechanisms by which these products are formed and their biological roles have not yet been fully described. It is possible that nitrosothiols are formed in vivo from reactive species produced from NO, such as peroxynitrite. Whether these nitrosothiol products function as a means to remove NO or to donate it is also unclear; they may constitute an elimination mechanism for sequestering and inactivating NO, or they may form a reservoir that can be used to provide NO. For these reasons, it is not possible at present to assess whether any of the nitrosothiols are likely to provide reliable indicators of NO production.

Low-mol-wt thiols and their *S*-nitrosated derivatives are difficult to analyze, but a method has been described by Stamler and Loscalzo using capillary electrophoresis *(30)*. The method is capable of separating thiols, their disulfide forms, and their *S*-nitrosated derivatives allowing distinction among cysteine, homocysteine, and glutathione with the rapidity and specificity that are characteristic of capillary electrophoresis. A disadvantage of the assay is the requirement to vary the polarity of the internal power supply and the buffer pH in order to analyze the different thiol derivatives. In addition, peak widths at half height for nitrosoglutathione (GSNO) are in the order of 20–30 s. Furthermore, the method lacks the sensitivity that is required for detection of physiological levels of *S*-nitrosothiols.

Recently, we have developed an assay for *S*-nitrosothiols that utilizes HPCE and the enhanced sensitivity provided by diode-array detection and extended light-path capillaries. This is capable of detecting a range of thiols, disulfides, and nitrosothiols with high sensitivity in a single direct analysis. An example of a 10 μ*M* GSNO standard with a peak width at half height of 3 s is shown in Fig. 3.

3.5. Electron Paramagnetic Resonance Spectroscopy

NO has an unpaired π orbital electron that can be excited by microwave and magnetic energy, giving rise to a characteristic spectrum on its return to the ground state; it is therefore detectable by electron paramagnetic resonance (EPR) spectroscopy. NO rapidly forms diamagnetic species by reaction with oxygen and radicals, including itself, so spin traps are necessary for stabilization before EPR analysis; nitroxides *(14)* and hemoglobin *(31)* are both effective. The complex formed between NO and reduced hem iron (Fe II) has well-defined chemical and spectroscopic properties *(14)*. The specificity of the hemoglobin NO (HbNO) assay is good because the characteristic three-line hyperfine

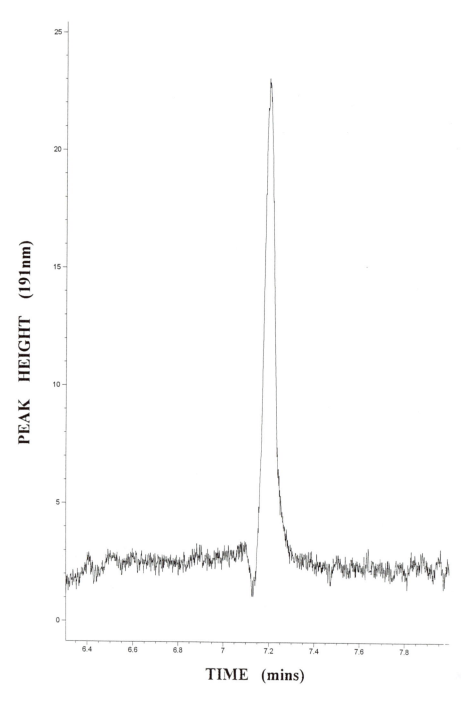

Fig. 3. A typical HPCE trace for a 10 μ*M* nitrosoglutathione standard.

pattern produced by HbNO is not seen with other nitrogen-containing adducts.

NO reacts readily with deoxy hemoglobin to form nitrosyl hemoglobin in vivo in the circulation, but it is not yet clear in which circumstances HbNO may provide a good index of NO production. Most indices are measured in basic units of moles per unit volume, and therefore, changes reflect not only production, but also metabolism. This is particularly important when assessing the meaning of changes in an index like HbNO, because the metabolic fate is not known. It is certainly important that it does not accumulate in the circulation, because the oxygen-carrying capacity of the blood would be rapidly compromised. A better understanding of the mechanisms of HbNO clearance is required before changes in HbNO concentration can be related with certainty to changes in NO production. A further problem is that in various animal models, it has not been possible to detect basal HbNO *(32)*. This is a real disadvantage; ideally, an assay for NO should be able to detect levels well below the physiological basal concentration, so that it can be applied to studies that involve inhibition of NO production, or to the investigation of pathologies that reduce constitutive NO production. A higher sensitivity is required for biological work than is available at present from EPR analyses. However, it has been possible to show increases in HbNO by EPR after inhalation of NO *(31)*, and in mouse and rat models of sepsis *(32,33)*. Lack of availability and technical complexity of EPR analyses are likely to be an additional hinderance at present to the usefulness of this index for many research groups.

4. Conclusion

Only a few methods, such as the porphyrinic microsensor, have been developed specifically for the direct detection of NO. Most of the techniques applied to the measurement of NO were available years before NO became recognized as an important cellular messenger. They represent the very wide range of analytical approaches that can be applied to a single problem in biology. The difficulties involved in measurement of NO have provided a thorough test for the biological analyst and they are likely to continue to present a challenge in the future.

References

1. Palmer, R. M. J., Ferrige, A. G., and Moncada, S. (1987) Nitric oxide release accounts for the biological activity of endothelium-derived relaxing factor. *Nature* **327,** 524–526.

2. Moncada, S., Palmer, R. M. J., and Higgs, E. A. (1991) Nitric oxide: physiology, pathophysiology and pharmacology. *Pharmacol. Rev.* **43,** 109–142.

3. Feelisch, M., te Poel, M., Zamora, R., Deussen, A., and Moncada, S. (1994) Understanding the controversy over the identity of EDRF. *Nature* **368,** 62–65.

4. Malinski, T. and Taha, Z. (1992) Nitric oxide release from a single cell measured in situ by a porphyrinic-based microsensor. *Nature* **358,** 676–678.

5. Shibuki, K. (1990) An electrochemical microprobe for detecting nitric oxide release in brain tissue. *Neurosci. Res.* **9,** 69–76.

6. Tsukahara, H., Gordienko, D. V., Tonshoff, B., Gelato, M. C., and Goligorsky, M. S. (1994) Direct demonstration of insulin-like growth factor-I-induced nitric oxide production by endothelial cells. *Kidney Int.* **45,** 598–604.

7. Kiechle, F. L. and Malinski, T. (1993) Nitric oxide, biochemistry, pathophysiology and detection. *Clin. Chem.* **100,** 567–575.

8. Malinski, T., Bailey, F., Zhang, Z. G., and Chopp, M. (1993) Nitric oxide measured by a porphyrinic microsensor in rat brain after transient middle cerebral artery occlusion. *J. Cereb. Blood Flow Metab.* **13,** 355–358.

9. Zafiriou, O. C. and McFarland, M. (1980) Determination of trace levels of nitric oxide in aqueous solution. *Anal. Chem.* **52,** 1662–1667.

10. Gustafsson, L. E., Leone, A. M., Persson, M. G., Wiklund, N. P., and Moncada, S. (1991) Endogenous nitric oxide is present in the exhaled air of rabbits, guinea pigs and humans. *Biochem. Biophys. Res. Commun.* **181,** 852–857.

11. Persson, M. G., Wiklund, P., and Gustafsson, L. E. (1993) Endogenous nitric oxide in single exhalations and the change during exercise. *Am. Rev. Respir. Dis.* **148,** 1210–1214.

12. Gerlach, H., Rossaint, R., Pappert, D., Knorr, M., and Falke, K. J. (1994) Autoinhalation of nitric oxide after endogenous synthesis in nasopharynx. *Lancet* **343,** 518,519.

13. Persson, M. G., Zetterstrom, O., Agrenius, V., Ihre, E., and Gustafsson, L. E. (1994) Single-breath nitric oxide measurements in asthmatic patients and smokers. *Lancet* **343,** 146,147.

14. Archer, S. (1993) Measurement of nitric oxide in biological model. *FASEB J.* **7,** 349–360.

15. Vallance, P., Collier, J., and Moncada, S. (1989) Effects of endothelium-derived nitric oxide on peripheral arteriolar tone in man. *Lancet* **2,** 997–999.

16. Murad, F., Arnold, W., Mittal, C. K., and Braughler, J. M. (1979) Properties and regulation of guanylate cyclase and some proposed functions for cyclic GMP. *Adv. Cyclic Nucleotide Res.* **11,** 175–204.

17. Wink, D. A., Darbyshire, J. F., Nims, R. W., Saavedra, J. E., and Ford, P. C. (1993) Reactions of the bioregulatory agent nitric oxide in oxygenated aqueous media: determination of the kinetics for oxidation and nitrosation by intermediates generated in the NO/O_2 reaction. *Chem. Res. Toxicol.* **6,** 23–27.

18. Kelm, M., Feelisch, M., Grube, R., Motz, W., and Strauer, B. E. (1992) Metabolism of endothelium-derived nitric oxide in human blood, in *The Biology of Nitric Oxide*, part 1 (Moncada, S. et al., ed.), Portland Press Proceedings, pp. 319–322.

19. Stamler, J. S., Simon, D. I., Osborne, J. A., Mullins, M. E., Jaraki, O., Michel, T., et al. (1992) *S*-nitrosylation of proteins with nitric oxide: synthesis and characterization of biologically active compounds. *Proc. Natl. Acad. Sci. USA* **89,** 444–448.

20. Wagner, D. A., Schultz, D. S., Deen, W. M., Young, V. R., and Tannenbaum, S. R. (1983) Metabolic fate of an oral dose of ^{15}N-labeled nitrate in humans: effect of diet supplementation with ascorbic acid. *Cancer Res.* **43,** 1921–1925.

21. Tesch, J. W., Rehg, W. R., and Sievers, R. E. (1976) Microdetermination of nitrates and nitrites in saliva, blood, water and suspended particulates in air by gas chromatography. *J. Chromatog.* **126,** 743–755.

22. Follett, M. J. and Ratcliff, P. W. (1963) Determination of nitrite and nitrate in cured meat products. *J. Sci. Food Agricul.* **14,** 138–144.

23. Romero, J. M., Lara, C., and Guerrero, M. G. (1989) Determination of intracellular nitrate. *Biochem. J.* **259,** 545–548.

24. Leone, A. M., Francis, P. L., Rhodes, P., and Moncada, S. (1994) A rapid and simple method for the measurement of nitrite and nitrate in plasma by high performance capillary electrophoresis. *Biochem. Biophys. Res. Commun.* **200,** 951–957.

25. Green, L. C., Wagner, D. A., Glogowski, J., Skipper, P. L., Wishnok, J. S., and Tannenbaum, S. R. (1982) Analysis of nitrate, nitrite and ^{15}N nitrate in biological fluids. *Anal. Biochem.* **126,** 131–138.

26. Farrell, A. J., Blake, D. R., Palmer, R. M. J., and Moncada, S. (1992) Increased concentrations of nitrite in synovial fluid and serum samples suggest increased nitric oxide synthesis in rheumatic diseases. *Ann. Rheum. Dis.* **51,** 1219–1222.

27. Wettig, K., Schulz, K. R., Scheiber, J., Broschinski, L., Diener, W., Fischer, G., and Namaschk, A. (1989) Nitrat-und Nitritgehalt in Spiechel, Urin, Blut, und Liquor von Patienten ciner Infectionsklinik. *Wiener klinische Wochenschrift* **101,** 386–388.

28. Ochoa, J. B., Udekwu, A. O., Billiar, T. R., Curran, R. D., Cerra, F. B., Simmons, R. L., and Peitzman, A. B. (1991) Nitrogen oxide levels in patients after trauma and during sepsis. *Ann. Surg.* **214,** 621–626.

29. Miles, D., Thomsen, L., Balkwill, F., Thavasu, P., and Moncada, S. (1994) Association between biosynthesis of nitric oxide and changes in immunological and vascular parameters in patients treated with interleukin-2. *Eur. J. Clin. Invest.* **24,** 27–31.

30. Stamler, J. S. and Loscalzo, J. (1992) Capillary zone electrophoretic detection of biological thiols and their *S*-nitrosated derivatives. *Anal. Chem.* **64,** 779–785.

31. Maples, K., Sandstrom, T., Su, Y., and Henderson, R. (1991) The nitric oxide/heme protein complex as a biological marker of exposure to nitrogen dioxide in humans, rats, and in vitro models. *Am. J. Respir. Cell Mol. Biol.* **4,** 538–543.

32. Wang, Q., Jacobs, J., DeLeo, J., Kruszyna, H., Kruszyna, R., Smith, R., and Wilcox, D. (1991) Nitric oxide haemoglobin in mice and rats in endotoxic shock. *Life Sci.* **49,** 55–60.

33. Westenberger, U., Thanner, S., Ruf, H., Gersonde, K., Sutter, G., and Trentz, O. (1990) Formation of free radicals and nitric oxide derivative of haemoglobin in rats during shock syndrome. *Free Radic. Res. Commun.* **11,** 167–178.

Index

Q

Quin-2, 203

R

Radioligand binding, 1–15
  choice of radioligand, 2
  data analysis, 14
  definition of receptors, 13
  depletion of radioligand, 10
  impure radioligands, 7
  incomplete separation of
    radioligand, 10
  ligand instability, 8
  radioligand dissociation, 10
  receptor instability, 9
  receptor preparation, 4
Reconstitution of receptors, 22
RNase, 47

S

Saponin, 217
Scintillation cocktail, 155

SDS-PAGE, 18, 23, 245
SH-SY5Y neuroblastoma cells, 215
Sodium dodecyl sulfate-polyacryla-
  mide gel electophoresis, *see*
  SDS-PAGE
Sodium fluoride, 73
Solubilization of receptors, 5, 17,19,
  21
Specific activity, 7
Specific binding, 6
Sulfydryl reagents, 147
Synapsin I, 242

T

TARDIS, 230
Terminal deoxynucleotidyl
  transferase, 42
Tyrosine hydroxylase, 242

X

Xenopus brain, 18, 19
Xenopus brain membranes, 19, 21

Methods in Molecular Biology™

Methods in Molecular Biology™ manuals are available at all medical bookstores. You may also order copies directly from Humana by filling in and mailing or faxing this form to: Humana Press, 999 Riverview Drive, Suite 208, Totowa, NJ 07512 USA, Phone: 201-256-1699/Fax: 201-256-8341.

Name _____

Department _____

Institution _____

Address _____

City/State/Zip _____

Country _____

Phone # _____ Fax # _____

denotes a tentative price. Prices listed are Humana Press prices, current as of October 1995, and do not reflect the prices at which books will be sold to you by suppliers other than Humana Press. All prices subject to change without notice.

Europe, Middle East, and Africa: Order directly from Chapman & Hall by faxing to: +44-171-522-9623.

Postage & Handling: *USA Prepaid (UPS):* Add $4.00 for the first book and $1.00 for each additional book. *Outside USA (Surface):* Add $5.00 for the first book and $1.50 for each additional book.

☐ **My check for $_____ is enclosed**
(Drawn on US funds from a US bank).

☐ Visa   ☐ MasterCard   ☐ American Express

Card # _____

Exp. date _____

Signature _____